VOLUME ONE HUNDRED AND THIRTY SIX

Progress in MOLECULAR BIOLOGY AND TRANSLATIONAL SCIENCE

Regulatory T Cells in Health and Disease

VOLUME ONE HUNDRED AND THIRTY SIX

PROGRESS IN MOLECULAR BIOLOGY AND TRANSLATIONAL SCIENCE

Regulatory T Cells in Health and Disease

Edited by

ADRIAN LISTON
Translational Immunology Laboratory,
Department of Microbiology and Immunology,
University of Leuven, Leuven, Belgium

ELSEVIER

AMSTERDAM • BOSTON • HEIDELBERG • LONDON
NEW YORK • OXFORD • PARIS • SAN DIEGO
SAN FRANCISCO • SINGAPORE • SYDNEY • TOKYO
Academic Press is an imprint of Elsevier

Academic Press is an imprint of Elsevier
225 Wyman Street, Waltham, MA 02451, USA
525 B Street, Suite 1800, San Diego, CA 92101-4495, USA
The Boulevard, Langford Lane, Kidlington, Oxford OX5 1GB, UK
125 London Wall, London, EC2Y 5AS, UK

First edition 2015

ISBN: 978-0-12-803415-6
ISSN: 1877-1173

For information on all Academic Press publications
visit our website at http://store.elsevier.com

CONTENTS

CONTRIBUTORS

Jakub Abramson
Faculty of Biology, Department of Immunology, Weizmann Institute of Science, Rehovot, Israel

Michael Delacher
Immune Tolerance, Tumor Immunology Program, German Cancer Research Center (DKFZ), Heidelberg, Germany

Maxime Dhainaut
Laboratory of Immunobiology, Department of Molecular Biology, Université Libre de Bruxelles, Brussel, Belgium

Darcy Ellis
Infection and Immunity Program, Monash Biomedicine Discovery Institute and Department of Biochemistry and Molecular Biology, Monash University, Clayton, Victoria, Australia

Markus Feuerer
Immune Tolerance, Tumor Immunology Program, German Cancer Research Center (DKFZ), Heidelberg, Germany

Thomas S. Fulford
Infection and Immunity Program, Monash Biomedicine Discovery Institute and Department of Biochemistry and Molecular Biology, Monash University, Clayton, Victoria, Australia

Steve Gerondakis
Infection and Immunity Program, Monash Biomedicine Discovery Institute and Department of Biochemistry and Molecular Biology, Monash University, Clayton, Victoria, Australia

Yael Goldfarb
Faculty of Biology, Department of Immunology, Weizmann Institute of Science, Rehovot, Israel

Ann-Cathrin Hofer
Immune Tolerance, Tumor Immunology Program, German Cancer Research Center (DKFZ), Heidelberg, Germany

Jochen Huehn
Department of Experimental Immunology, Helmholtz Centre for Infection Research, Braunschweig, Germany

Axel Kallies
Walter and Eliza Hall Institute of Medical Research, and Department of Medical Biology, University of Melbourne, Parkville, Victoria, Australia

Danny Kägebein
Immune Tolerance, Tumor Immunology Program, German Cancer Research Center (DKFZ), Heidelberg, Germany

Yohko Kitagawa
Department of Experimental Immunology, Immunology Frontier Research Center, Osaka University, Suita, Osaka, Japan

Noriko Komatsu
Department of Immunology, Graduate School of Medicine and Faculty of Medicine, The University of Tokyo, Bunkyo-ku, Tokyo, Japan

Adrian Liston
Translational Immunology Laboratory, VIB, and Department of Microbiology and Immunology, University of Leuven, Leuven, Belgium

Matthias Lochner
Institute of Infection Immunology, TWINCORE, Centre for Experimental and Clinical Infection Research: A Joint Venture Between the Medical School Hannover (MHH) and the Helmholtz Centre for Infection Research (HZI), Hannover, Germany

Jennifer M. Lund
Vaccine and Infectious Disease Division, Fred Hutchinson Cancer Research Center, and Department of Global Health, University of Washington, Seattle, Washington, USA

Muriel Moser
Laboratory of Immunobiology, Department of Molecular Biology, Université Libre de Bruxelles, Brussel, Belgium

Vitalijs Ovcinnikovs
Institute of Immunity & Transplantation, Division of Infection & Immunity, University College London, London, United Kingdom

Maria Pasztoi
Department of Experimental Immunology, Helmholtz Centre for Infection Research, Braunschweig, Germany

Joern Pezoldt
Department of Experimental Immunology, Helmholtz Centre for Infection Research, Braunschweig, Germany

David M. Richards
Immune Tolerance, Tumor Immunology Program, German Cancer Research Center (DKFZ), Heidelberg, Germany

Laura E. Richert-Spuhler
Vaccine and Infectious Disease Division, Fred Hutchinson Cancer Research Center, Seattle, Washington, USA

Shimon Sakaguchi
Department of Experimental Immunology, Immunology Frontier Research Center, Osaka University, Suita, Osaka, Japan

Tim Sparwasser
Institute of Infection Immunology, TWINCORE, Centre for Experimental and Clinical Infection Research: A Joint Venture Between the Medical School Hannover (MHH) and the Helmholtz Centre for Infection Research (HZI), Hannover, Germany

Hiroshi Takayanagi
Department of Immunology, Graduate School of Medicine and Faculty of Medicine, The University of Tokyo, and Japan Science and Technology Agency, Exploratory Research for Advanced Technology Program, Takayanagi Osteonetwork Project, Bunkyo-ku, Tokyo, Japan

Peggy P. Teh
Walter and Eliza Hall Institute of Medical Research, and Department of Medical Biology, University of Melbourne, Parkville, Victoria, Australia

Annemarie van Nieuwenhuijze
Translational Immunology Laboratory, VIB, and Department of Microbiology and Immunology, University of Leuven, Leuven, Belgium

Ajithkumar Vasanthakumar
Walter and Eliza Hall Institute of Medical Research, and Department of Medical Biology, University of Melbourne, Parkville, Victoria, Australia

Lucy S.K. Walker
Institute of Immunity & Transplantation, Division of Infection & Immunity, University College London, London, United Kingdom

Zuobai Wang
Institute of Infection Immunology, TWINCORE, Centre for Experimental and Clinical Infection Research: A Joint Venture Between the Medical School Hannover (MHH) and the Helmholtz Centre for Infection Research (HZI), Hannover, Germany

James Badger Wing
Department of Experimental Immunology, Immunology Frontier Research Center, Osaka University, Suita, Osaka, Japan

PREFACE

Regulatory T cells (or Tregs) are a unique subpopulation of T cells with suppressive properties, acting to counter the immunogenic function of other T cells. This function is critical for the prevention of autoimmune disease and also has profound impacts on other aspects of the mammalian immune system, leading to an intensive effort to harness the power of Tregs as a novel therapeutic strategy across multiple immune diseases.

This volume takes a broad and comprehensive look at Tregs in health and disease states. We have expert chapters on the generation of Tregs, with contributions by Sakaguchi, Huehn, Feuerer, and Abramson on the processes by which Tregs are generated in the thymus and peripheral organs such as the gut. Complementing these chapters, we have articles by Gerondakis, van Nieuwenhuijze, and Kallies, which dissect the molecular pathways that control the induction and differentiation of Tregs. Sparwasser and Moser discuss the cellular dynamics Tregs share with Th17 cells and dendritic cells. Finally, we have an assessment of the physiological impact on Tregs in disease, with expert chapters by Takayanagi, Lund, and Walker on the role of Tregs in arthritis, infection, and diabetes.

Adrian Liston

CHAPTER ONE

Transcriptional and Epigenetic Control of Regulatory T Cell Development

Yohko Kitagawa, James Badger Wing, Shimon Sakaguchi[1]
Department of Experimental Immunology, Immunology Frontier Research Center, Osaka University, Suita, Osaka, Japan
[1]Corresponding author: e-mail address: shimon@ifrec.osaka-u.ac.jp

Contents

Abstract

The control of immune responses against self and nonharmful environmental antigens is of critical importance to the immune homeostasis. Regulatory T (Treg) cells are the key players of such immune regulation and their deficiency and dysfunction are associated with various immune disorders, such as autoimmunity and allergy. It is therefore essential to understand the molecular mechanisms that make up Treg cell characteristics; that is, how their unique gene expression profile is regulated at transcriptional and

Progress in Molecular Biology and Translational Science, Volume 136
ISSN 1877-1173
http://dx.doi.org/10.1016/bs.pmbts.2015.07.011

epigenetic levels. In this chapter, we focus on the components of molecular features of Treg cells and discuss how they are introduced during their development.

1. INTRODUCTION

Treg cells are a subset of $CD4^+$ T cells, specialized in the maintenance of immune tolerance and prevention of autoimmunity. Treg cells are unique in that their primary function is to suppress aberrant or excessive immune responses harmful to the host by counteracting the effects of conventional T cells. This property of Treg cells is particularly important in the establishment of self-tolerance. Discrimination between self and nonself is required for the immune system to avoid attacking self-tissues and organs and causing autoimmune diseases. Along with deletion of self-reactive T cells during their development and induction of an anergic state in self-reactive T cells in peripheral lymphoid organs, thymic production of Treg cells, and their immune suppression in the periphery are a critical mechanism of self-tolerance. In addition, conventional T cells can give rise to Treg cells under certain conditions, contributing to immune homeostasis in the periphery.

The production of suppressive cells in the thymus was initially noted in experiments where the removal of thymus from neonatal mice led to severe autoimmunity.[1] However, it was not until 1995 that Treg cells were definitively identified by specific expression of the alpha chain of the IL-2 receptor (CD25),[2] which enabled the finding that Treg cells constituted around 10% of $CD4^+$ T cells and clearly demonstrating that they have a critical role in self-tolerance. This was then further confirmed with the discovery of the lineage defining transcription factor Foxp3.[3,4] Foxp3 is essential for the function of Treg cells, as loss-of-function mutations of Foxp3 in either the scurfy mouse strain or IPEX (immunodysregulation, polyendocrinopathy, enteropathy, X-linked) syndrome leads to severe autoimmunity including Type-1 diabetes (T1D), immunopathology such as inflammatory bowel disease, and allergy accompanying hyperproduction of IgE.[5–7] Furthermore, depletion of Treg cells in adults also leads to similar autoimmune pathology, demonstrating that Treg cells are needed not just for the establishment, but also the lifelong maintenance, of immune self-tolerance and homeostasis.[8]

In addition to severe acute autoimmunity seen in the complete absence of Treg cells, more subtle defects in Treg cell function have been implicated in the development of a wide range of chronic autoimmune diseases. Partial loss of Treg cell function or reduction in Treg cell numbers has been associated with a range of human autoimmune disorders such as T1D, rheumatoid arthritis, systemic lupus erythematosus, thyroiditis, hepatic disease, and dermatitis (reviewed in Ref. 9). These finding are confirmed in a number of mouse models of autoimmunity. In nonobese diabetes mice, a model of T1D, defective IL-2 signaling is associated with low Treg cell numbers in the pancreas and the development of diabetes. Conversely, treatment with IL-2 expands Treg cells and prevents the development of diabetes.[10] In the case of colitis, transfer of naïve ($CD45RB^{high}$) $CD4^+$ T cells into T cell-deficient mice leads to the development of colitis; while cotransfer of Treg cells is able to prevent the disease.[11] Treg cells also play a critical role in the regulation of humoral immunity and prevention of allergy, as evidenced by the characteristically high levels of IgE production seen in scurfy mice and IPEX patients.[12] Another aspect of Treg cell-mediated suppression of self-reactive T cells is that Treg cells are able to suppress antitumor immune responses. The presence of Treg cells in tumors is often inversely correlated with survival in both mice and humans. This indicates that depletion of Treg cells and targeting of their suppressive functions can be an important tool in antitumor immunotherapy.[13]

A wide range of Treg cell-mediated suppressive mechanisms have been described, suggesting that they may have context-specific roles at different sites.[14] To date, CTLA4, IL-10, TGFβ, ITGβ8, micro-RNA containing exosomes, IL-35, granzyme, perforin, CD39, CD73, and TIGIT have all been demonstrated to have a role in Treg suppressive function. In particular, CTLA4 expression by Treg cells is crucial for Treg cell-mediated immune suppression. CTLA4 downregulates the expression of the costimulatory molecules CD80 and CD86 on the surface of antigen presenting cells, thereby influencing their ability to activate conventional T cells.[15] Treg cell-specific loss of CTLA4 leads to the development of fatal autoimmunity and dysregulated humoral immunity, similar to that seen in scurfy or Treg-depleted mice.[16–18] Further information on the critical role of CTLA4 in humans has been revealed by the finding that haploinsufficiency of CTLA4 leads to a severe autoimmune syndrome, similar to that seen in IPEX, albeit with variable penetrance and age of onset.[19,20]

Another key feature of Treg cells is their inability to produce IL-2, despite their high dependency on IL-2 for survival and proliferation. IL-2 is also a driving factor for conventional T cell proliferation and some effector T cell differentiation. In this competition for IL-2, high expression of the high-affinity IL-2 receptor even at the resting state gives Treg cells an advantage and IL-2 deprivation by Treg cells from other T cells is one mechanism of immune suppression. Indeed, overexpression of CTLA4 and repression of IL-2 in conventional T cells enable them to behave like Treg cells.[21] Conversely, failure to repress IL-2 in Treg cells is associated with the development of autoimmunity.[22]

These molecular features are regulated at both the transcriptional and epigenetic levels. Foxp3-dependent transcriptional programs, which often involve interaction with other transcription factors, control some Treg cell-type gene expression, while Foxp3-independent epigenetic modifications also contribute to the generation of Treg cell characteristics. There is dynamic cross talk between transcriptional and epigenetic regulation in a cooperative manner, which enables stable maintenance of Treg cell characteristics throughout multiple divisions, regardless of environmental changes. Given the critical and wide-ranging roles of Treg cells in autoimmunity, allergy, infection, and tumor immunology, it is vital to understand the molecular mechanisms underlying the development and maintenance of Treg cells in order to develop more sophisticated strategies to either enhance or dampen the function of Treg cells in clinical settings. Here, we review the current understanding of transcriptional and epigenetic regulation in Treg cells and discuss how these molecular changes occur during Treg cell development.

2. TRANSCRIPTIONAL REGULATION IN TREG CELLS

Treg cells have a distinct gene expression profile. Foxp3 regulates some gene expression directly and others in cooperation with its cofactors, while there is also a set of gene expression that is controlled independently of Foxp3.

2.1 Foxp3-Dependent Transcriptional Regulation

2.1.1 Foxp3 as a Master Regulator

Foxp3 is a transcription factor that is specifically expressed by Treg cells. As its deletion impairs the suppressive function of Treg cells and causes similar autoimmune diseases to Treg cell depletion, Foxp3 is indispensable for Treg cell function and is considered as the master regulator of Treg cells. Indeed, Foxp3 is able to upregulate or downregulate about half of the genes that are overexpressed or underexpressed, respectively, in Treg cells, compared to conventional T cells.[23] Importantly, such transcriptional changes induced by overexpression of Foxp3 in conventional $CD4^+$ T cells are sufficient to provide suppressive function similar to that of Treg cells.[4] Moreover, overexpression of Foxp3 and certain transcription factors, such as Irf4, Eos, and Gata1, generates almost complete Treg cell-type transcription profile in conventional CD4$^+$ T cells.[24] Taken together, these findings demonstrate that Foxp3, solely or cooperatively with other transcription factors, regulates the majority of gene transcription in Treg cells.

At the molecular level, Foxp3 mainly functions as a transcriptional repressor and contributes to some of the key characteristics of Treg cells.[25,26] The direct targets of Foxp3 are predominantly those that are normally upregulated by TCR stimulation in conventional CD4$^+$ T cells. A large fraction of them are involved in signaling pathways, such as *Zap70*, *Ptpn22*, and *Itk*.[27] Foxp3 also represses the expression of IL-2.[28] This repression and high dependence on paracrine IL-2 enable Treg cells to suppress conventional T cell proliferation by IL-2 deprivation. Furthermore, Foxp3 directly represses Satb1 by binding to its promoter and inducing microRNAs that target Satb1, to prevent the expression of proinflammatory cytokines that are normally produced by effector T helper cells.[29] Thus, one function of Foxp3 is to repress genes that are activated by T cell activation, and Foxp3 targets genes that serve as regulators of many other genes, thereby efficiently maintaining Treg cell characteristics.

Foxp3 is also involved in upregulation some genes. Hallmarks of Treg cells such as *Il2ra*, *Ctla4*, and *Tnfrsf18* are all bound by Foxp3 and positively regulated.[27] However, Foxp3-null Treg cells, analyzed using mouse models that express a fluorescent marker instead of Foxp3, still express these genes, as well as most of the genes upregulated in Treg cells, but at a lower level than in wild-type Treg cells.[30] These findings illustrate the role of Foxp3 in amplification of pre-existing molecular features.

In terms of the regions that Foxp3 binds to, only a subset of Foxp3-bound genes showed differential expression between Foxp3$^+$ and Foxp3$^-$ T cell hybridomas, suggesting that Foxp3 requires cofactors for its transcription.[27] Consistently, many of the Foxp3-binding sites overlap with other transcription factor binding sites.[31] Therefore, Foxp3, as a master regulator of Treg cells, is able to directly regulate some characteristics of Treg cells, but is insufficient for the generation of full Treg cell-type gene expression, which may require other transcription factors and epigenetic regulation.

2.1.2 Foxp3 and Its Cofactors

As with most transcription factors, Foxp3 interacts with a number of other transcription factors: some being general transcriptional regulators and others being T cell or Treg cell-specific ones. Some of the proteins currently reported to be capable of interacting with Foxp3 are NFκB,[32] NFAT,[22] Runx1,[28] Eos, CtBP1,[33] CBFb, Gata3, Ash2l, Bcl11b, Ikzf3, Foxp1, Smarcc1, Smarce1, Smarca4, Smarca5, Chd4, Hdac2, Rcor1, Lsd1,[34] HIF-1α, IRF-4,[35] p300, TIP60,[36] and Ezh2.[26] Though Foxp3 is likely to exist in a large protein complex, not all these cofactors are always found in the same complex. There are two features determined by the interaction with particular cofactors: effects of binding on target gene transcription and location of Foxp3 binding.

First, Foxp3 can serve as both transcriptional activator and repressor and these modes of action are determined by the recruitment of coactivators or corepressors. For example, human FOXP3 protein is capable of interacting with the coactivators p300 and TIP60 and such interaction promotes the transcriptional activity of FOXP3, while Treg cell-specific deletion of p300 and TIP60 results in loss of Treg function.[36] In contrast, Foxp3 recruits Eos and the corepressor CtBP1 to repress the expression of genes such as Il2. Since IL-2 repression is critical for Treg cell-mediated immune regulation, silencing Eos in Treg cells abrogates their suppressive function.[33] Notably, some of the factors that Foxp3 interacts with, such as Smarca4, Hdac2, and Ezh2 are known as epigenetic regulators, suggesting that Foxp3 recruits these factors to modulate epigenetic features for long-term control of gene expression (discussed in Section 4). Thus, Foxp3 interacts with appropriate cofactors in a locus-specific manner in order to generate Treg cell-type gene expression (Fig. 1).

Second, Foxp3 is dependent on other transcription factors for binding guidance in some loci, meaning that cofactors alter the targets of its gene regulation. Some interactions are fundamentally required for generating Treg cell phenotypes in physiological conditions. For example, NFκB and NFAT transcription factors have been shown to interact with Foxp3 and cooperatively repress the expression of proinflammatory cytokine genes such as *Il2*, *Il4*, and *Ifng*.[22,32] Mutations at the interface of Foxp3 and NFAT interaction resulted in the production of IL-2 by Treg cells and failure to prevent the manifestation of type I diabetes.[22] Other interactions are utilized for particular purposes, such as regulation of specific effector T cell subsets during inflammation. For example, during Th2-type inflammation, Foxp3 interacts with IRF4, which is a transcription factor essential for Th2 cell differentiation program, and enables Treg cells to efficiently control Th2-type inflammation.[37] Importantly, in addition to the variety of Foxp3 complexes at different genomic loci, the repertoire of Foxp3–cofactor complexes

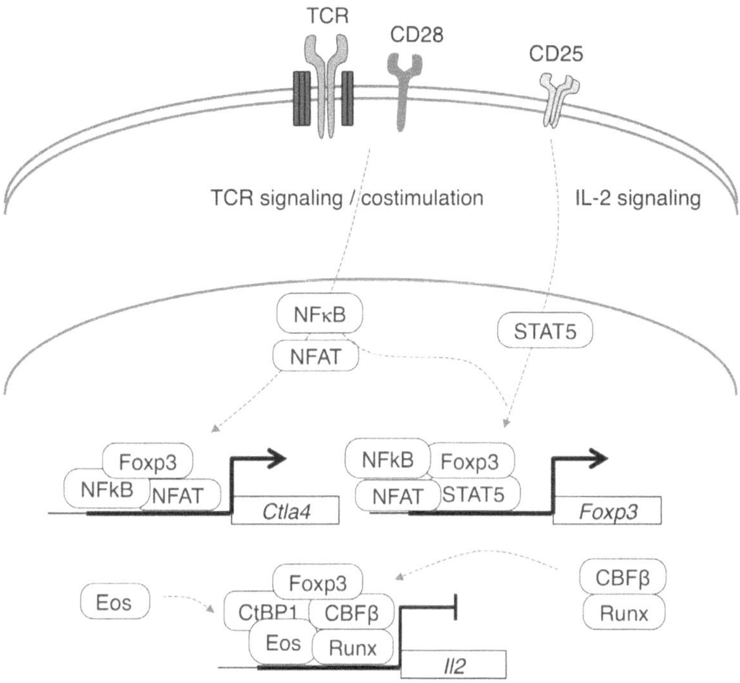

Figure 1 Foxp3-dependent gene expression. Some Foxp3-dependent gene regulation is mediated by the interaction of Foxp3 with transcription factors downstream of TCR/costimulation and IL-2, which are also required for the induction of Foxp3 expression. Others involve the interaction of Foxp3 with T cell-specific or Treg cell-specific transcription factors, such as Runx and Eos.

within a cell may vary depending on the differentiation stage of Treg cells and the environmental conditions they are exposed to. In this sense, the balance among Foxp3 cofactors may be an important determinant of what Foxp3 interacts with. When a fluorescent marker is fused to the N-terminus of Foxp3, it impaired the interaction of Foxp3 with HIF-1α and instead recruited IRF4.[35] Consequently, some gene regulation is altered with particular upregulation of IRF4 signature genes, and these mutant Treg cells alleviated rheumatoid arthritis, but exacerbated type I diabetes.[35] The cause of cofactor change may be due to the competition between HIF-1α and IRF4, or due to the alteration in posttranslational modification of Foxp3. Nevertheless, selection of partners for Foxp3 can serve as a molecular switch for Foxp3-dependent transcription and consequent Treg cell function.

The requirement of Foxp3 to interact with its cofactors indicates that these cofactors also need to be expressed in Treg cells for Foxp3-dependent transcription. Interestingly, a large proportion of these cofactors are direct targets of Foxp3.[34] This notion indicates that Foxp3 directly upregulates the minimum targets by itself, and then regulate the rest of the gene expression in cooperation with these Foxp3 targets that now serve as cofactors. Furthermore, some cofactors such as Runx1, NFAT, and Bcl11b are known to promote Foxp3 transcription, suggesting that Foxp3 and some cofactors positively regulate each other to achieve stable gene regulation.[38–40] There are also cofactors that are independently expressed from Foxp3. For example, NFκB and NFAT are transcription factors activated upon TCR/costimulation. The requirement of these factors for Foxp3-dependent transcriptional regulation suggests that Treg cell specification and maintenance requires TCR signaling in addition to Foxp3 expression. In fact, a large part of Foxp3 targets are coregulated by TCR/costimulation and the number of genes regulated by Foxp3 increase dramatically, as Treg cells become activated.[23,25] Consistent with this, genetic ablation of TCR in mature Treg cells results in a loss of 25% of activated Treg cell signature.[41] Therefore, while some cofactors are upregulated by Foxp3, others are independently expressed, possibly under limited conditions in which Treg cell lineage specification occurs.

Finally, there are "quintet" factors that have been shown to redundantly cooperate with Foxp3 to generate most of the Treg-type gene expression: Eos, Gata1, IRF4, Satb1, and Lef1.[24] Notably, these factors and Foxp3 were retrovirally transduced in conventional $CD4^+$ T cells in this experimental setting, suggesting that TCR stimulation required for retroviral transduction may contribute to some of the Treg cell-type transcriptional regulation. However, even so, coexpression of at least one of the quintet factors with Foxp3 enabled the much more efficient induction of the Treg up- and downregulated gene expression profile than the overexpression of Foxp3 alone. Not all of these quintet factors have been shown to physically interact with Foxp3 protein yet, but they are certainly the coregulators of Foxp3-dependent transcription. How they maximize the transcriptional capacity of Foxp3 remains to be elucidated and it is particularly puzzling that two of the quintet factors, Satb1 and Lef1, are downregulated in Treg cells. One speculation is that coexpression of quintet factors and Foxp3 turns on the molecular switch to build and activate the protein complex around Foxp3. The redundancy among quintet factors, despite belonging to different families and having different functions, may be a mechanism to allow the generation of Treg cell-type gene expression, once Foxp3 is expressed, in various settings where only one of the quintet factors may be expressed.

2.1.3 Foxp3 Posttranslational Modification

For protein interaction and activity of each protein, posttranslational modifications are crucial. Foxp3 is also subjected to such modification. In particular, acetylation of lysine residues is a key determinant of Foxp3 stability and transcriptional activity. Histone acetyltransferases p300 and TIP60, acetylate Foxp3, whereas histone deacetylases SIRT1, HDAC7, and HDAC9 reverse this process.[42] When acetylated, Foxp3 has higher DNA-binding capacity, thereby enhancing transcriptional activity and becomes more resistant to polyubiquitination and proteasomal degradation.[43] This accords with the result that deleting SIRT1 does not have much effect on conventional T cell function and proliferation, but increases Foxp3 expression and Treg cell suppressive activity. These positive effects on Foxp3 function make SIRT1 a promising target for the induction of transplantation tolerance. Indeed, T cell-specific deletion of SIRT1 or administration of pharmacological SIRT1 inhibitors in mice prevented allograft rejection.[44]

Another posttranslational modification that regulates Foxp3 transcriptional activity is the phosphorylation of a serine residue (Ser418 in humans). Lack of this modification results in impaired Foxp3 function as indicated by the failure to repress IL-2 production.[45] Ser418 can be dephosphorylated by protein phosphatase 1 (PP1), and during rheumatoid arthritis, induction of PP1 by the proinflammatory cytokine, TNFα, in inflamed synovium dephosphorylates Foxp3 protein, impairs Treg cell function and contributes to disease pathogenesis. This demonstrates that posttranslational modifications of Foxp3 serve as a key regulator of Treg cell-mediated immune suppression.

2.2 Foxp3-Independent Transcriptional Regulation

Though Foxp3 is the master regulator of Treg cells, Treg cell-type gene regulation also includes Foxp3-independent features.[30,46] This is evident from the fact that Foxp3-null Treg cells retain a large portion of Treg-type gene expression.[30,47,48] This finding can be partly explained by the fact that TCR, IL-2, and TGFβ signaling also regulate the majority of Foxp3 target genes and the number of genes that are solely controlled by Foxp3 is limited.[23] However, there is still a significant fraction (more than 25%) of Treg-type gene expression that is not regulated by Foxp3, TCR, IL-2, or TGFβ signaling.[30,46] Some are regulated by other transcription factors coexpressed in

Treg cells. For example, Foxo1, which is highly expressed and activated by phosphorylation in Treg cells, controls a subset of Treg cell-type gene expression, independently of Foxp3.[49] Others, such as Eos and Helios, are associated with Treg cell-type epigenetic modifications. This suggests that the permissive chromatin status of these genes enables constitutively expressed transcription factors to induce their expression, rather than specifically expressed transcription factors being responsible for their expression.[48]

3. EPIGENETIC REGULATION IN TREG CELLS

To understand the mechanisms of cell type-specific transcriptional regulation, in addition to the activity of transcription factors, the status of target gene loci is another factor that needs to be considered. That is, there are two requirements for the activation of gene transcription: (1) the responsible transcription factors (*trans*-regulatory factors) are expressed and (2) the chromatin configuration of the target gene locus (*cis*-regulatory elements) is permissive so that the transcription factors can bind. The latter is regulated by various epigenetic modifications of chromatin, such as DNA methylation, histone modification, and nucleosome positioning (Fig. 2). These basic criteria need to be met at least at the gene promoters. In addition, such requirements extend to enhancers for stabilizing high gene expression.

Epigenetic modifications of *cis*-regulatory elements have been implicated in lineage determination. There is a close association among cell differentiation, permissive epigenetic modifications at gene loci associated with the cell lineage, and repressive epigenetic modifications at gene loci related to the alternative cell fate. For example, as multipotent progenitors differentiate into common lymphoid progenitors, they show DNA demethylation in lymphoid lineage-specific genes, while undergoing DNA methylation at myeloid lineage-specific genes.[50] These lineage-specific epigenetic modifications are thought to assist irreversible lineage specification by ensuring the stable expression of key regulator genes. This concept is also applicable to Treg cells, which are indeed characterized by distinct epigenetic modifications.

3.1 Stability of the Treg Cell Lineage

The gene expression regulation by Foxp3 and its cofactors is required not only during the Treg cell development but also for their functional maintenance. Ablation of Foxp3 in mature Treg cells resulted in the reversal of Foxp3-dependent gene expression program and consequently these cells lost

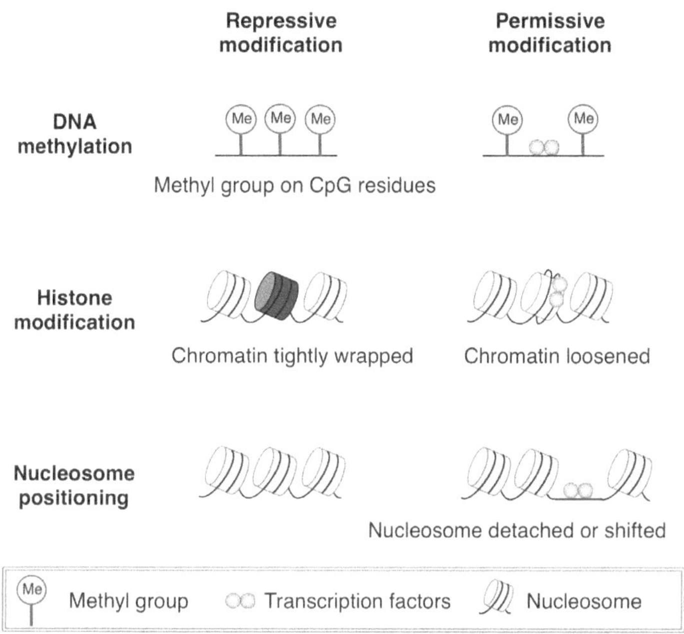

Figure 2 Alteration in transcription factor accessibility by epigenetic modifications. Accessibility of transcription factor to target regions can be determined by various epigenetic modifications: the removal of methyl group on CpG residues by DNA demethylation, loosening of chromatin around histone octamer by permissive histone modification, and detachment or sliding of nucleosome that facilitates transcription factor binding to DNA.

suppressive function.[47] Thus, maintaining Treg cell identity requires continuous expression of Foxp3. This raises an important question regarding the stability of Treg cells. If Treg cells lose Foxp3 under certain conditions, such as during inflammation where effector T cell polarizing stimuli are abundant, they can lose suppressive function and behave like effector T cells. Because Treg cells possess a relatively self-reactive TCR repertoire, Treg cell plasticity indicates a potential hazard of eliciting immune responses against self. Several studies showed Treg cells could be plastic when they receive chronic antigenic stimulation.[51,52] Moreover, once they lose Foxp3, they secrete proinflammatory cytokines such as IFNγ, possibly contributing to the amplification of inflammation. In contrast, there are also reports demonstrating that Treg cells are stable regardless of environmental conditions.[53,54] This controversy remains unsolved, but may be explained by the definition of function-competent Treg cells and different experimental systems utilized.

There are some Foxp3$^+$ T cells that are not committed to Treg cell lineage. TCR stimulation induces Foxp3 expression in some murine conventional T cells and these activation-induced Foxp3$^+$ T cells do not possess Treg cell characteristics except low expression of Foxp3.[54] In humans, Foxp3 is more loosely regulated and can be transiently upregulated by *in vitro* stimulation of CD4$^+$CD25$^-$ non-Treg cells, while there is a clear fraction of Foxp3$^+$ T cells with no suppressive function *in vivo*.[55,56] Foxp3 expression can also be induced *in vitro* by stimulating both murine and human conventional T cells in the presence of TGFβ and IL-2.[57] These cells have some suppressive function but are unable to maintain prolonged Foxp3 expression upon removal of such stimulation, indicating that they are not fully committed to Treg cell lineage. Therefore, maintaining Foxp3 expression involves additional mechanisms to those that activate Foxp3 promoter activity. This suggests that whatever that ensures the stable expression of Foxp3 is the true indicator of Treg cell lineage.

3.2 *cis*-Regulatory Elements of the *Foxp3* Gene

There are four *cis*-regulatory elements in the *Foxp3* locus, important for the regulation of gene expression: the promoter and three enhancers. The promoter contains response elements for transcription factors downstream of TCR/costimulation, such as NFAT, AP-1, and Nr4a family members and for STAT5, which is activated by IL-2 signaling. This explains the induction of low, transient Foxp3 expression by these signals. In order to achieve stable Foxp3 expression at high level, however, appropriate enhancers need to be activated and looped to *Foxp3* promoter. The *Foxp3* gene has three conserved noncoding regions that serve as enhancers. These are referred to as conserved noncoding sequence (CNS) 1, CNS2, and CNS3.[58]

CNS1 is an enhancer within intron 1 of the *Foxp3* locus, shown to be required for peripheral Treg cell differentiation. It contains a TGFβ signaling response element with Smad3-binding site. CNS2 is another enhancer located in the intron 1 with binding sites for Ets1, Foxp3, and CREB and is associated with Foxp3 expression stability. Its deletion results in gradual loss of Foxp3 expression as cells divide.[58] CNS3 is considered as an enhancer required for efficient thymic Treg cell development, as its deletion leads to a severe reduction in Treg cells in the thymus. In this way, these enhancers are activated at different stages of Treg cell development and maintenance.

3.3 DNA Demethylation

Of currently known Treg cell-specific epigenetic modifications, DNA demethylation of *Foxp3* CNS2 region most highly correlates with the stability of Treg cells. When CpG residues are methylated, the methyl group interferes with transcription factor binding, whereas demethylation increases the accessibility for transcription factors by revealing their consensus sequence. Indeed, the transcription factors CREB, Ets1, and *Foxp3* bind to CNS2 in a demethylation-dependent manner.[58–60] Treg cell-specific DNA demethylated regions (TSDRs) are present not only at *Foxp3* locus but also at other Treg signature gene loci, such as *Ctla4*, *Ikzf4*, *Tnfrsf18*, and *Ikzf2* and their demethylated status is highly stable and specific in Treg cells, suggesting that TSDR DNA demethylation ensures the stable expression of genes closely associated with Treg cell function.[48] DNA demethylation at lineage-specific gene loci has also been observed in many other cell types,[50] indicating that this is a common mechanism of lineage specification.

One possible mechanism for the key role of CNS2 demethylation in determining stable expression of Foxp3 is the constitutive expression of transcription factors bound to CNS2. Unlike Smad3 that is activated upon TGFβ signaling, CNS2-bound transcription factors such as Runx, Gata3, and Ets1 are constitutively present in T cells, enabling stable enhancer activation regardless of changes to the cell environment. Indeed, Runx/Cbfb deletion results in gradual loss of Foxp3 expression in mature Treg cells.[61] Similarly, Gata3 binds to CNS2 in Treg cells and Treg cell-specific Gata3 deletion leads to failure to maintain Foxp3 expression.[62] Furthermore, Foxp3 binds to CNS2 by interacting with CNS2-bound Runx1/Cbfb complex and amplifies its own expression, forming a positive feedback loop.[58] These mechanisms may contribute to the stable inheritance of Foxp3 expression as cells divide. However, in order for Treg cells to stably exert their suppressive function even during inflammation, they also need mechanisms to counteract the effects of helper T cell polarizing stimuli. Recent evidence demonstrates that TCR activation in Treg cells facilitates the binding of downstream transcription factors, NFAT and NFκB to promoter and CNS2, which are looped to ensure stable expression of Foxp3.[63] STAT5 activated by IL-2 signaling also binds to CNS2 and protects Treg cells from losing Foxp3 expression.[64] These results suggest that DNA demethylation at Foxp3 CNS2 is a key determinant of Treg cell lineage stability and TCR stimulation and IL-2 signaling further lock their identity under inflammatory conditions.

DNA demethylation of TSDRs may also contribute to the regulation of Foxp3-dependent and -independent gene expression in Treg cells. First,

TSDR demethylation is observed at *Ikzf4* and *Ikzf2* gene loci and these genes are not upregulated by Foxp3, TCR stimulation, or TGFβ signaling.[23,48] With no signals or transcription factors identified to induce and maintain these genes in Treg cells, it is conceivable that TSDR demethylation, followed by binding of constitutively expressed transcription factors induces the expression of some Foxp3-independent genes. Second, TSDR demethylation is also observed at genes upregulated by Foxp3 and TCR stimulation, such as *Ctla4* and *Tnfrsf18*.[48] This may explain how Treg cells maintain high level of these molecules, even at a resting state. Taken together, these findings suggest that TSDR demethylation facilitates Foxp3-dependent transcriptional regulation by stabilizing Foxp3 expression as well as assisting with both Foxp3-dependent and -independent gene regulation.

3.4 Histone Modification

Histone modification is another relatively well-studied epigenetic feature. Histones form an octomeric nucleosome core, consisting of two copies each of the core histones H2A, H2B, H3, and H4, and together with DNA wrapped around it, make up the nucleosome. Posttranslation modifications of histones have a critical role in the control of transcription as they alter the accessibility of a particular genomic region to transcription factors. There are various modification types, each associated with a permissive or repressive effect on transcription. Well-studied modifications are monomethylation, dimethylation, and trimethylation of histone H3 at Lys4 (H3K4me1, H3K4me2, and H3K4me3, respectively), acetylation and trimethylation of histone H3 at Lys27 (H3K27ac, H3K27me3), and acetylation of histone H3 at Lys9 (H3K9ac). Studying these histone codes can reveal the status of gene transcription and enhancer activity. For example, H3K4me3 and H3K9ac are found in actively transcribed promoters, whereas H3K4me1 and H3K27ac indicate poised and active enhancers, respectively.[65]

In Treg cells, promoters of Treg cell-associated genes, such as Foxp3, are marked with permissive H3K4me3 modification, strongly correlating with gene expression.[66] Moreover, DNA demethylation at TSDRs was found to correlate with increased H3K4me3 modification, suggesting that Treg cell-specific DNA demethylation and H3K4me3 modification have similar roles in the maintenance of Treg cell lineage.[67] However, H3K4me3 modification at the *Foxp3* promoter is more easily induced, compared to DNA demethylation of CNS2 region; whereas DNA demethylation does not occur after TCR/costimulation, IL-2 signaling or TGFβ signaling,

H3K4me3 modification at transcription start site increases when naïve $CD4^+$ T cells are stimulated.[48] This correlates with the temporary induction of Foxp3 expression, but as Foxp3 expression is lost, H3K4me3 modification also decreases.[68] Thus, this particular type of histone modification may be merely an indicator of active transcription.

Treg cells are also characterized by unique patterns of H3K4me1 and H3K27ac modifications. With these modifications serving as the markers for poised and active enhancers, their comparison in human conventional T cells and Treg cells revealed the presence of Treg cell-specific enhancers and there was a correlation between the activation of cell-specific enhancers and neighboring gene expression.[69] Moreover, Treg cell-specific enhancers were enriched with STAT5 binding, whereas conventional T cell-specific enhancers were frequently bound by Runx and Ets1, indicating that Treg cells are more dependent on the IL-2-STAT5 pathway for their enhancer activation and/or gene regulation at activated enhancers. It will be important to address whether enhancer activation is required for binding of these transcription factors, or vice versa.

While permissive histone modifications are found near genes upregulated in Treg cells, repressive histone modification, H3K27me3, is found near genes downregulated in Treg cells and is controlled at least partially in a Foxp3-dependent manner.[26] Foxp3 itself does not possess the ability to directly modify epigenetic features, but it interacts with a number of components of nucleosome remodeling deacetylase complex and SWI/SNF complex.[34] These complexes are known to modulate chromatin organization and histone modification, suggesting that Foxp3 utilizes these complexes to stably control gene expression. A recent study has revealed that many Foxp3 target genes are characterized by H3K27me3 modification in activated Treg cells and their expression is epigenetically repressed.[26] The loss of H3K27me3 pattern at some of these locations in Foxp3-deficient Treg cells demonstrates that Foxp3 and its partner proteins induce epigenetic repression. PRC2 (polycomb repressive complex 2) is one of the partner protein complexes recruited by Foxp3 for this purpose. Foxp3 has been shown to interact with a PRC2 component, Ezh2 in activated Treg cells and when Ezh2 was specifically ablated in Treg cells, certain genes were upregulated in a similar manner to that found when *Foxp3* was deleted in Treg cells, and a large fraction of them were characterized by H3K27me3 modification in wild-type activated Treg cells.[70] These findings suggest that Foxp3 and Ezh2 cooperatively repress some genes by induction of repressive histone modifications (Fig. 3).

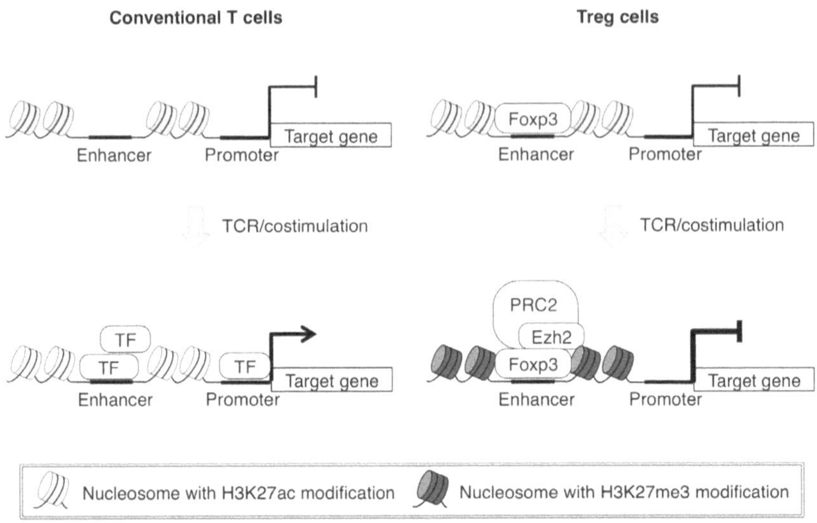

Figure 3 Foxp3-mediated induction of repressive histone modification. Foxp3 mainly serves as a transcription repressor, targeting genes that are normally upregulated by TCR/costimulation. One mechanism of Foxp3-dependent gene repression in activated Treg cells is the induction of repressive histone modification, H3K27me3 by recruiting Ezh2-containing polycomb complex PRC2.

3.5 Nucleosome Positioning

Nucleosomes are a subunit of chromatin made up of a histone octamer and DNA wrapped around it. Chromatin remodeling enzymes can slide nucleosomes, remove the histone octamer or loosen the DNA around histones, in an ATP-dependent manner.[71] The consequence of these epigenetic events is the alteration in the exposed region of the genome, which changes the accessibility to transcription factors. One method to assess nucleosome positioning is the examination of DNase I hypersensitivity (DHS) sites, taking advantage of the fact that nucleosome-free regions can be cleaved by DNase I. For example, active enhancers are bound by a number of transcription factors and are characterized by high DHS.

The nucleosome positioning in naïve $CD4^+$ T cells and Treg cells are largely similar, but there are some limited DHS regions specific to Treg cells (less than 1% of all DHS sites).[31] These regions are located near Treg signature genes, such as *Foxp3*, *Ctla4*, and *Ikzf2*, suggesting that the key molecules that define Treg cell lineage are marked with Treg cell-specific epigenetic modifications to ensure their stable expression. The overlap of gene sets with TSDRs and Treg cell-specific DHS regions suggest that a common mechanism may regulate these two epigenetic processes. Given that these genes are highly associated with Treg cell function and identity, these epigenetic modifications may ensure their stable expression by promoting the binding of transcription factors.

In terms of the interaction between transcription factors and nucleosome positioning, Foxp3 does not have a profound effect. Foxp3 binding sites mostly show an open chromatin structure both in naïve $CD4^+$ T cells and Treg cells, indicating that Foxp3 does not have to modulate chromatin structure in order to bind to its targets.[31] Instead, Foxp3 binds to regions that are already bound by other transcription factors; in some regions, Foxp3 binds to where Runx1 is bound and cooperatively regulate the gene expression, while in other regions, it replaces Foxo1 and initiates Treg cell-type gene regulation.

4. CROSS TALK BETWEEN FOXP3-DEPENDENT GENE REGULATION AND TREG CELL-TYPE EPIGENETIC MODIFICATIONS

Treg cell-type gene regulation involves both Foxp3-dependent transcriptional programs and epigenetic modifications. Both factors contribute cooperatively to the regulation of some genes, while in other cases Foxp3 is required but epigenetic modifications are not, and vice versa. As examples of the former, *Foxp3* and *Ctla4* gene upregulation is ensured by DNA demethylation and open chromatin structure of their enhancers, to which Foxp3 binds and promotes transcription. In contrast, as examples of the latter scenario, *Ikzf4* and *Ikzf2* gene upregulation in Treg cells occurs independently of Foxp3 expression but are associated with DNA demethylation and DHS, and the repression of *Il2* is dependent on Foxp3 expression but not associated with Foxp3-independent epigenetic modification.[28,31,48,59]

Genome-wide analyses of gene expression, Foxp3-binding sites, and TSDR demethylation further revealed that indeed there is a division of labor between Foxp3 and TSDR demethylation. Foxp3 acts predominantly as a transcriptional repressor after TCR stimulation, whereas TSDR demethylation is associated with gene upregulation before activation.[25,26] This is consistent with the observation that Foxp3-deficient Treg cells express most of the Treg hallmark genes at steady state, yet they cannot repress the expression of proinflammatory cytokines such as IFNγ and IL-17, especially under inflammatory conditions.[30,48] This suggests that in a simplified model, Treg cell-type chromatin landscape sets the environment in which general or

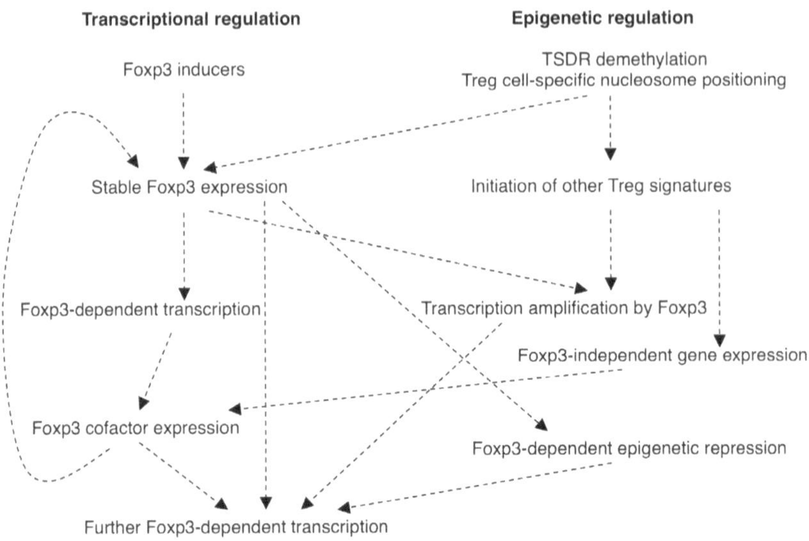

Figure 4 Cross talk between transcriptional and epigenetic regulation for the generation of Treg cell-type gene expression. Foxp3 expression is induced and maintained by both transcription factors and epigenetic modifications. Foxp3 then generates some Treg cell-type gene expression, which includes the upregulation of Foxp3 cofactors. There are also genes expressed independently of Foxp3, but associated with Treg cell-specific epigenetic modifications. Some of these genes are further upregulated by Foxp3, and serve as Foxp3 cofactors. Foxp3 and its cofactors then cooperatively regulate more gene expression.

T cell-specific transcription factors can induce the expression of genes upregulated in Treg cells, while Foxp3 acts later, predominantly preventing the activation of effector T cell differentiation programs and maintains Treg cell identity (Fig. 4).

5. TREG CELL DEVELOPMENT

The transcriptional and epigenetic features of Treg cells described above are introduced during the development of Treg cells. The highly specific and stable nature of Treg cell-type epigenetic features, interacting with the transcriptional networks allows the irreversible commitment of progenitor cells into the Treg cell lineage. Then the important questions are what kind of signals trigger Treg cell development and what kind of molecular mechanisms are involved in interpreting such signals and coordinating transcriptional and epigenetic changes?

Treg cells are broadly divided into two subpopulations, based on their site of origin. The majority develops in the thymus and is referred to as thymus-derived Treg (tTreg) cells, while some Treg cells also differentiate from conventional $CD4^+$ T cells in the peripheral lymphoid organs as peripherally induced Treg (pTreg) cells (Fig. 5). tTreg cells, particularly those that develop during neonatal period, are nonredundantly required for the establishment of self-tolerance.[72] In contrast, pTreg cells are predominantly found in mucosa-associated lymphoid tissues such as Peyers' patches and lamina propria of small and large intestines, and are involved in the induction of immune tolerance to commensal microbes and nonpathogenic environmental antigens, such as food antigens. Moreover, pTreg cells, which exist only in placental mammals, appear to play roles in the establishment of maternal–fetal tolerance.[73] Therefore, tTreg and pTreg cells have a division of labor in some scenarios. However, in terms of their

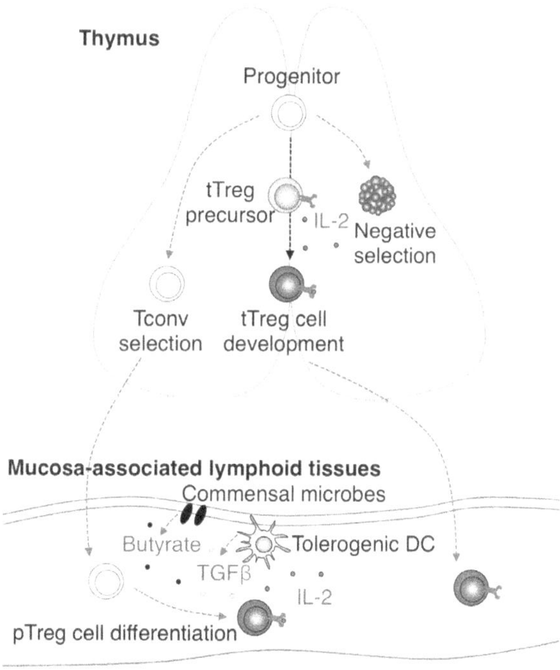

Figure 5 Thymic and peripheral development of Treg cells. Thymus-derived Treg (tTreg) cells develop from progenitors to tTreg cells, through tTreg precursor cells in the thymus, dependently on IL-2 availability. In contrast, peripherally induced Treg (pTreg) cells differentiate from conventional T cells at mucosa-associated lymphoid tissues, and this is facilitated by TGFβ and butyrate produced by tolerogenic DC and commensal microbes, respectively.

transcriptional and epigenetic profiles, these two subpopulations of Treg cells show high levels of similarity, with some exceptions such as *Nrp1*, *Itgb8*, and *Ikzf2*, which are expressed primarily in tTreg cells.[74–77] Both of them are stably committed to the Treg cell lineage with similar suppressive capacity, yet they may exert immune regulatory function at different locations and/or timing, which could explain the need for two distinct developmental pathways. As expected from the different environments that they develop in, different, but overlapping molecular mechanisms are involved in these two developmental systems.

5.1 Signals Involved in Treg Cell Development

Current understanding of tTreg cell development is based on the two-step model where the first step generates tTreg precursor cells ($CD4^+CD8^-CD25^+Foxp3^-$ thymocytes) by agonistic TCR/costimulation and the second step converts them to tTreg cells by IL-2 stimulation.[78] During thymocyte selection, interaction with self-antigens presented in the medulla measures the self-reactivity of TCRs and determines the fate of individual thymocytes. In general, weak interaction with self-antigens selects conventional thymocytes, strong interaction, indicative of thymocytes being highly self-reactive, causes apoptosis, and tTreg cell development comes in between. The requirement of agonistic TCR/costimulation for tTreg cell generation is clearly demonstrated by the lack of tTreg cells in foreign antigen-specific TCR transgenic mice, but the enhanced tTreg cell development in double transgenic mice where cognate antigens and corresponding TCRs are transgenically expressed (reviewed in Ref. 79). These findings indicate that development of tTreg cells requires stronger TCR/costimulation than that positively selects conventional thymocytes. In support of this notion, blockade of CD28-mediated costimulation by CD28 deletion severely reduces tTreg numbers but not so much conventional CD4SP thymocytes.[80,81] Consequently, tTreg cells possess relatively self-reactive TCR repertoire, which enables them to efficiently suppress immune response against self-antigens. However, given the relatively high level of TCR/costimulation that developing Treg cells are subject to and the proapoptotic nature of Foxp3 protein, tTreg cell development occurs at the verge of death and requires other signals to diverge from apoptosis.[82] Particularly, IL-2 signaling is critical for survival and expansion of Treg cells and the expression of high-affinity IL-2 receptor, CD25, prior to Foxp3 expression in tTreg precursors is necessary to be protected from the proapoptotic effect of Foxp3.[82]

In contrast, pTreg cell differentiation is more dependent on the exposure to nonpathogenic environmental antigens. First, persistent low dose-antigenic stimulation, instead of strong TCR/costimulation required for tTreg cell development, is favorable for pTreg cell generation, as evident from the induction of pTreg cells when cognate antigens were orally administered in corresponding TCR transgenic mice but not when antigens were injected with adjuvant.[83] This oral tolerance is a pivotal mechanism to prevent food allergy. Second, pTreg cell generation in the gut-associated lymphoid tissues is dependent on the interaction of immune system with commensal microbes. This is evident from the fact that maintaining mice under germ-free conditions reduces colonic Treg cell numbers and inoculation of a particular species of microbiota increases them.[84] At mucosa-associated lymphoid tissues, the immune system requires appropriate immune regulation, so that immune responses are elicited toward pathogenic microbes but are suppressed to tolerate commensal microbes and nonpathogenic environmental antigens. As a mechanism of the latter, commensal bacteria, such as Clostridia, generate a fermentation product, butyrate, which induces pTreg cells.[85,86] Furthermore, antigens found in the mucosa modulate the function of dendritic cells (DCs) to tolerogenic type and this then facilitates pTreg cell development. Tolerogenic DCs express a lower level of costimulatory molecules such as CD80 and CD86 compared to mature DCs, making it less likely to drive effector T cell differentiation programs. They also express immunosuppressive cytokines such as TGFβ, IL-10, and retinoic acids, which potentiate pTreg cell conversion.[87] While there are distinct molecular pathways involved in tTreg and pTreg cell development, the common signals include IL-2 signaling and TCR/costimulation.[38,88] Thus, Treg cell development occurs by combination of signals that are available in different settings.

5.2 Transcription Factors Involved in Foxp3 Induction

Given the indispensable roles of Foxp3 in Treg cell function, Foxp3 induction is the central event during Treg cell development and therefore whatever induces Foxp3 expression holds the key for Treg cell lineage determination. However, if there should be one transcription factor specifically appointed to induce Foxp3 expression, the next question is what induces this transcription factor and then what upstream factors induce this further factor and so on, a question that might continue until identifying the initial lineage-specifying factor. Given the highly specific expression of Foxp3 in Treg cells and the close association of Foxp3 with Treg cell function, it is plausible that Foxp3 is the ultimate lineage-specifying factor for Treg cells. In this case, multiple transcription factors may be involved in Foxp3 transcription in a nonredundant manner, reflecting the need for various requirements, such as appropriate TCR/costimulation and IL-2 availability, to be met for Treg cell development. In this section, transcription factors currently known to be involved in Foxp3 transcription are discussed.

First, several transcription factors downstream of TCR/costimulation are required for Foxp3 induction. This is expected as agonistic TCR stimulation is essential for Treg cell development and the deletion of CD28 also severely diminishes the efficiency of thymic Treg cell development.[79] cRel is a transcription factor of NFκB family, downstream of TCR/costimulation. It is involved in the initiation of Foxp3 transcription and often referred to as the pioneer factor for tTreg cell development. cRel knockout mice show a severely reduced frequency of Treg cells in the thymus, whereas enhanced NFκB signaling pathway in mice that transgenically express constitutively active IKK-β kinase have an increased number of Foxp3$^+$ cells among both CD4SP and CD8SP thymocytes.[89,90] Mechanistically, one study shows that cRel is capable of binding to the *Foxp3* enhancer, CNS3. CNS3 shows permissive histone modification, H3K4me1, indicating that the enhancer is at a poised state, at least from the double positive (DP) thymocyte stage. A possible scenario is that cRel activation and nuclear translocation upon agonistic TCR/costimulation leads to activation of CNS3.[58] In fact, CNS3 knockout mice shows impaired tTreg cell development, similar to cRel knockout mice, supporting this scenario. Another study observed cRel binding to CNS2, even when its CpG residues are methylated, in T cell lines and primary CD4$^+$ T cells, but not in non-T cell lines or DP thymocyte cell lines, suggesting that cRel may also be involved in DNA demethylation of CNS2 and enhancement of Foxp3 transcription.[90] However, cRel is not exclusively expressed in tTreg precursor cells and tTreg cells, and it is unclear at present whether cRel binding to *Foxp3* enhancers occurs specifically in developing tTreg cells, and if so, what directs cRel to these regions and whether cRel binding is sufficient to activate *Foxp3* transcription. Another report has demonstrated the ability of cRel and another subunit of the NFκB family, p65, to bind to the *Foxp3* promoter during *in vitro* TGFβ-dependent *Foxp3* induction in peripheral naïve CD4$^+$ T cells.[89] cRel ablation also hampers *in vitro* induction of Foxp3 expression in this system, partly by impairing endogenous IL-2 production and by preventing the formation of *Foxp3* enhanceosome at the *Foxp3* promoter.[89,91] This *Foxp3* enhanceosome contains transcription factors involved in *Foxp3* transcription such as Smad3, CREB, and NFAT, in addition to cRel/p65 and because some of these factors first bind to enhancers CNS1 and CNS2, cRel/p65 may serve as the base of enhanceosome formation and local chromatin looping to connect enhancers and promoter.[89]

Orphan nuclear receptors, Nr4a family members, are also involved in Foxp3 induction. These transcription factors are upregulated upon TCR stimulation and Nr4a1 and Nr4a3 are redundantly essential for Foxp3 induction, as clearly shown by the complete lack of Foxp3$^+$ cells both in the thymus and periphery of Nr4a1/3 double-knockout mice.[92] There is a binding site for Nr4a family members in the *Foxp3* core promoter region and at least Nr4a2 has been shown to bind to *Foxp3* promoter and CNS1.[93] *In vitro* Nr4a2 overexpression alone is sufficient to induce Foxp3 expression in conventional CD4$^+$ T cells, indicating it has the ability to activate Foxp3 promoter. However, Nr4a family members are expressed immediately after general TCR stimulation in conventional CD4$^+$ T cells and such expression is insufficient to activate *Foxp3* transcription. Interestingly, Nr4a1 is not only required for Foxp3 expression but also involved in promoting negative selection and its expression level reflects TCR stimulation strength.[94,95] Thus, as a sensitive interpreter of TCR stimulation during thymocyte selection, the expression level and the activity of Nr4a family members may contribute to the fate decision among conventional T cell selection, Treg cell development, and negative selection.

Foxo1 and Foxo3 are another set of transcription factors regulated by TCR/costimulation and redundantly required for Foxp3 expression. Foxo1/3 double-knockout mice have severely reduced Treg cell percentages and the remaining Treg cells are not functional.[96] These factors bind to *Foxp3* promoter, CNS2 and CNS3, regulating *Foxp3* transcription. Notably, Foxo proteins are exported out of the nucleus when phosphorylated by PI3K–Akt signal, which is activated by prolonged TCR/costimulation.[96] This suggests that Foxo proteins act as another sensor of TCR/costimulation, setting a window of TCR/costimulation duration that allows Treg cell differentiation.

IL-2 signaling is also required for both tTreg and pTreg cell development, and the responsible downstream transcription factor is STAT5. *Stat5* deletion results in severely impaired development of Foxp3$^+$ Treg cells both in the thymus and periphery, while transgenic expression of a constitutively active form of STAT5 increases Treg cell percentages.[97] Furthermore, the ability of IL-2 signaling to induce Foxp3 expression from tTreg precursor cells *in vitro* indicates the direct involvement of IL-2–STAT5 pathway in

Foxp3 transcription.[78] Mechanistically, STAT5 binds to *Foxp3* promoter and CNS2 and likely activates these *cis*-regulatory elements.[97,98]

Furthermore, there is a group of transcription factors that are expressed in the T cell lineage and are involved in *Foxp3* transcription. Ets1, Bcl11b, and Bach2 fall into this category; these are not specifically induced during Treg cell development but binds to *cis*-regulatory elements of *Foxp3* gene and their genetic ablation results in impaired generation of Foxp3$^+$ Treg cells.[40,99,100] Given their roles in the development and maintenance of Treg progenitors, it is possible that these factors are involved in the pre-establishment of chromatin landscape in which transcription factors downstream of TCR and IL-2 signaling can work.

In addition, some transcription factors are required only for pTreg cell development. Reflecting the involvement of TGFβ signaling in pTreg cell development, deletion of either Smad2, or Smad3, transcription factors downstream of TGFβ signaling, does not affect tTreg cell development efficiency, but impairs TGFβ-dependent Foxp3 induction from peripheral naïve CD4$^+$ T cells.[101] In concordance with this finding, the deletion of *Foxp3* CNS1 region or Smad3 response element within CNS1 results in the specific reduction in Treg cells within gut-associated lymphoid tissues, which are enriched with pTreg cells.[101,102] Although TGFβ signaling itself is also required for tTreg cell development, these results indicate that the activation of CNS1 enhancer by Smad3 is required for pTreg, but not tTreg cell development.[103]

In summary, there are a number of transcription factors reported to be required for Foxp3 induction. Most of them fit with the required signals, such as TCR/costimulation and IL-2 signaling, suggesting that only when all the required signals are provided, the downstream transcription factors cooperatively activate Foxp3 transcription.

5.3 Induction of Epigenetic Modification During Treg Cell Development

Along with Foxp3 induction, Treg cell-specific epigenetic modifications occur during Treg cell development. CD25$^+$Foxp3$^+$ CD4SP thymocytes are already fully committed to Treg cell lineage and are capable of maintaining stable expression of Foxp3.[54] This is at least partially due to the DNA demethylation of CNS2, taking place at this stage.[104] pTreg cells *in vivo* are similarly demethylated at the CNS2 region.[74] In terms of the molecules that induce these changes, it is not as well understood as transcription factors that regulate *Foxp3* expression. It was only in 2009 that enzymes capable of demethylating 5-methylcytosines in CpG residues were identified.[105] These enzymes, Tet family proteins, actively convert 5-methylcytosines to 5-hydroxymethylcytosines, which are then reverted to cytosines by TDG-mediated basic excision repair or through DNA replication.[106] Alternatively, DNA demethylation can also occur passively, by the loss of methyl group as DNA replication takes place without the DNA methylation maintenance machinery. Regarding the mechanism of Treg cell-specific DNA demethylation, it has not been reported yet whether Tet family members are involved; however, DNA demethylation during tTreg cell development has been shown to proceed even when cell division is inhibited, indicating the involvement of an active process.[104] As DNA demethylation of lineage-specific genes is a common mechanism regulating lineage commitment, it is also important to address how DNA demethylating enzymes are recruited to TSDRs during Treg cell development.

Mechanisms of other epigenetic modifications involved in Treg cell development are even less clearly understood. Enzymes responsible for each type of histone modifications have been identified, such as p300 for H3K27ac and SETD7 for H3K4me1, but similarly to the case with DNA demethylation, how they are recruited to specific loci in developing Treg cells are not known. Histone modifications are dynamically regulated, and inhibition of repressive histone modifying enzymes also leads to the induction of permissive H3K27ac modification. A recent report shows that butyrate, which is produced by the colonic commensal microbes, acts as a histone deacetylase inhibitor and enhances *Foxp3* transcription, at least in part by facilitating H3K27ac modification at *Foxp3* promoter and enhancers.[86]

5.4 Coordination of Transcriptional and Epigenetic Changes During Treg Cell Development

Both the induction of Foxp3 for Foxp3-dependent transcriptional regulation and DNA demethylation at TSDRs to ensure the stable expression of Foxp3 and other Treg signature genes are required for generation of Treg cells. The outstanding questions are which comes first and what is the prerequisite for Treg cell development.

Foxp3 induction and TSDR demethylation are independent, simultaneous processes during tTreg cell development. TSDR demethylation at *Foxp3* CNS2 is not required for Foxp3 induction, and TSDR demethylation also occurs even in the absence of Foxp3.[48,58,104] Yet, tTreg precursor cells acquire both Foxp3 expression and TSDR demethylation upon receiving TCR/costimulation and IL-2 signaling.[104] Similarly, recently developed Treg cells identified in the periphery of fate mapping reporter mice show partial demethylation.[54,74] These findings suggest that Foxp3 induction and DNA demethylation at CNS2 occur in parallel, likely initiated by a common upstream mechanism.

However, some epigenetic modifications required for efficient Treg cell development take place at the progenitor stage, prior to the transduction of Treg cell-inducing stimuli. *Foxp3* enhancer CNS3 already shows H3K4me1 modification in DP and CD4SP thymocytes, but not in B cells.[58] This poised status means the enhancer is ready for use by late-arriving transcription factors such as cRel, and indicates the potential of progenitors to

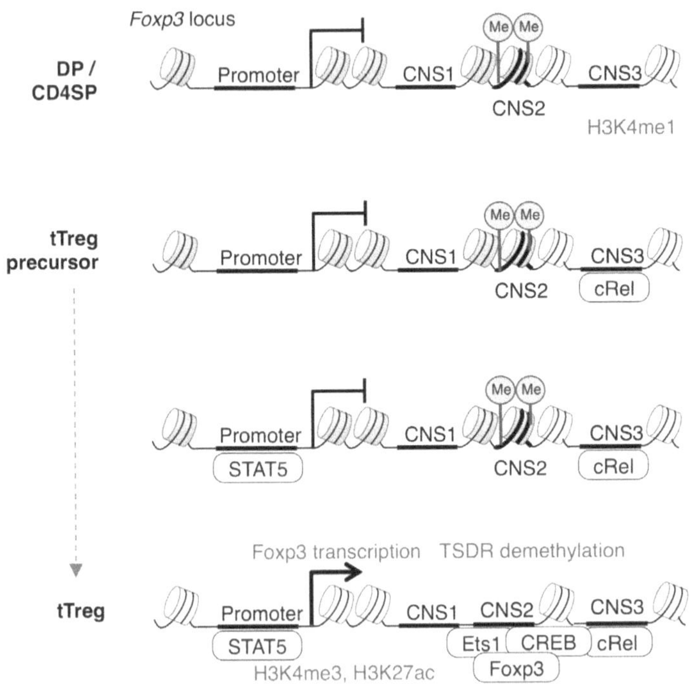

Figure 6 Transcriptional and epigenetic changes at *Foxp3* locus during thymic Treg cell development. In Treg progenitor cells, CNS2 is methylated but CNS3 shows a poised state with H3K4me1 modification. As they receive Treg cell-differentiating stimuli, transcription factors such as cRel binds to CNS3 in tTreg precursor cells. IL-2 signaling and further TCR stimulation then converts precursor cells into tTreg cells by allowing STAT5-mediated transcription activation, DNA demethylation at CNS2, and H3K4me3 and H3K27ac modifications at the promoter.

develop into Treg cells, while commitment into other lineages leads to the loss of this potential. In this sense, a large part of nucleosome positioning is also set in progenitor cells, so that Foxp3 can utilize the pre-established chromatin landscape for gene regulation.[31] It is therefore likely that such epigenetic modifications are initiated by transcription factors that are expressed in progenitors and have the capacity to recruit appropriate enzymes. Alternatively, these enzymes may be guided to specific loci by long non-coding RNAs, whose involvement in lineage determination is becoming clear.[107] Thus, there are many layers of molecular events, which need to accumulate to allow Treg cell development upon appropriate signal transduction (Fig. 6). The precise characterization of each molecular event may help us find new targets for Treg cell manipulation and methods for stable Treg cell generation *in vitro*.

6. CONCLUSION

Given the critical role of Treg cells in the maintenance of immune homeostasis and the potential clinical benefits of being able to convert pathogenic effector T cells into truly stable and functional Treg cells, understanding the mechanisms by which Treg cells develop and are maintained is an issue of vital importance. In the relatively short period of time since the discovery of Treg cells, great leaps in our understanding of these cells have been made; such as the essential role of Foxp3, the interaction of Foxp3 with other transcription factors for Foxp3-dependent transcription regulation, and the epigenetic modifications that ensure stable commitment to the Treg cell lineage. Despite these advances, however, key questions remain unanswered, such as the exact nature of the initial events that control both Foxp3 expression and epigenetic modifications at the Treg precursor stage. It is hoped that the ongoing search for the molecules that determine the Treg cell lineage will provide new therapeutic targets for treatment of various immunological disorders.

ACKNOWLEDGMENT

We thank Hannah Hume for proofreading the chapter.

REFERENCES

1. Nishizuka Y, Sakakura T. Thymus and reproduction: sex-linked dysgenesia of the gonad after neonatal thymectomy in mice. *Science*. 1969;166(3906):753–755.
2. Sakaguchi S, Sakaguchi N, Asano M, Itoh M, Toda M. Immunologic self-tolerance maintained by activated T cells expressing IL-2 receptor alpha-chains (CD25).

Breakdown of a single mechanism of self-tolerance causes various autoimmune diseases. *J Immunol.* 1995;155(3):1151–1164.
3. Fontenot JD, Gavin MA, Rudensky AY. Foxp3 programs the development and function of CD4+CD25+ regulatory T cells. *Nat Immunol.* 2003;4(4):330–336.
4. Hori S, Nomura T, Sakaguchi S. Control of regulatory T cell development by the transcription factor Foxp3. *Science.* 2003;299(5609):1057–1061.
5. Brunkow ME, Jeffery EW, Hjerrild KA, et al. Disruption of a new forkhead/winged-helix protein, scurfin, results in the fatal lymphoproliferative disorder of the scurfy mouse. *Nat Genet.* 2001;27(1):68–73.
6. Bennett CL, Christie J, Ramsdell F, et al. The immune dysregulation, polyendocrinopathy, enteropathy, X-linked syndrome (IPEX) is caused by mutations of FOXP3. *Nat Genet.* 2001;27(1):20–21.
7. Wildin RS, Smyk-Pearson S, Filipovich AH. Clinical and molecular features of the immunodysregulation, polyendocrinopathy, enteropathy, X linked (IPEX) syndrome. *J Med Genet.* 2002;39(8):537–545.
8. Kim JM, Rasmussen JP, Rudensky AY. Regulatory T cells prevent catastrophic autoimmunity throughout the lifespan of mice. *Nat Immunol.* 2007;8(2):191–197.
9. Grant CR, Liberal R, Mieli-Vergani G, Vergani D, Longhi MS. Regulatory T-cells in autoimmune diseases: challenges, controversies and—yet—unanswered questions. *Autoimmun Rev.* 2015;14(2):105–116.
10. Tang Q, Adams JY, Penaranda C, et al. Central role of defective interleukin-2 production in the triggering of islet autoimmune destruction. *Immunity.* 2008;28(5):687–697.
11. Powrie F, Leach MW, Mauze S, Caddle LB, Coffman RL. Phenotypically distinct subsets of CD4+ T cells induce or protect from chronic intestinal inflammation in C. B-17 scid mice. *Int Immunol.* 1993;5(11):1461–1471.
12. Wing JB, Sakaguchi S. Foxp3(+) T(reg) cells in humoral immunity. *Int Immunol.* 2014;26(2):61–69.
13. Nishikawa H, Sakaguchi S. Regulatory T cells in cancer immunotherapy. *Curr Opin Immunol.* 2014;27:1–7.
14. Yamaguchi T, Wing JB, Sakaguchi S. Two modes of immune suppression by Foxp3(+) regulatory T cells under inflammatory or non-inflammatory conditions. *Semin Immunol.* 2011;23(6):424–430.
15. Qureshi OS, Zheng Y, Nakamura K, et al. Trans-endocytosis of CD80 and CD86: a molecular basis for the cell-extrinsic function of CTLA-4. *Science.* 2011;332(6029):600–603.
16. Wing K, Onishi Y, Prieto-Martin P, et al. CTLA-4 control over Foxp3+ regulatory T cell function. *Science.* 2008;322(5899):271–275.
17. Wing JB, Ise W, Kurosaki T, Sakaguchi S. Regulatory T cells control antigen-specific expansion of Tfh cell number and humoral immune responses via the coreceptor CTLA-4. *Immunity.* 2014;41(6):1013–1025.
18. Sage PT, Paterson AM, Lovitch SB, Sharpe AH. The coinhibitory receptor CTLA-4 controls B cell responses by modulating T follicular helper, T follicular regulatory, and T regulatory cells. *Immunity.* 2014;41(6):1026–1039.
19. Kuehn HS, Ouyang W, Lo B, et al. Immune dysregulation in human subjects with heterozygous germline mutations in CTLA4. *Science.* 2014;345(6204):1623–1627.
20. Schubert D, Bode C, Kenefeck R, et al. Autosomal dominant immune dysregulation syndrome in humans with CTLA4 mutations. *Nat Med.* 2014;20(12):1410–1416.
21. Yamaguchi T, Kishi A, Osaki M, et al. Construction of self-recognizing regulatory T cells from conventional T cells by controlling CTLA-4 and IL-2 expression. *Proc Natl Acad Sci USA.* 2013;110(23):E2116–E2125.
22. Wu Y, Borde M, Heissmeyer V, et al. FOXP3 controls regulatory T cell function through cooperation with NFAT. *Cell.* 2006;126(2):375–387.

23. Hill JA, Feuerer M, Tash K, et al. Foxp3 transcription-factor-dependent and -independent regulation of the regulatory T cell transcriptional signature. *Immunity*. 2007;27(5):786–800.
24. Fu W, Ergun A, Lu T, et al. A multiply redundant genetic switch 'locks in' the transcriptional signature of regulatory T cells. *Nat Immunol*. 2012;13(10):972–980.
25. Morikawa H, Ohkura N, Vandenbon A, et al. Differential roles of epigenetic changes and Foxp3 expression in regulatory T cell-specific transcriptional regulation. *Proc Natl Acad Sci USA*. 2014;111(14):5289–5294.
26. Arvey A, van der Veeken J, Samstein RM, Feng Y, Stamatoyannopoulos JA, Rudensky AY. Inflammation-induced repression of chromatin bound by the transcription factor Foxp3 in regulatory T cells. *Nat Immunol*. 2014;15(6):580–587.
27. Marson A, Kretschmer K, Frampton GM, et al. Foxp3 occupancy and regulation of key target genes during T-cell stimulation. *Nature*. 2007;445(7130):931–935.
28. Ono M, Yaguchi H, Ohkura N, et al. Foxp3 controls regulatory T-cell function by interacting with AML1/Runx1. *Nature*. 2007;446(7136):685–689.
29. Beyer M, Thabet Y, Muller RU, et al. Repression of the genome organizer SATB1 in regulatory T cells is required for suppressive function and inhibition of effector differentiation. *Nat Immunol*. 2011;12(9):898–907.
30. Gavin MA, Rasmussen JP, Fontenot JD, et al. Foxp3-dependent programme of regulatory T-cell differentiation. *Nature*. 2007;445(7129):771–775.
31. Samstein RM, Arvey A, Josefowicz SZ, et al. Foxp3 exploits a pre-existent enhancer landscape for regulatory T cell lineage specification. *Cell*. 2012;151(1):153–166.
32. Bettelli E, Dastrange M, Oukka M. Foxp3 interacts with nuclear factor of activated T cells and NF-kappa B to repress cytokine gene expression and effector functions of T helper cells. *Proc Natl Acad Sci USA*. 2005;102(14):5138–5143.
33. Pan F, Yu H, Dang EV, et al. Eos mediates Foxp3-dependent gene silencing in CD4+ regulatory T cells. *Science*. 2009;325(5944):1142–1146.
34. Rudra D, deRoos P, Chaudhry A, et al. Transcription factor Foxp3 and its protein partners form a complex regulatory network. *Nat Immunol*. 2012;13(10):1010–1019.
35. Darce J, Rudra D, Li L, et al. An N-terminal mutation of the Foxp3 transcription factor alleviates arthritis but exacerbates diabetes. *Immunity*. 2012;36(5):731–741.
36. Xiao Y, Nagai Y, Deng G, et al. Dynamic interactions between TIP60 and p300 regulate FOXP3 function through a structural switch defined by a single lysine on TIP60. *Cell Reports*. 2014;7(5):1471–1480.
37. Zheng Y, Chaudhry A, Kas A, et al. Regulatory T-cell suppressor program co-opts transcription factor IRF4 to control T(H)2 responses. *Nature*. 2009;458(7236):351–356.
38. Tone Y, Furuuchi K, Kojima Y, Tykocinski ML, Greene MI, Tone M. Smad3 and NFAT cooperate to induce Foxp3 expression through its enhancer. *Nat Immunol*. 2008;9(2):194–202.
39. Bruno L, Mazzarella L, Hoogenkamp M, et al. Runx proteins regulate Foxp3 expression. *J Exp Med*. 2009;206(11):2329–2337.
40. Vanvalkenburgh J, Albu DI, Bapanpally C, et al. Critical role of Bcl11b in suppressor function of T regulatory cells and prevention of inflammatory bowel disease. *J Exp Med*. 2011;208(10):2069–2081.
41. Levine AG, Arvey A, Jin W, Rudensky AY. Continuous requirement for the TCR in regulatory T cell function. *Nat Immunol*. 2014;15(11):1070–1078.
42. Xiao Y, Li B, Zhou Z, Hancock WW, Zhang H, Greene MI. Histone acetyltransferase mediated regulation of FOXP3 acetylation and Treg function. *Curr Opin Immunol*. 2010;22(5):583–591.
43. van Loosdregt J, Vercoulen Y, Guichelaar T, et al. Regulation of Treg functionality by acetylation-mediated Foxp3 protein stabilization. *Blood*. 2010;115(5):965–974.

44. Beier UH, Wang L, Bhatti TR, et al. Sirtuin-1 targeting promotes Foxp3+ T-regulatory cell function and prolongs allograft survival. *Mol Cell Biol.* 2011;31(5):1022–1029.
45. Nie H, Zheng Y, Li R, et al. Phosphorylation of FOXP3 controls regulatory T cell function and is inhibited by TNF-alpha in rheumatoid arthritis. *Nat Med.* 2013;19(3):322–328.
46. Sugimoto N, Oida T, Hirota K, et al. Foxp3-dependent and -independent molecules specific for CD25+CD4+ natural regulatory T cells revealed by DNA microarray analysis. *Int Immunol.* 2006;18(8):1197–1209.
47. Lin W, Haribhai D, Relland LM, et al. Regulatory T cell development in the absence of functional Foxp3. *Nat Immunol.* 2007;8(4):359–368.
48. Ohkura N, Hamaguchi M, Morikawa H, et al. T cell receptor stimulation-induced epigenetic changes and Foxp3 expression are independent and complementary events required for Treg cell development. *Immunity.* 2012;37(5):785–799.
49. Ouyang W, Liao W, Luo CT, et al. Novel Foxo1-dependent transcriptional programs control T(reg) cell function. *Nature.* 2012;491(7425):554–559.
50. Bock C, Beerman I, Lien WH, et al. DNA methylation dynamics during in vivo differentiation of blood and skin stem cells. *Mol Cell.* 2012;47(4):633–647.
51. Zhou X, Bailey-Bucktrout SL, Jeker LT, et al. Instability of the transcription factor Foxp3 leads to the generation of pathogenic memory T cells in vivo. *Nat Immunol.* 2009;10(9):1000–1007.
52. Bailey-Bucktrout SL, Martinez-Llordella M, Zhou X, et al. Self-antigen-driven activation induces instability of regulatory T cells during an inflammatory autoimmune response. *Immunity.* 2013;39(5):949–962.
53. Rubtsov YP, Niec RE, Josefowicz S, et al. Stability of the regulatory T cell lineage in vivo. *Science.* 2010;329(5999):1667–1671.
54. Miyao T, Floess S, Setoguchi R, et al. Plasticity of Foxp3(+) T cells reflects promiscuous Foxp3 expression in conventional T cells but not reprogramming of regulatory T cells. *Immunity.* 2012;36(2):262–275.
55. Miyara M, Yoshioka Y, Kitoh A, et al. Functional delineation and differentiation dynamics of human CD4+ T cells expressing the FoxP3 transcription factor. *Immunity.* 2009;30(6):899–911.
56. Allan SE, Crome SQ, Crellin NK, et al. Activation-induced FOXP3 in human T effector cells does not suppress proliferation or cytokine production. *Int Immunol.* 2007;19(4):345–354.
57. Chen W, Jin W, Hardegen N, et al. Conversion of peripheral CD4+CD25- naive T cells to CD4+CD25+ regulatory T cells by TGF-beta induction of transcription factor Foxp3. *J Exp Med.* 2003;198(12):1875–1886.
58. Zheng Y, Josefowicz S, Chaudhry A, Peng XP, Forbush K, Rudensky AY. Role of conserved non-coding DNA elements in the Foxp3 gene in regulatory T-cell fate. *Nature.* 2010;463(7282):808–812.
59. Polansky JK, Schreiber L, Thelemann C, et al. Methylation matters: binding of Ets-1 to the demethylated Foxp3 gene contributes to the stabilization of Foxp3 expression in regulatory T cells. *J Mol Med.* 2010;88(10):1029–1040.
60. Kim HP, Leonard WJ. CREB/ATF-dependent T cell receptor-induced FoxP3 gene expression: a role for DNA methylation. *J Exp Med.* 2007;204(7):1543–1551.
61. Rudra D, Egawa T, Chong MM, Treuting P, Littman DR, Rudensky AY. Runx-CBFbeta complexes control expression of the transcription factor Foxp3 in regulatory T cells. *Nat Immunol.* 2009;10(11):1170–1177.
62. Wang Y, Su MA, Wan YY. An essential role of the transcription factor GATA-3 for the function of regulatory T cells. *Immunity.* 2011;35(3):337–348.

63. Li X, Liang Y, LeBlanc M, Benner C, Zheng Y. Function of a Foxp3 cis-element in protecting regulatory T cell identity. *Cell*. 2014;158(4):734–748.
64. Feng Y, Arvey A, Chinen T, van der Veeken J, Gasteiger G, Rudensky AY. Control of the inheritance of regulatory T cell identity by a cis element in the Foxp3 locus. *Cell*. 2014;158(4):749–763.
65. Bannister AJ, Kouzarides T. Regulation of chromatin by histone modifications. *Cell Res*. 2011;21(3):381–395.
66. Wei G, Wei L, Zhu J, et al. Global mapping of H3K4me3 and H3K27me3 reveals specificity and plasticity in lineage fate determination of differentiating CD4+ T cells. *Immunity*. 2009;30(1):155–167.
67. Schmidl C, Klug M, Boeld TJ, et al. Lineage-specific DNA methylation in T cells correlates with histone methylation and enhancer activity. *Genome Res*. 2009;19(7):1165–1174.
68. Sauer S, Bruno L, Hertweck A, et al. T cell receptor signaling controls Foxp3 expression via PI3K, Akt, and mTOR. *Proc Natl Acad Sci USA*. 2008;105(22):7797–7802.
69. Schmidl C, Hansmann L, Lassmann T, et al. The enhancer and promoter landscape of human regulatory and conventional T-cell subpopulations. *Blood*. 2014;123(17):e68–e78.
70. DuPage M, Chopra G, Quiros J, et al. The chromatin-modifying enzyme Ezh2 is critical for the maintenance of regulatory T cell identity after activation. *Immunity*. 2015;42(2):227–238.
71. Jiang C, Pugh BF. Nucleosome positioning and gene regulation: advances through genomics. *Nat Rev Genet*. 2009;10(3):161–172.
72. Yang S, Fujikado N, Kolodin D, Benoist C, Mathis D. Regulatory T cells generated early in life play a distinct role in maintaining self-tolerance. *Science*. 2015;348(6234):589–594.
73. Samstein RM, Josefowicz SZ, Arvey A, Treuting PM, Rudensky AY. Extrathymic generation of regulatory T cells in placental mammals mitigates maternal-fetal conflict. *Cell*. 2012;150(1):29–38.
74. Haribhai D, Williams JB, Jia S, et al. A requisite role for induced regulatory T cells in tolerance based on expanding antigen receptor diversity. *Immunity*. 2011;35(1):109–122.
75. Weiss JM, Bilate AM, Gobert M, et al. Neuropilin 1 is expressed on thymus-derived natural regulatory T cells, but not mucosa-generated induced Foxp3+ T reg cells. *J Exp Med*. 2012;209(10):1723–1742. S1721.
76. Yadav M, Louvet C, Davini D, et al. Neuropilin-1 distinguishes natural and inducible regulatory T cells among regulatory T cell subsets in vivo. *J Exp Med*. 2012;209(10):1713–1722. S1711–1719.
77. Edwards JP, Thornton AM, Shevach EM. Release of active TGF-beta1 from the latent TGF-beta1/GARP complex on T regulatory cells is mediated by integrin beta8. *J Immunol*. 2014;193(6):2843–2849.
78. Lio CW, Hsieh CS. A two-step process for thymic regulatory T cell development. *Immunity*. 2008;28(1):100–111.
79. Hsieh CS, Lee HM, Lio CW. Selection of regulatory T cells in the thymus. *Nat Rev Immunol*. 2012;12(3):157–167.
80. Lio CW, Dodson LF, Deppong CM, Hsieh CS, Green JM. CD28 facilitates the generation of Foxp3(−) cytokine responsive regulatory T cell precursors. *J Immunol*. 2010;184(11):6007–6013.
81. Tai X, Cowan M, Feigenbaum L, Singer A. CD28 costimulation of developing thymocytes induces Foxp3 expression and regulatory T cell differentiation independently of interleukin 2. *Nat Immunol*. 2005;6(2):152–162.
82. Tai X, Erman B, Alag A, et al. Foxp3 transcription factor is proapoptotic and lethal to developing regulatory T cells unless counterbalanced by cytokine survival signals. *Immunity*. 2013;38(6):1116–1128.

83. Curotto de Lafaille MA, Lafaille JJ. Natural and adaptive foxp3+ regulatory T cells: more of the same or a division of labor? *Immunity*. 2009;30(5):626–635.
84. Atarashi K, Tanoue T, Shima T, et al. Induction of colonic regulatory T cells by indigenous Clostridium species. *Science*. 2011;331(6015):337–341.
85. Furusawa Y, Obata Y, Fukuda S, et al. Commensal microbe-derived butyrate induces the differentiation of colonic regulatory T cells. *Nature*. 2013;504(7480):446–450.
86. Arpaia N, Campbell C, Fan X, et al. Metabolites produced by commensal bacteria promote peripheral regulatory T-cell generation. *Nature*. 2013;504(7480):451–455.
87. Maldonado RA, von Andrian UH. How tolerogenic dendritic cells induce regulatory T cells. *Adv Immunol*. 2010;108:111–165.
88. Curotto de Lafaille MA, Lino AC, Kutchukhidze N, Lafaille JJ. CD25- T cells generate CD25+Foxp3+ regulatory T cells by peripheral expansion. *J Immunol*. 2004;173(12): 7259–7268.
89. Ruan Q, Kameswaran V, Tone Y, et al. Development of Foxp3(+) regulatory T cells is driven by the c-Rel enhanceosome. *Immunity*. 2009;31(6):932–940.
90. Long M, Park SG, Strickland I, Hayden MS, Ghosh S. Nuclear factor-kappaB modulates regulatory T cell development by directly regulating expression of Foxp3 transcription factor. *Immunity*. 2009;31(6):921–931.
91. Visekruna A, Huber M, Hellhund A, et al. c-Rel is crucial for the induction of Foxp3 (+) regulatory CD4(+) T cells but not T(H)17 cells. *Eur J Immunol*. 2010;40(3):671–676.
92. Sekiya T, Kashiwagi I, Yoshida R, et al. Nr4a receptors are essential for thymic regulatory T cell development and immune homeostasis. *Nat Immunol*. 2013;14(3): 230–237.
93. Sekiya T, Kashiwagi I, Inoue N, et al. The nuclear orphan receptor Nr4a2 induces Foxp3 and regulates differentiation of CD4+ T cells. *Nat Commun*. 2011;2:269.
94. Fassett MS, Jiang W, D'Alise AM, Mathis D, Benoist C. Nuclear receptor Nr4a1 modulates both regulatory T-cell (Treg) differentiation and clonal deletion. *Proc Natl Acad Sci USA*. 2012;109(10):3891–3896.
95. Moran AE, Holzapfel KL, Xing Y, et al. T cell receptor signal strength in Treg and iNKT cell development demonstrated by a novel fluorescent reporter mouse. *J Exp Med*. 2011;208(6):1279–1289.
96. Ouyang W, Beckett O, Ma Q, Paik JH, DePinho RA, Li MO. Foxo proteins cooperatively control the differentiation of Foxp3+ regulatory T cells. *Nat Immunol*. 2010;11(7):618–627.
97. Burchill MA, Yang J, Vogtenhuber C, Blazar BR, Farrar MA. IL-2 receptor beta-dependent STAT5 activation is required for the development of Foxp3+ regulatory T cells. *J Immunol*. 2007;178(1):280–290.
98. Zorn E, Nelson EA, Mohseni M, et al. IL-2 regulates FOXP3 expression in human CD4+CD25+ regulatory T cells through a STAT-dependent mechanism and induces the expansion of these cells in vivo. *Blood*. 2006;108(5):1571–1579.
99. Mouly E, Chemin K, Nguyen HV, et al. The Ets-1 transcription factor controls the development and function of natural regulatory T cells. *J Exp Med*. 2010;207(10): 2113–2125.
100. Roychoudhuri R, Hirahara K, Mousavi K, et al. BACH2 represses effector programs to stabilize T(reg)-mediated immune homeostasis. *Nature*. 2013;498(7455):506–510.
101. Takimoto T, Wakabayashi Y, Sekiya T, et al. Smad2 and Smad3 are redundantly essential for the TGF-beta-mediated regulation of regulatory T plasticity and Th1 development. *J Immunol*. 2010;185(2):842–855.
102. Schlenner SM, Weigmann B, Ruan Q, Chen Y, von Boehmer H. Smad3 binding to the foxp3 enhancer is dispensable for the development of regulatory T cells with the exception of the gut. *J Exp Med*. 2012;209(9):1529–1535.

103. Liu Y, Zhang P, Li J, Kulkarni AB, Perruche S, Chen W. A critical function for TGF-beta signaling in the development of natural CD4+CD25+Foxp3+ regulatory T cells. *Nat Immunol.* 2008;9(6):632–640.
104. Toker A, Engelbert D, Garg G, et al. Active demethylation of the Foxp3 locus leads to the generation of stable regulatory T cells within the thymus. *J Immunol.* 2013;190(7):3180–3188.
105. Tahiliani M, Koh KP, Shen Y, et al. Conversion of 5-methylcytosine to 5-hydroxymethylcytosine in mammalian DNA by MLL partner TET1. *Science.* 2009;324(5929):930–935.
106. Kohli RM, Zhang Y. TET enzymes, TDG and the dynamics of DNA demethylation. *Nature.* 2013;502(7472):472–479.
107. Lee JT. Epigenetic regulation by long noncoding RNAs. *Science.* 2012;338(6113):1435–1439.

CHAPTER TWO

Microenvironment Matters: Unique Conditions Within Gut-Draining Lymph Nodes Favor Efficient *De Novo* Induction of Regulatory T Cells

Maria Pasztoi[1], Joern Pezoldt[1], Jochen Huehn[1,2]
Department of Experimental Immunology, Helmholtz Centre for Infection Research, Braunschweig, Germany
[2]Corresponding author: e-mail address: jochen.huehn@helmholtz-hzi.de

Contents

Abstract

The gastrointestinal tract constitutes the largest surface of the body and thus has developed multitude mechanisms to either prevent pathogen entry or to efficiently eliminate invading pathogens. At the same time, the gastrointestinal system has to avoid unwanted immune responses against self and harmless nonself antigens, such as nutrients and commensal microbiota. Therefore, it is somewhat not unexpected that the gastrointestinal mucosa serves as the largest repository of immune cells throughout the body, harboring both potent pro- as well as anti-inflammatory properties. One additional key element of this regulatory machinery is created by trillions of symbiotic commensal bacteria in the gut. The microbiota not only simply contribute to the breakdown

[1] All authors contributed equally to this work.

Progress in Molecular Biology and Translational Science, Volume 136
ISSN 1877-1173
http://dx.doi.org/10.1016/bs.pmbts.2015.07.012

of nutrients, but are essential in limiting the expansion of pathogens, directing the development of the intestinal immune system, and establishing mucosal tolerance by fostering the induction of regulatory T cells (Tregs). In this review, we will discuss our current understanding about the microenvironmental factors fostering the *de novo* generation of Tregs within the gastrointestinal immune system, focusing on unique properties of antigen-presenting cells, tolerogenic cytokines, commensal-derived metabolites and the contribution of lymph node stromal cells.

1. INTRODUCTION: KEY TREG CHARACTERISTICS

1.1 The Lineage Specification Factor Foxp3

Tregs are a subset of $CD4^+$ T cells having not only fundamental functions in the maintenance of immune homeostasis and peripheral tolerance but also in the prevention of overwhelming immune responses against invading pathogens.[1] Originally, they were described as $CD4^+$ T cells constitutively expressing CD25, the α-chain of the IL-2 receptor.[2] Later, the transcription factor Forkhead box protein 3 (Foxp3) was identified as the lineage specification factor of Tregs,[3,4] being of utmost importance for the Tregs' suppressive properties through the maintenance of a Treg-specific gene expression signature.[5,6] The importance of Foxp3 is further illustrated by mutations in the *Foxp3* gene, resulting in the development of fatal autoimmune diseases, like the lymphoproliferative scurfy phenotype in mice or the IPEX syndrome (immune dysregulation, polyendocrinopathy, enteropathy, and X-linked inheritance) in humans.[7–9] For the maintenance of their suppressive phenotype, Tregs require permanent expression of Foxp3.[10] Work by others and us has demonstrated that epigenetic mechanisms contribute to the stabilization of Foxp3 expression within the Treg lineage.[11,12] Particularly, the selective demethylation of a CpG-rich conserved noncoding sequence 2 (CNS2) in the *Foxp3* locus, also known as Treg-specific demethylated region (TSDR), contributes to the fixation of the Treg lineage identity.[13–17]

1.2 Thymic-Derived Versus Peripherally Induced Tregs

The vast majority of $Foxp3^+$ Tregs is generated during thymic development (thymus-derived, tTregs) and seems to be selected for recognition of self-antigens.[18] However, in the periphery the Treg repertoire can be complemented by specificities directed against harmless nonself antigens, including commensal microbiota and food antigens, and this expansion of the $Foxp3^+$

Treg pool is achieved by peripheral conversion of conventional $Foxp3^-CD4^+$ T cells into $Foxp3^+$ Tregs (peripherally induced, pTregs).[19–23] Finally, $Foxp3^+$ Tregs can also be generated *in vitro* by stimulating conventional $Foxp3^-CD4^+$ T cells in the presence of tumor growth factor-β (TGF-β),[24] generating so-called *in vitro*-induced Tregs (iTregs).[24]

Despite intense research, reliable marker distinguishing between tTregs and pTregs are missing. None of the canonical Treg markers such as CD25, Foxp3, inducible costimulator (ICOS), CD103, cytotoxic T-lymphocyte antigen-4 (CTLA-4), and glucocorticoid-induced TNFR family-related protein (GITR) is specifically expressed on either tTregs or pTregs. Helios, a member of the Ikaros transcription factor family, had been suggested as a marker for tTregs[25,26]; however, Helios expression is also associated with T-cell activation and differentiation, and under certain conditions can be induced in pTregs.[20,27,28] More recently, the cell surface receptor neuropilin-1 (Nrp1) was reported to identify tTregs, but only under steady-state conditions, precluding its use under infectious or inflammatory settings.[29,30] Finally, tTregs and pTregs can also not be distinguished by their epigenetic signature since both subsets are selectively demethylated at the *Foxp3* TSDR and the other recently identified Treg-specific epigenetic signature genes.[12,14,15,17] Thus, a specific marker for the discrimination of tTregs and pTregs is presently missing.

1.3 Suppressive Mechanisms

An extensive body of evidence underlines that Tregs possess myriads of mediators and mechanisms to suppress numerous types of immune responses. First of all, it is well known that Tregs release several anti-inflammatory cytokines including IL-10, TGF-β, and IL-35.[31] Tregs can also exert their suppressive capacity through direct contact with effector T cells and/or antigen-presenting cells (APCs) via receptor–ligand interactions such as CTLA-4, GITR, Nrp1, programmed death-1 (PD-1), lymphocyte activation gene 3 (LAG-3), fibrinogen-like protein-2 (FGL-2), and the TNFR superfamily member 4 (also known as OX40).[32,33] Moreover, the high CD25 expression levels on Tregs enables them to compete with conventional T cells for IL-2, resulting in consumption of this T-cell growth factor and finally in Bim-mediated apoptosis of the target cells.[34] Another way of suppression includes the release of the anti-inflammatory β-galactoside binding galectin-1,[35] cytolitic molecules such as granzyme B and perforin or cAMP.[36,37] Interestingly, it has recently been shown that Tregs have the ability to release exosomes, thereby delivering immunosuppression due to their microRNA,[38] or ecto-5-nucleotide CD73 content.[39]

Most importantly, accumulating evidence suggests that Tregs are highly responsive to their local environment. The differentiation of polarized tissue-specific Treg subsets endows them with special properties optimally relieving inflammatory responses under which they have been generated.[40–42] Although differences in suppressive mechanisms depending on Foxp3 expression or between tTregs and pTregs have not been studied extensively, the possible variability in the suppressing apparatus might have an essential role while adapting to the unique microenvironment of the gut.

2. PERIPHERAL *DE NOVO* INDUCTION OF FOXP3$^+$ TREGS WITHIN THE GASTROINTESTINAL IMMUNE SYSTEM

Accumulating evidence suggests that gut-draining lymph nodes (LNs), including mesenteric LNs (mLNs) and the liver-draining celiac LN (cLN), represent a site of superior pTreg generation when compared to skin-draining LNs.[23,43–45] Further studies made clear that the generation of pTregs is a stepwise process.[46] The initial step of peripheral tolerance induction includes the CCR7-dependent migration of CD103$^+$ dendritic cells (DCs) from the lamina propria (LP) into the mLN,[47] where they induce Foxp3 and gut-homing molecule expression of naive CD4$^+$ T cells (Fig. 1).[44,48] The establishment of mucosal tolerance requires subsequent homing of pTregs from the mLN into the small intestine, where the pTregs then locally expand. This process depends on IL-10 released from the fractalkine receptor CX3C chemokine receptor 1 (CX3CR1) expressing gut-resident macrophages, since mice deficient in CX3CR1 failed to develop oral tolerance.[49] Therefore, the latent state of tolerance established in mLN is converted into an irreversible tissue-located state of tolerance through the relocation of the Treg pool from the mLN into the intestine where Tregs re-encounter their respective antigen and suppress local immune responses.[46]

Although pTregs were originally thought to be the dominant population responsible for establishment of peripheral tolerance, it is highly probable that the establishment of an effective Treg response requires the mutual interaction of both tTregs and pTregs widening inhibitory tools and the T-cell receptor (TCR) repertoire in the intestine.[20,50–52]

In order to establish peripheral tolerance, the tolerogenic microenvironment of the gut plays a critical role as recently emphasized by a study

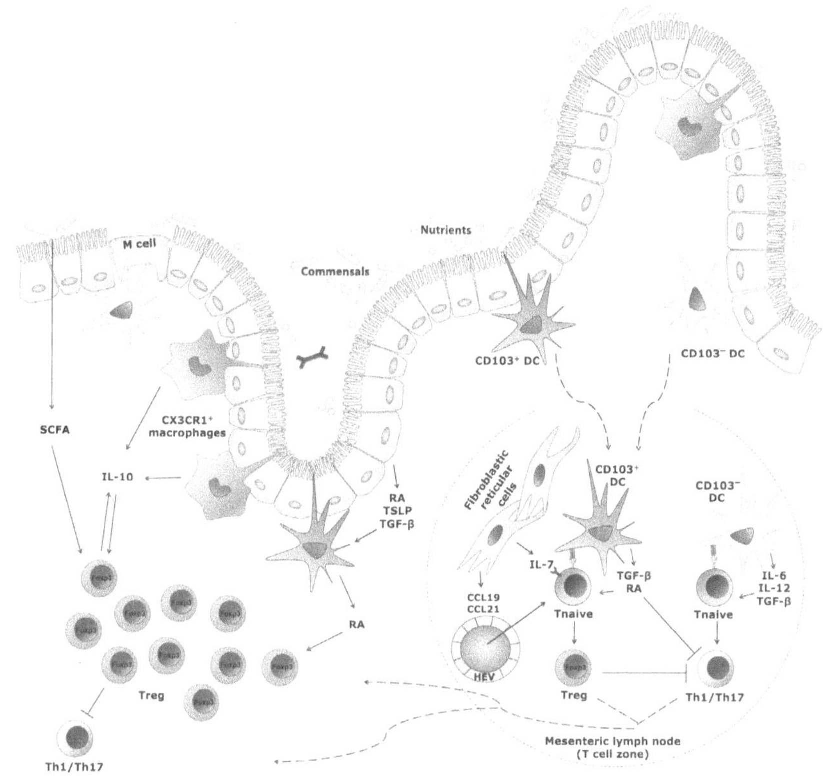

Figure 1 The tolerogenic microenvironment of the gut favors Treg induction and expansion. Peripheral Treg generation in the gut is a two-step process. First, conversion of naive CD4$^+$ T cells into Foxp3$^+$ Tregs is efficiently induced in the mLN by CD103$^+$ DCs migrating from the LP into the mLN and acting in concert with FRCs. Subsequently, Tregs migrate back to the LP where they further expand. This expansion step is strongly supported by CX3CR1$^+$ gut-resident macrophages due to their constant IL-10 production. Within this tolerogenic environment Tregs re-encounter their respective antigens and can efficiently suppress local immune responses. HEV, high endothelial venules; RA, retinoic acid; TGF-β, tumor growth factor-β; TSLP, thymic stromal lymphopoietin; SCFA, short-chain fatty acids. Migration events are indicated by dashed arrows.

investigating steady-state migrating lymph DCs. It was shown that tolerogenic DC subsets *in vitro* induce strong proliferation and proinflammatory cytokine production from naive T cells, instead of exerting anti-inflammatory responses, but still had the ability to stimulate Foxp3$^+$ Tregs.[53] Although these data show that intestinal tolerogenic DCs can stimulate naive T cells in an inflammatory manner, it is very likely that *in vivo* numerous other factors might influence this phenotype, as addition of steady-state lymph to the cultures abrogated DC-induced T-cell proliferation.[53] These data suggest that the lymph contains tolerogenic factors having essential role in connecting LP and its draining LNs and thereby inducing immune tolerance.

Next to lymph-derived molecules, a range of other microenvironmental factors in the intestinal mucosa favor pTreg generation, thereby contributing to the maintenance of intestinal homeostasis and establishment of oral tolerance against food-borne and microbiota-derived antigens. This intricate balance is achieved by an interlocking system of unique APCs, soluble immunoregulatory factors, and further local features of the intestine. Therefore, in this review we first discuss how special properties of intestinal APCs drive pTreg formation. Then, we focus on tolerogenic cytokines released into the intestinal microenvironment. Our review also considers the influence of signals from commensal microbiota on Treg function and epigenetic modulation. Finally, we highlight recent advances with regard to LN stromal cells as key players of tolerogenic immune responses.

2.1 The Role of Antigen-Presenting Cells in pTreg Generation

Antigen-presentation by DCs is critical for efficient pTreg induction. This has been emphasized by studies from others and us demonstrating that depletion of $CD11c^+$ DCs leads to decreased pTreg generation, Foxp3 expression and Treg homeostasis.[23,54] Importantly, the make-up of costimulatory molecules on the DCs, which is influenced by their priming conditions,[55] determines the Treg induction rate. For example, strong costimulation via CD28 inhibits pTreg generation,[56] whereas the coinhibitory molecule CTLA-4 has a stimulatory effect on pTreg induction.[57]

2.1.1 Intestinal Antigen-Presenting Cell Subsets

In the intestine, DCs and macrophages create the major pool of APCs, but their precise characterization still remains a major challenge. The routine phenotyping of conventional DCs includes detection of CD11c and MHCII coexpression, however, most of the macrophages in the periphery also express CD11c and show comparably high MHCII levels.[58] Therefore, further markers are needed for proper discrimination, such as the integrin α_E CD103, which, among myeloid cells, is exclusively expressed by DCs, whereas high expression of CX3CR1 is characteristic for the $F4/80^+CD64^+$ intestinal macrophage subset.[59]

In light of these observations, intestinal DCs can be further subdivided into four subsets based on the expression of CD103 and CD11b. In the

LP and gut-draining LNs a majority of the DCs is $CD103^+$, although they still represent a heterogeneous population. Most of them are $CD103^+CD11b^+$ DCs, but they also include a minor but functionally different $CD103^+CD11b^-CD8\alpha^+$ population.[58,60]

$CD103^+$ DCs have a unique tolerogenic phenotype supported by local factors, such as bile and dietary retinoids in the intestinal lumen, vasoactive intestinal peptide (VIP), and prostaglandin (PG)E2.[58] After migrating from the intestinal mucosa into the mLN,[47] they promote pTreg generation in a TGF-β-, retinoic acid- (RA), indoleamine 2,3-dioxygenase- (IDO), and thymic stromal lymphopoietin (TSLP)-dependent manner (Fig. 1).[43,61–63] Among these factors, RA is produced during vitamin A metabolism by the retinal dehydrogenase (RALDH) enzyme expressed at high levels in $CD103^+$ DCs.[61,64,65] On one hand, RA has a direct effect on pTreg generation by mediating permissive histone acetylation in the *Foxp3* promoter leading to enhanced transcriptional activity.[65] On the other hand, it exerts an indirect effect on Tregs via reducing cytokine production of effector T cells, which have a negative impact on pTreg generation.[66] Finally, RA has a key role in balancing between Tregs and Th17 cells, via potentiating TGF-β-induced pTreg generation and opposing IL-6-driven Th17 differentiation.[67] The involvement of both $CD103^+$ DC populations in establishing intestinal tolerance has been shown recently, as deletion of either $CD11b^+$ or $CD11b^-$ subsets did not affect $Foxp3^+$ Treg homeostasis, only deletion of the whole $CD103^+$ subset resulted in reduced Treg numbers.[68] In contrast to $CD103^+$ DCs, *$CD103^-$ DC subsets* such as $CD103^-CD11b^+CX3CR1^{int}$ and $CD103^-CD11b^-$ DCs have been shown to be mainly involved in the priming of Th1/Th17 cells.[69]

Intestinal macrophages predominantly represent a gut-resident, nonmigratory population as a result of their lack of CCR7 expression. They play an essential role in professional sensing and early combatting microbial invaders based on their transendothelial dendrites.[59,70,71] Under homeostatic settings, they represent an anti-inflammatory subset constitutively producing IL-10, which also helps to maintain their hyporesponsiveness to TLR ligands.[72] Although gut-resident macrophages were initially thought not to be involved in T-cell priming processes, this concept has been changed, especially due to their high MHCII expression levels. Indeed, $CD11b^+F4/80^+CD11c^-CX3CR1^+$ macrophages of the small intestinal LP have Treg-inducing capacity dependent on IL-10, RA, and exogenous TGF-β (Fig. 1) and are able to reverse IL-17-inducing effect of $CD11b^+$ DCs.[73] Since $CX3CR1^+$ mononuclear phagocytes are the main producers of IL-10 in the intestine, it is not surprising that they are also responsible for the maintenance of Foxp3 expression and suppressive capacity of Tregs.[74]

Next to DCs and macrophages, *intestinal epithelial cells* are also implicated in the maintenance of intestinal homeostasis due to the release of soluble factors like TGF-β, RA, and TSLP driving the conversion of DCs into tolerogenic CD103$^+$ subsets (Fig. 1).[75,76]

Under steady-state conditions, DC- or macrophage-like APCs of the intestine and its draining LNs are equipped with unique functions to promote pTreg induction. The characterization of tolerogenic APCs via surface markers such as CD103 and CX3CR1 is challenging in the context of inflammation, where previously "tolerogenic" CD103$^+$ DCs lose their Treg-inducing capacity.[77]

2.2 Tolerogenic Cytokines Involved in pTreg Induction

Various cytokines have been reported to contribute to a tolerogenic environment within the intestine, thereby favoring the induction of pTregs. In this regard, *IL-10* plays a particular role. Initially described as a Th2 cytokine inhibiting Th1 immune responses, IL-10 is now known as a major effector molecule of Tregs mediating the suppression of proinflammatory cytokine and chemokine production of different APCs and their expression of MHCII and costimulatory molecules. Furthermore, IL-10 can act on proinflammatory T helper subsets such as Th17 or Th1 cells and block their proliferation.[31] Upon Treg-specific IL-10 ablation, mice spontaneously develop signs of colitis,[78] illustrating the key role of Treg-derived IL-10 for the maintenance of intestinal tolerance. Additionally, IL-10 produced by CD103$^+$ DCs or CX3CR1$^+$ macrophages also provides a unique microenvironment sustaining Foxp3 expression and Treg stability,[49,74] and inhibiting proliferation of Th17 cells.[79]

The central role of *TGF-β* in immune tolerance is long known.[80] More recently, it has been demonstrated that TGF-β-induced Smad3 can bind to CNS1 in the *Foxp3* locus and induces Foxp3 expression within naive conventional CD4$^+$ T cells.[16,22,81] In the tolerogenic intestinal environment, TGF-β is mainly released from CD103$^+$ DCs and acts synergistically with RA while inducing Tregs.[43,56,61,65,81,82] TGF-β is produced as an inactive protein complex, called latent TGF-β1, which has to undergo further maturation steps. Previous studies described that the integrin $\alpha_v\beta_8$ on DCs is involved in processing and activating latent TGF-β1.[83,84] Very recently, the expression of integrin $\alpha_v\beta_8$ on Tregs was shown to allow the conversion of latent TGF-β and suppression of inflammatory responses.[85]

In addition to IL-10 and TGF-β, *IL-35* has also been recognized as an anti-inflammatory cytokine. It belongs to the IL-12 family and is mainly produced by Tregs and regulatory B cells (Bregs), but can also be released by vascular endothelial cells, smooth muscle cells and monocytes.[31,86] IL-35 induces conventional $CD4^+$ T cells to differentiate into Tregs, which in turn express IL-35 and can reverse the development of autoimmune diseases.[87] More recently, IL-35 was reported to be produced by DCs, hence contribute to their tolerogenic phenotype.[88]

IL-37 is a novel IL-1 family cytokine being produced by hematopoietic cells and acts in an IL-10-dependent manner.[31] It plays an essential role in downregulating inflammatory responses and protects against colitis development.[89] Its importance in tolerance induction has been suggested recently as IL-37 was shown to be a potent inhibitor of adaptive immune responses having an indirect effect on Tregs via inducing tolerogenic properties of DCs.[90] In this study, contact hypersensitivity was diminished upon adoptive transfer of hapten-sensitized DCs from IL-37 transgenic mice via impairment of effector T-cell responses and induction of Tregs.

2.3 Intestinal Commensals and Their Impact on pTreg Differentiation

LNs are subjected to their surrounding local environment including tissue-dependent composition of the commensal microflora. Several studies have focused on how the intestinal flora is influencing the immune system and in turn how the immune system influences the composition of the gut microbiota.[91] Moreover, the composition of commensal microbiota is a decisive factor for the development of intestinal inflammatory disorders, autoimmune diseases, allergy, and cancer,[92–94] underlining the importance of commensal microbiota in shaping different types of immune responses.

2.3.1 Intestinal Commensal Metabolites Shape T Cell Migration and Differentiation

The importance of the interaction between commensals and pTreg generation has been suggested by a study showing that germ free (GF) mice harbor less $Helios^-$ Tregs, especially in the colon.[26] The phenotype could be reversed upon colonization with commensal microbiota,[26] an effect that was already achievable using a cocktail of 17 human-derived *Clostridiales* strains.[95] The robust development of Tregs upon colonization with *Clostridiales* strains is independent of MyD88-signaling, indicating that sensing of bacterial material via TLRs is not required for Treg induction and/or expansion.[26] Instead, the immune-shaping via the *Clostridiales* strains has been attributed to short-chain fatty acids (SCFAs), including acetate, propionate, and butyrate.[96–98] The highest concentration of commensals producing SCFAs is reached in the caecum and colon, where fermenting bacteria of the *Bacteroidetes* phylum break down dietary fibers and produce SCFAs (Fig. 1).[96,99,100] The impact of SCFAs on the immune system highly depends on their origin or type of administration. Feeding with butyrylated high-amylose maize starch led to an increased number of colonic Tregs.[96] However, administration of SCFAs in drinking water seemed to have a systemic impact on immune homeostasis, as feeding of antibiotics-treated mice with propionate or butyrate resulted in an increase in Treg frequencies in the spleen.[97] These data underline that SCFAs can directly impact the immune system, not only locally but also systemically. However, it is still unclear whether SCFAs only induce Treg expansion in the intestinal LP or already impact T-cell differentiation in the draining LNs.

SCFAs can influence T-cell differentiation at different levels, including modulation of intestinal DCs, regulation of Treg migration, and epigenetic imprinting of Tregs in order to fix their tolerogenic phenotype. For example, butyrate binds to the Gpr109a, a high-affinity butyrate and niacin receptor of colonic immune cells. Although intestinal DCs of Gpr109a-deficient mice show elevated levels of IL-17 and IL-6, reduced IL-10, and RALDH1, and concomitantly reduced Foxp3$^+$ Treg frequencies in the colonic LP,[101] it remains unclear whether this is an effect on Tregs in the LP or on the DCs migrating to the LNs. In addition, SCFAs, namely propionate, can directly modulate the homing phenotype of Tregs. This function was dependent on Gpr43, a receptor for SCFAs, as propionate-treated GF and specific pathogen free (SPF) Gpr43-deficient mice were incapable of upregulating chemoattractant-receptor Gpr15,[98] which is required for migration of Tregs to the colonic LP.[102] Finally, several studies have implicated the role of SCFAs in stabilizing the Treg lineage.[103] The most intense SCFA-mediated impact on Treg induction *in vitro* was observed with butyrate.[96] Butyrate has also been shown to inhibit class I and IIa histone deacetylases.[104] While it elevates the permissive histone acetylation within CNS1 and CNS3 of the *Foxp3* locus, thereby favoring increased Foxp3 expression, it does not affect other Th cell lineage specification factors like RORγt, T-bet, and Gata3.[96,97] Whether butyrate promotes the stability of pTregs by contributing to the demethylation of the TSDR is currently unknown.

Although SCFAs have been shown to modulate different aspects of pTreg generation, stabilization, and expansion, there are no concise studies that distinguish the function of commensal-derived metabolites with regard to the priming of the immune response in the gut-draining LNs and expansion of Tregs within the LP. A defined understanding and dissection of both processes and how they are influenced by commensals and their metabolites is required to precisely shape intestinal immune responses.

2.4 Lymph Node Stromal Cells in pTreg Induction

To allow for effective priming of adaptive immune responses, including pTreg induction, naive antigen-specific T cells need to encounter their cognate antigen. The gathering place for DCs and lymphocytes is located in fully functional LNs, which allows for constant provision of antigens and provides the infrastructure for migration and interaction.

Lymphatic endothelial cells (LEC) establish a network of vessels transporting lymph from the tissue to and around the LN. Blood endothelial cells (BEC) comprise the high endothelial venules (HEV) interspersing the LN cortex and serving as an entry gate for circulating lymphocytes into the LN. The cortex of the LN is made up of a dense network of follicular dendritic cells (FDC) and fibroblastic reticular cells (FRCs), dominating in the B cell follicles and T-cell zones, respectively.[105,106] Here, we will merely discuss recent findings that implicate FRCs as modulators of T-cell differentiation including pTreg induction.

2.4.1 FRCs and Their Modulatory Function on T Cell Responses

FRCs have been linked to several immune modulatory functions, ranging from limiting T-cell expansion to regulating peripheral tolerance.[107–111] Several studies have assessed the function of MHCII expression on FRCs and potential antigen-specific modulation of T-cell responses. Although FRCs are not considered as professional APCs, they endogenously express MHCII at low levels under homeostatic conditions and upregulate MHCII expression under inflammatory settings in an IFNγ-dependent manner.[112,113] Utilizing LN transplantation experiments, Baptista *et al.* have shown that endogenous expression of MHCII on LN stromal cells is required for homeostatic maintenance of Tregs, preventing immune activation and ameliorating skin graft rejection in a K14-mOVA model.[107]

Despite the expression of MHCII in FRCs, a recent study has suggested that the majority of MHCII molecules on FRCs are derived from DCs that transfer MHCII complexes in a contact-dependent manner to LECs and FRCs.[112] When activated Ova-specific OTII T cells were pre-cultured with FRCs containing Ova-peptide-MHCII complexes gained from DCs, OTII proliferation was attenuated.[112] Thus, FRCs have the ability to express and present self-antigens, potentially filtered from the conduits or directly processed by DCs from the periphery and interact with cognate T cells in order to attenuate T-cell proliferation and contribute to Treg maintenance.

2.4.2 Lymph Node Infrastructure and Its Implications for pTreg Induction

Tissue-derived APCs and soluble antigens are delivered into the LN via afferent lymphatics, reflecting the immunological state of the drained tissue. This conduit network made up of LECs collects lymph from the tissue and transports the fluid in a unidirectional way to the subcapsular sinus.[114] Within the subcapsular sinus the constituents of the lymph are separated according to their size. Larger particles, such as viruses are captured in the medulla by subcapsular sinus macrophages and translocated toward underlying B cell follicles in the cortex.[115] Smaller particles, with a molecular weight of less than 70 kDa, including chemokines, peptides, cytokines, and small metabolites, enter a conduit system of collagen cores sheathed by FRCs.[116] The sharp separation is enforced by the tight plasmalemma vesicle-associated protein (PLVAP) diaphragm, which is directly connected to the FRC-collagen conduit.[117,118] The content of the conduit network is sampled by DCs adjoined to the FRCs,[118,119] although in which context DCs present the material without being primed in the tissue is unresolved.

As the FRC network resides in close proximity to the content of the lymph, they are subjected to the flood of immune-modulatory signals derived from the drained tissue. Thus, under steady-state conditions, FRCs are extraordinarily well positioned to sample or support sampling of constituents of the lymph and therefore might have an outstanding role in modulating pTreg generation in concert with DCs.

2.4.3 Lymph Node Stromal Cells Retain Tissue-Specific Functional Properties

LNs reside at distinct anatomical localizations and are the starting points to mount new or to reinitiate adaptive immune responses from the pool of circulating naive and memory lymphocytes, respectively. The content of the lymph resembles site-dependent information of the drained tissue and therefore modulates the immune response depending on the tissue type and immunological state.[120] The different tissues are constantly sampled by DCs. Depending on the environmental cues, DCs are primed to promote respective immune responses to maintain tolerance or optimally target pathogens. DCs that reach the LN are retained, but have a short functional life span of several days and are continuously replaced in the LN.[69,121] In contrast to DCs in the LN, stromal cells have a low turn-over under steady-state conditions and thus could provide a constant microenvironmental framework for the LN. Several studies have shown that DCs can be influenced by the environment of the LNs and even change their respective tissue-derived priming,[122,123] indicating that the LNs, including stromal cells, provide a microenvironment that shapes T-cell responses by modulating incoming DCs.

To dissect the major cellular compartments behind the LN-location-specific functions, LN transplantation experiments have been utilized.[124] Surgical resection of endogenous intestinal or skin-draining LN enables the engraftment of a transplanted LN from a distinct site.[125] During engraftment, the hematopoietic compartment is replaced by cells from the draining tissue; however, LN-resident cells, predominantly stromal cells of donor origin, remain.[123] The transplanted LNs retain functional properties, such as MAdCAM-1 expression on HEVs originated from the mLN, whereas pLN stromal cells do not gain MAdCAM-1 expression if transplanted to the mesenteries.[123,125] Furthermore, adoptively transferred naive OTII T cells readily upregulate $\alpha_4\beta_7$ but not CCR9 expression in mLN transplanted into the skin-draining *popliteal fossa* in a DC-dependent manner.[126] Thus, stromal cells not only retain the functional property to express location-specific adhesion molecules but also to modulate T-cell differentiation in a location-specific manner. These findings indicate that in contrast to DCs, stromal cells stably retain their functional properties.

The conservation of functional properties with regard to expression of MAdCAM-1 is only observed if the whole intact LN is transplanted. When single-cell suspensions of stromal cells from skin-draining LNs were transplanted into the mesenteries using a collagen sponge, the newly establishing LN contains HEVs acquiring location-specific properties and expressing MAdCAM-1.[127] These findings indicate that a newly forming LN functionally adapts to the new tissue-draining site and can be influenced by the draining environment. Therefore, it is very likely that this process also takes place during LN development in a tissue-dependent manner.

2.4.4 Stromal Cells Shaped by Their Local Microenvironment Contribute to the Treg-Inducing Properties of Their LN

The tissue-specific adhesion molecules expressed on HEVs, such as MAdCAM-1 in mLN, allows circulating lymphocytes to recirculate back to the tissue and LN, where they were initially primed. It is becoming increasingly evident that each tissue demands and shapes its own adaptive immune response tailored to its environmental stimuli.[120] Thus, it is reasonable to conclude that mesenchymal stromal cells as "scaffold" within LNs are shaped by their respective environment and modulate tissue-specific immune responses. One example is the high expression of RALDH in FRCs from mLN,[123,128,129] which together with high RALDH expression in $CD103^+$ DCs contributes to the high RA levels found in mLN. Hence, FRCs significantly contribute to create a specific environment within LNs.

The role of stromal cells for mucosal tolerance was first shown in an intranasal model of an Ova-induced delayed-type hypersensitivity reaction. In this study, the surgical removal of cervical LNs and their replacement by skin-draining LNs did not allow for tolerance induction, whereas replacement by cervical LNs enabled the maintenance of functional tolerance.[130] Furthermore, removal of mLNs results in an abrogation of oral tolerance,[47] underlining the importance of the mLNs for intestinal tolerance. However, these studies did neither assess the Treg compartment nor pTreg induction.

Therefore, we initiated a study and performed LN transplantation experiments in gut- and skin-draining sites to investigate the role of LN stromal cells to modulate *de novo* Treg induction. Transplantation and engraftment of cLN and mLN into the skin-draining nontolerogenic *popliteal fossa* generated an environment favoring Treg induction from ovalbumin (OVA)-specific, TCR-transgenic (DO11.10) naive $Foxp3^-CD4^+$ T cells upon i.v. injection of OVA peptide. In contrast, transplantation of skin-draining LNs into the gut mesenteries led to reduced Treg induction as compared to transplanted mLN.[23] Together, these data suggest that LNs stably retain their distinct Treg-inducing capacities upon transplantation into different sites.

The gut-draining LNs, such as the cLN and mLN are subject to specific tolerogenic environmental factors. Whereas the liver-draining cLN is potentially subject to high levels of vitamin A,[131] the mLN is exposed to the entity of commensal microbiota. Importantly, if either cLNs from vitamin A-deficient or mLNs from GF mice were transplanted into the *popliteal fossa* a reduced Treg-inducing capacity was observed as compared to cLN from normally fed or mLN from SPF mice, respectively.[23] Thus, gut-draining LNs harbor a stably imprinted high pTreg induction capacity, which is dependent on respective tissue-specific environmental cues such as the availability of vitamin A or presence of commensals during ontogeny. These findings indicate that LN stromal cells have key roles in shaping immune responses, memorizing their tissue-specific imprinting and impacting T-cell differentiation due to their respective environmental cues.[23] As DCs have been shown to be indispensable for pTreg generation[23] and are known to abundantly interact with FRCs,[132] one could conclude that FRCs tweak incoming DCs toward an imprinted homeostatic condition to influence subsequent T-cell differentiation. As FRCs are fixed location-wise and can memorize functional properties in a tissue-specific manner, it is tempting to speculate that they significantly contribute to the shaping of antigen-specific T-cell responses under steady-state conditions, thereby promoting induction of tolerance toward commensals and food-borne antigens.

3. OUTLOOK

pTregs are essential to shape peripheral tolerance and maintain intestinal homeostasis. Besides Foxp3-dependent epigenetic characteristics allowing for Treg lineage commitment and suppressive functions, the intestinal microenvironment plays a central role in pTreg generation to account for food- and commensal-derived antigens. Although it is known that pTreg generation primarily takes places in gut-draining LNs followed by the expansion of Tregs within the LP, the precise mechanisms driving stable tolerogenic lineage commitment need to be dissected to unravel the mediators required for efficient pTreg induction.

In addition to unique DC subsets being present in the LP and gut-draining LNs, residential LN stromal cells are also key contributors in shaping tissue-specific tolerance by fostering the induction of pTreg. The stromal cells' phenotype is strongly influenced and stably imprinted by microenvironmental factors, including vitamin A and commensal-derived metabolites among others. This raises the question, whether during LN ontogeny stromal cells require tissue-specific microenvironmental cues to acquire their location-dependent functional properties.

As LN transplantation experiments have shown, LN stromal cells maintain the expression of tissue-specific homing molecules and respectively influence pTreg induction. The maintenance of pTreg induction in mLN transplanted to skin-draining sites indicates underlying epigenetic mechanisms allowing for stable LN properties. However, during inflammation the LN architecture is dynamically changing, resulting in fast volume expansion without FRC proliferation dependent on the interplay between FRCs and activated DCs.[133,134] The loosening of the dense FRC network potentially ameliorates the impact of the tolerogenic properties of mLN stromal cells during the acute phase of infections. During and after clearance of infection FRCs proliferate and re-establish the former infrastructure.[113] As to whether infection and inflammation can modulate the tolerogenic capacity of mLN stromal cells in the long-run needs to be elucidated.

Taken together, this complex network responsible for intestinal pTreg generation is maintained in the unique microenvironment of the gut in the face of continuously changing antigenic challenges and immunomodulatory signals. Only a precise understanding of the numerous factors and mechanisms responsible for pTreg generation and subsequent expansion will allow us to manipulate the Treg compartment into the desired direction, and efficiently treat allergy, autoimmune disorders and chronic infections in the future.

REFERENCES

1. Smigiel KS, Srivastava S, Stolley JM, Campbell DJ. Regulatory T-cell homeostasis: steady-state maintenance and modulation during inflammation. *Immunol Rev.* 2014;259(1):40–59.
2. Sakaguchi S, Sakaguchi N, Asano M, Itoh M, Toda M. Immunologic self-tolerance maintained by activated T cells expressing IL-2 receptor alpha-chains (CD25). Breakdown of a single mechanism of self-tolerance causes various autoimmune diseases. *J Immunol.* 1995;155(3):1151–1164.
3. Fontenot JD, Gavin MA, Rudensky AY. Foxp3 programs the development and function of $CD4^+CD25^+$ regulatory T cells. *Nat Immunol.* 2003;4(4):330–336.
4. Hori S, Nomura T, Sakaguchi S. Control of regulatory T cell development by the transcription factor Foxp3. *Science.* 2003;299(5609):1057–1061.
5. Hill JA, Feuerer M, Tash K, et al. Foxp3 transcription-factor-dependent and -independent regulation of the regulatory T cell transcriptional signature. *Immunity.* 2007;27(5):786–800.
6. Gavin MA, Rasmussen JP, Fontenot JD, et al. Foxp3-dependent programme of regulatory T-cell differentiation. *Nature.* 2007;445(7129):771–775.
7. Bennett CL, Christie J, Ramsdell F, et al. The immune dysregulation, polyendocrinopathy, enteropathy, X-linked syndrome (IPEX) is caused by mutations of FOXP3. *Nat Genet.* 2001;27(1):20–21.
8. Brunkow ME, Jeffery EW, Hjerrild KA, et al. Disruption of a new forkhead/winged-helix protein, scurfin, results in the fatal lymphoproliferative disorder of the scurfy mouse. *Nat Genet.* 2001;27(1):68–73.
9. Bacchetta R, Passerini L, Gambineri E, et al. Defective regulatory and effector T cell functions in patients with FOXP3 mutations. *J Clin Invest.* 2006;116(6):1713–1722.
10. Williams LM, Rudensky AY. Maintenance of the Foxp3-dependent developmental program in mature regulatory T cells requires continued expression of Foxp3. *Nat Immunol.* 2007;8(3):277–284.
11. Huehn J, Polansky JK, Hamann A. Epigenetic control of FOXP3 expression: the key to a stable regulatory T-cell lineage? *Nat Rev Immunol.* 2009;9(2):83–89.
12. Ohkura N, Hamaguchi M, Morikawa H, et al. T cell receptor stimulation-induced epigenetic changes and Foxp3 expression are independent and complementary events required for Treg cell development. *Immunity.* 2012;37(5):785–799.

13. Kim HP, Leonard WJ. CREB/ATF-dependent T cell receptor-induced FoxP3 gene expression: a role for DNA methylation. *J Exp Med.* 2007;204(7):1543–1551.
14. Floess S, Freyer J, Siewert C, et al. Epigenetic control of the foxp3 locus in regulatory T cells. *PLoS Biol.* 2007;5(2):e38.
15. Polansky JK, Kretschmer K, Freyer J, et al. DNA methylation controls Foxp3 gene expression. *Eur J Immunol.* 2008;38(6):1654–1663.
16. Zheng Y, Josefowicz S, Chaudhry A, Peng XP, Forbush K, Rudensky AY. Role of conserved non-coding DNA elements in the Foxp3 gene in regulatory T-cell fate. *Nature.* 2010;463(7282):808–812.
17. Toker A, Engelbert D, Garg G, et al. Active demethylation of the Foxp3 locus leads to the generation of stable regulatory T cells within the thymus. *J Immunol.* 2013;190(7):3180–3188.
18. Hsieh CS, Lee HM, Lio CW. Selection of regulatory T cells in the thymus. *Nat Rev Immunol.* 2012;12(3):157–167.
19. Curotto de Lafaille MA, Lafaille JJ. Natural and adaptive foxp3$^+$ regulatory T cells: more of the same or a division of labor? *Immunity.* 2009;30(5):626–635.
20. Haribhai D, Williams JB, Jia S, et al. A requisite role for induced regulatory T cells in tolerance based on expanding antigen receptor diversity. *Immunity.* 2011;35(1):109–122.
21. Lathrop SK, Bloom SM, Rao SM, et al. Peripheral education of the immune system by colonic commensal microbiota. *Nature.* 2011;478(7368):250–254.
22. Bilate AM, Lafaille JJ. Induced CD4$^+$Foxp3$^+$ regulatory T cells in immune tolerance. *Annu Rev Immunol.* 2012;30:733–758.
23. Cording S, Wahl B, Kulkarni D, et al. The intestinal micro-environment imprints stromal cells to promote efficient Treg induction in gut-draining lymph nodes. *Mucosal Immunol.* 2014;7(2):359–368.
24. Chen W, Jin W, Hardegen N, et al. Conversion of peripheral CD4$^+$CD25$^-$ naive T cells to CD4$^+$CD25$^+$ regulatory T cells by TGF-beta induction of transcription factor Foxp3. *J Exp Med.* 2003;198(12):1875–1886.
25. Thornton AM, Korty PE, Tran DQ, et al. Expression of Helios, an Ikaros transcription factor family member, differentiates thymic-derived from peripherally induced Foxp3$^+$ T regulatory cells. *J Immunol.* 2010;184(7):3433–3441.
26. Atarashi K, Tanoue T, Shima T, et al. Induction of colonic regulatory T cells by indigenous Clostridium species. *Science.* 2011;331(6015):337–341.
27. Akimova T, Beier UH, Wang L, Levine MH, Hancock WW. Helios expression is a marker of T cell activation and proliferation. *PLoS One.* 2011;6(8):e24226.
28. Serre K, Benezech C, Desanti G, et al. Helios is associated with CD4 T cells differentiating to T helper 2 and follicular helper T cells in vivo independently of Foxp3 expression. *PLoS One.* 2011;6(6):e20731.
29. Weiss JM, Bilate AM, Gobert M, et al. Neuropilin-1 is expressed on thymus-derived natural regulatory T cells, but not mucosa-generated induced Foxp3$^+$ Treg cells. *J Exp Med.* 2012;209(10):1723–1742.
30. Yadav M, Louvet C, Davini D, et al. Neuropilin-1 distinguishes natural and inducible regulatory T cells among regulatory T cell subsets in vivo. *J Exp Med.* 2012;209(10): 1713–1722.
31. Banchereau J, Pascual V, O'Garra A. From IL-2 to IL-37: the expanding spectrum of anti-inflammatory cytokines. *Nat Immunol.* 2012;13(10):925–931.
32. Shevach EM. Mechanisms of foxp3$^+$ T regulatory cell-mediated suppression. *Immunity.* 2009;30(5):636–645.
33. Park HJ, Park JS, Jeong YH, et al. PD-1 upregulated on regulatory T cells during chronic virus infection enhances the suppression of CD8$^+$ T cell immune response via the interaction with PD-L1 expressed on CD8$^+$ T cells. *J Immunol.* 2015;194(12):5801–5811.

34. Pandiyan P, Zheng L, Ishihara S, Reed J, Lenardo MJ. $CD4^+CD25^+Foxp3^+$ regulatory T cells induce cytokine deprivation-mediated apoptosis of effector $CD4^+$ T cells. *Nat Immunol.* 2007;8(12):1353–1362.
35. Garin MI, Chu CC, Golshayan D, Cernuda-Morollon E, Wait R, Lechler RI. Galectin-1: a key effector of regulation mediated by $CD4^+CD25^+$ T cells. *Blood.* 2007;109(5):2058–2065.
36. Cao X, Cai SF, Fehniger TA, et al. Granzyme B and perforin are important for regulatory T cell-mediated suppression of tumor clearance. *Immunity.* 2007;27(4):635–646.
37. Bopp T, Becker C, Klein M, et al. Cyclic adenosine monophosphate is a key component of regulatory T cell-mediated suppression. *J Exp Med.* 2007;204:1303–1310.
38. Okoye IS, Czieso S, Ktistaki E, et al. Transcriptomics identified a critical role for Th2 cell-intrinsic miR-155 in mediating allergy and antihelminth immunity. *Proc Natl Acad Sci USA.* 2014;111(30):E3081–E3090.
39. Smyth LA, Ratnasothy K, Tsang JY, et al. CD73 expression on extracellular vesicles derived from $CD4^+$ $CD25^+$ $Foxp3^+$ T cells contributes to their regulatory function. *Eur J Immunol.* 2013;43(9):2430–2440.
40. Cretney E, Kallies A, Nutt SL. Differentiation and function of $Foxp3^+$ effector regulatory T cells. *Trends Immunol.* 2013;34(2):74–80.
41. Delacher M, Schreiber L, Richards DM, Farah C, Feuerer M, Huehn J. Transcriptional control of regulatory T cells. *Curr Top Microbiol Immunol.* 2014;381:83–124.
42. Huehn J, Beyer M. Epigenetic and transcriptional control of Foxp3 regulatory T cells. *Semin Immunol.* 2015;27(1):10–18.
43. Sun CM, Hall JA, Blank RB, et al. Small intestine lamina propria dendritic cells promote de novo generation of Foxp3 Treg cells via retinoic acid. *J Exp Med.* 2007;204(8):1775–1785.
44. Siewert C, Menning A, Dudda J, et al. Induction of organ-selective $CD4^+$ regulatory T cell homing. *Eur J Immunol.* 2007;37(4):978–989.
45. Feuerer M, Hill JA, Kretschmer K, von Boehmer H, Mathis D, Benoist C. Genomic definition of multiple ex vivo regulatory T cell subphenotypes. *Proc Natl Acad Sci USA.* 2010;107(13):5919–5924.
46. Pabst O, Bernhardt G. On the road to tolerance—generation and migration of gut regulatory T cells. *Eur J Immunol.* 2013;43(6):1422–1425.
47. Worbs T, Bode U, Yan S, et al. Oral tolerance originates in the intestinal immune system and relies on antigen carriage by dendritic cells. *J Exp Med.* 2006;203(3):519–527.
48. Agace WW. Tissue-tropic effector T cells: generation and targeting opportunities. *Nat Rev Immunol.* 2006;6(9):682–692.
49. Hadis U, Wahl B, Schulz O, et al. Intestinal tolerance requires gut homing and expansion of $FoxP3^+$ regulatory T cells in the lamina propria. *Immunity.* 2011;34(2):237–246.
50. Cebula A, Seweryn M, Rempala GA, et al. Thymus-derived regulatory T cells contribute to tolerance to commensal microbiota. *Nature.* 2013;497(7448):258–262.
51. Paiva RS, Lino AC, Bergman ML, et al. Recent thymic emigrants are the preferential precursors of regulatory T cells differentiated in the periphery. *Proc Natl Acad Sci USA.* 2013;110(16):6494–6499.
52. Huang YJ, Haist V, Baumgartner W, et al. Induced and thymus-derived $Foxp3^+$ regulatory T cells share a common niche. *Eur J Immunol.* 2014;44(2):460–468.
53. Milling SW, Jenkins CD, Yrlid U, et al. Steady-state migrating intestinal dendritic cells induce potent inflammatory responses in naive $CD4^+$ T cells. *Mucosal Immunol.* 2009;2(2):156–165.
54. Darrasse-Jeze G, Deroubaix S, Mouquet H, et al. Feedback control of regulatory T cell homeostasis by dendritic cells in vivo. *J Exp Med.* 2009;206(9):1853–1862.
55. Scott CL, Aumeunier AM, Mowat AM. Intestinal $CD103^+$ dendritic cells: master regulators of tolerance? *Trends Immunol.* 2011;32(9):412–419.

56. Benson MJ, Pino-Lagos K, Rosemblatt M, Noelle RJ. All-trans retinoic acid mediates enhanced T reg cell growth, differentiation, and gut homing in the face of high levels of co-stimulation. *J Exp Med.* 2007;204(8):1765–1774.
57. Zheng SG, Wang JH, Stohl W, Kim KS, Gray JD, Horwitz DA. TGF-beta requires CTLA-4 early after T cell activation to induce FoxP3 and generate adaptive $CD4^+CD25^+$ regulatory cells. *J Immunol.* 2006;176(6):3321–3329.
58. Cerovic V, Bain CC, Mowat AM, Milling SW. Intestinal macrophages and dendritic cells: what's the difference? *Trends Immunol.* 2014;35(6):270–277.
59. Schulz O, Jaensson E, Persson EK, et al. Intestinal $CD103^+$, but not $CX3CR1^+$, antigen sampling cells migrate in lymph and serve classical dendritic cell functions. *J Exp Med.* 2009;206(13):3101–3114.
60. Fujimoto K, Karuppuchamy T, Takemura N, et al. A new subset of $CD103^+CD8alpha^+$ dendritic cells in the small intestine expresses TLR3, TLR7, and TLR9 and induces Th1 response and CTL activity. *J Immunol.* 2011;186(11):6287–6295.
61. Coombes JL, Siddiqui KR, Arancibia-Carcamo CV, et al. A functionally specialized population of mucosal $CD103^+$ DCs induces $Foxp3^+$ regulatory T cells via a TGF-beta and retinoic acid-dependent mechanism. *J Exp Med.* 2007;204(8):1757–1764.
62. Matteoli G, Mazzini E, Iliev ID, et al. Gut $CD103^+$ dendritic cells express indoleamine 2,3-dioxygenase which influences T regulatory/T effector cell balance and oral tolerance induction. *Gut.* 2010;59(5):595–604.
63. Spadoni I, Iliev ID, Rossi G, Rescigno M. Dendritic cells produce TSLP that limits the differentiation of Th17 cells, fosters Treg development, and protects against colitis. *Mucosal Immunol.* 2012;5(2):184–193.
64. Raverdeau M, Mills KH. Modulation of T cell and innate immune responses by retinoic Acid. *J Immunol.* 2014;192(7):2953–2958.
65. Kang SG, Lim HW, Andrisani OM, Broxmeyer HE, Kim CH. Vitamin A metabolites induce gut-homing $foxP3^+$ regulatory T cells. *J Immunol.* 2007;179(6):3724–3733.
66. Hill JA, Hall JA, Sun CM, et al. Retinoic acid enhances Foxp3 induction indirectly by relieving inhibition from $CD4^+CD44^{hi}$ Cells. *Immunity.* 2008;29(5):758–770.
67. Mucida D, Park Y, Kim G, et al. Reciprocal TH17 and regulatory T cell differentiation mediated by retinoic acid. *Science.* 2007;317(5835):256–260.
68. Welty NE, Staley C, Ghilardi N, Sadowsky MJ, Igyarto BZ, Kaplan DH. Intestinal lamina propria dendritic cells maintain T cell homeostasis but do not affect commensalism. *J Exp Med.* 2013;210(10):2011–2024.
69. Cerovic V, Houston SA, Scott CL, et al. Intestinal $CD103^-$ dendritic cells migrate in lymph and prime effector T cells. *Mucosal Immunol.* 2013;6(1):104–113.
70. Farache J, Zigmond E, Shakhar G, Jung S. Contributions of dendritic cells and macrophages to intestinal homeostasis and immune defense. *Immunol Cell Biol.* 2013;91:232–239.
71. Aychek T, Mildner A, Yona S, et al. IL-23-mediated mononuclear phagocyte crosstalk protects mice from Citrobacter rodentium-induced colon immunopathology. *Nat Commun.* 2015;6:6525.
72. Hirotani T, Lee PY, Kuwata H, et al. The nuclear IkappaB protein IkappaBNS selectively inhibits lipopolysaccharide-induced IL-6 production in macrophages of the colonic lamina propria. *J Immunol.* 2005;174(6):3650–3657.
73. Denning TL, Wang YC, Patel SR, Williams IR, Pulendran B. Lamina propria macrophages and dendritic cells differentially induce regulatory and interleukin 17-producing T cell responses. *Nat Immunol.* 2007;8(10):1086–1094.
74. Murai M, Turovskaya O, Kim G, et al. Interleukin 10 acts on regulatory T cells to maintain expression of the transcription factor Foxp3 and suppressive function in mice with colitis. *Nat Immunol.* 2009;10(11):1178–1184.
75. Iliev ID, Spadoni I, Mileti E, et al. Human intestinal epithelial cells promote the differentiation of tolerogenic dendritic cells. *Gut.* 2009;58(11):1481–1489.

76. Bimczok D, Kao JY, Zhang M, et al. Human gastric epithelial cells contribute to gastric immune regulation by providing retinoic acid to dendritic cells. *Mucosal Immunol.* 2015;8(3):533–544.
77. Laffont S, Siddiqui KR, Powrie F. Intestinal inflammation abrogates the tolerogenic properties of MLN CD103^{+} dendritic cells. *Eur J Immunol.* 2010;40(7):1877–1883.
78. Rubtsov YP, Rasmussen JP, Chi EY, et al. Regulatory T cell-derived interleukin-10 limits inflammation at environmental interfaces. *Immunity.* 2008;28(4):546–558.
79. Chaudhry A, Samstein RM, Treuting P, et al. Interleukin-10 signaling in regulatory T cells is required for suppression of Th17 cell-mediated inflammation. *Immunity.* 2011;34(4):566–578.
80. Shull MM, Ormsby I, Kier AB, et al. Targeted disruption of the mouse transforming growth factor-beta 1 gene results in multifocal inflammatory disease. *Nature.* 1992;359(6397):693–699.
81. Wang J, Huizinga TW, Toes RE. De novo generation and enhanced suppression of human CD4^{+}CD25^{+} regulatory T cells by retinoic acid. *J Immunol.* 2009;183(6): 4119–4126.
82. Mucida D, Pino-Lagos K, Kim G, et al. Retinoic acid can directly promote TGF-beta-mediated Foxp3^{+} Treg cell conversion of naive T cells. *Immunity.* 2009;30(4):471–472. author reply 472–473.
83. Travis MA, Reizis B, Melton AC, et al. Loss of integrin alpha(v)beta8 on dendritic cells causes autoimmunity and colitis in mice. *Nature.* 2007;449(7160):361–365.
84. Worthington JJ, Czajkowska BI, Melton AC, Travis MA. Intestinal dendritic cells activate transforming growth factor-beta and induce foxp3^{+} T regulatory cells via integrin alphavbeta8. *Gastroenterology.* 2011;141(5):1802–1812.
85. Worthington JJ, Kelly A, Smedley C, et al. Integrin alphavbeta8-mediated TGF-beta activation by effector regulatory T cells is essential for suppression of T-cell-mediated inflammation. *Immunity.* 2015;42(5):903–915.
86. Li X, Mai J, Virtue A, et al. IL-35 is a novel responsive anti-inflammatory cytokine—a new system of categorizing anti-inflammatory cytokines. *PLoS One.* 2012;7(3):e33628.
87. Collison LW, Chaturvedi V, Henderson AL, et al. IL-35-mediated induction of a potent regulatory T cell population. *Nat Immunol.* 2010;11(12):1093–1101.
88. Dixon KO, van der Kooij SW, Vignali DA, van Kooten C. Human tolerogenic dendritic cells produce IL-35 in the absence of other IL-12 family members. *Eur J Immunol.* 2015;45(6):1736–1747.
89. McNamee EN, Masterson JC, Jedlicka P, et al. Interleukin 37 expression protects mice from colitis. *Proc Natl Acad Sci USA.* 2011;108(40):16711–16716.
90. Luo Y, Cai X, Liu S, et al. Suppression of antigen-specific adaptive immunity by IL-37 via induction of tolerogenic dendritic cells. *Proc Natl Acad Sci USA.* 2014;111(42): 15178–15183.
91. Brestoff JR, Artis D. Commensal bacteria at the interface of host metabolism and the immune system. *Nat Immunol.* 2013;14(7):676–684.
92. Round JL, Mazmanian SK. The gut microbiota shapes intestinal immune responses during health and disease. *Nat Rev Immunol.* 2009;9(5):313–323.
93. Kosiewicz MM, Zirnheld AL, Alard P. Gut microbiota, immunity, and disease: a complex relationship. *Front Microbiol.* 2011;2:180.
94. Belkaid Y, Hand TW. Role of the microbiota in immunity and inflammation. *Cell.* 2014;157(1):121–141.
95. Atarashi K, Tanoue T, Oshima K, et al. Treg induction by a rationally selected mixture of Clostridia strains from the human microbiota. *Nature.* 2013;500(7461):232–236.
96. Furusawa Y, Obata Y, Fukuda S, et al. Commensal microbe-derived butyrate induces the differentiation of colonic regulatory T cells. *Nature.* 2013;504(7480):446–450.

97. Arpaia N, Campbell C, Fan X, et al. Metabolites produced by commensal bacteria promote peripheral regulatory T-cell generation. *Nature*. 2013;504(7480):451–455.
98. Smith PM, Howitt MR, Panikov N, et al. The microbial metabolites, short-chain fatty acids regulate colonic Treg cell homeostasis. *Science*. 2013;341:569–573.
99. Kuhn KA, Stappenbeck TS. Peripheral education of the immune system by the colonic microbiota. *Semin Immunol*. 2013;25(5):364–369.
100. Mowat AM, Agace WW. Regional specialization within the intestinal immune system. *Nat Rev Immunol*. 2014;14(10):667–685.
101. Singh N, Gurav A, Sivaprakasam S, et al. Activation of Gpr109a, receptor for niacin and the commensal metabolite butyrate, suppresses colonic inflammation and carcinogenesis. *Immunity*. 2014;40(1):128–139.
102. Kim SV, Xiang WV, Kwak C, et al. GPR15-mediated homing controls immune homeostasis in the large intestine mucosa. *Science*. 2013;340:1456–1459.
103. Kamada N, Seo SU, Chen GY, Nunez G. Role of the gut microbiota in immunity and inflammatory disease. *Nat Rev Immunol*. 2013;13(5):321–335.
104. Licciardi PV, Ververis K, Karagiannis TC. Histone deacetylase inhibition and dietary short-chain fatty acids. *ISRN Allergy*. 2011;2011:869647.
105. Bajenoff M. Stromal cells control soluble material and cellular transport in lymph nodes. *Front Immunol*. 2012;3:304.
106. Fletcher AL, Acton SE, Knoblich K. Lymph node fibroblastic reticular cells in health and disease. *Nat Rev Immunol*. 2015;15:350–361.
107. Baptista AP, Roozendaal R, Reijmers RM, et al. Lymph node stromal cells constrain immunity via MHC class II self-antigen presentation. *eLife*. 2014;3:e04433.
108. Fletcher AL, Lukacs-Kornek V, Reynoso ED, et al. Lymph node fibroblastic reticular cells directly present peripheral tissue antigen under steady-state and inflammatory conditions. *J Exp Med*. 2010;207(4):689–697.
109. Lukacs-Kornek V, Malhotra D, Fletcher AL, et al. Regulated release of nitric oxide by nonhematopoietic stroma controls expansion of the activated T cell pool in lymph nodes. *Nat Immunol*. 2011;12(11):1096–1104.
110. Siegert S, Huang HY, Yang CY, et al. Fibroblastic reticular cells from lymph nodes attenuate T cell expansion by producing nitric oxide. *PLoS One*. 2011;6(11):e27618.
111. Khan O, Headley M, Gerard A, Wei W, Liu L, Krummel MF. Regulation of T cell priming by lymphoid stroma. *PLoS One*. 2011;6(11):e26138.
112. Dubrot J, Duraes FV, Potin L, et al. Lymph node stromal cells acquire peptide-MHCII complexes from dendritic cells and induce antigen-specific $CD4^+$ T cell tolerance. *J Exp Med*. 2014;211(6):1153–1166.
113. Abe J, Shichino S, Ueha S, et al. Lymph node stromal cells negatively regulate antigen-specific $CD4^+$ T cell responses. *J Immunol*. 2014;193(4):1636–1644.
114. Forster R, Braun A, Worbs T. Lymph node homing of T cells and dendritic cells via afferent lymphatics. *Trends Immunol*. 2012;33(6):271–280.
115. Junt T, Moseman EA, Iannacone M, et al. Subcapsular sinus macrophages in lymph nodes clear lymph-borne viruses and present them to antiviral B cells. *Nature*. 2007;450(7166):110–114.
116. Gretz JE, Norbury CC, Anderson AO, Proudfoot AE, Shaw S. Lymph-borne chemokines and other low molecular weight molecules reach high endothelial venules via specialized conduits while a functional barrier limits access to the lymphocyte microenvironments in lymph node cortex. *J Exp Med*. 2000;192(10):1425–1440.
117. Rantakari P, Auvinen K, Jappinen N, et al. The endothelial protein PLVAP in lymphatics controls the entry of lymphocytes and antigens into lymph nodes. *Nat Immunol* 2015;16(4):386–396.

118. Sixt M, Kanazawa N, Selg M, et al. The conduit system transports soluble antigens from the afferent lymph to resident dendritic cells in the T cell area of the lymph node. *Immunity*. 2005;22(1):19–29.
119. Roozendaal R, Mempel TR, Pitcher LA, et al. Conduits mediate transport of low-molecular-weight antigen to lymph node follicles. *Immunity*. 2009;30(2):264–276.
120. Matzinger P, Kamala T. Tissue-based class control: the other side of tolerance. *Nat Rev Immunol*. 2011;11(3):221–230.
121. Kamath AT, Henri S, Battye F, Tough DF, Shortman K. Developmental kinetics and lifespan of dendritic cells in mouse lymphoid organs. *Blood*. 2002;100(5):1734–1741.
122. Dudda JC, Lembo A, Bachtanian E, et al. Dendritic cells govern induction and reprogramming of polarized tissue-selective homing receptor patterns of T cells: important roles for soluble factors and tissue microenvironments. *Eur J Immunol*. 2005;35(4):1056–1065.
123. Hammerschmidt SI, Ahrendt M, Bode U, et al. Stromal mesenteric lymph node cells are essential for the generation of gut-homing T cells in vivo. *J Exp Med*. 2008;205(11):2483–2490.
124. Mebius RE, Breve J, Kraal G, Streeter PR. Developmental regulation of vascular addressin expression: a possible role for site-associated environments. *Int Immunol*. 1993;5(5):443–449.
125. Ahrendt M, Hammerschmidt SI, Pabst O, Pabst R, Bode U. Stromal cells confer lymph node-specific properties by shaping a unique microenvironment influencing local immune responses. *J Immunol*. 2008;181(3):1898–1907.
126. Molenaar R, Greuter M, van der Marel AP, et al. Lymph node stromal cells support dendritic cell-induced gut-homing of T cells. *J Immunol*. 2009;183(10):6395–6402.
127. Buettner M, Dittrich-Breiholz O, Falk CS, et al. Stromal cells as trend-setters for cells migrating into the lymph node. *Mucosal Immunol*. 2015;8(3):640–649.
128. Malhotra D, Fletcher AL, Astarita J, et al. Transcriptional profiling of stroma from inflamed and resting lymph nodes defines immunological hallmarks. *Nat Immunol*. 2012;13(5):499–510.
129. Molenaar R, Knippenberg M, Goverse G, et al. Expression of retinaldehyde dehydrogenase enzymes in mucosal dendritic cells and gut-draining lymph node stromal cells is controlled by dietary vitamin A. *J Immunol*. 2011;186(4):1934–1942.
130. Wolvers DA, Coenen-de Roo CJ, Mebius RE, et al. Intranasally induced immunological tolerance is determined by characteristics of the draining lymph nodes: studies with OVA and human cartilage gp-39. *J Immunol*. 1999;162(4):1994–1998.
131. Winau F, Quack C, Darmoise A, Kaufmann SH. Starring stellate cells in liver immunology. *Curr Opin Immunol*. 2008;20(1):68–74.
132. Acton SE, Astarita JL, Malhotra D, et al. Podoplanin-rich stromal networks induce dendritic cell motility via activation of the C-type lectin receptor CLEC-2. *Immunity*. 2012;37:276–289.
133. Acton SE, Farrugia AJ, Astarita JL, et al. Dendritic cells control fibroblastic reticular network tension and lymph node expansion. *Nature*. 2014;514(7523):498–502.
134. Astarita JL, Cremasco V, Fu J, et al. The CLEC-2-podoplanin axis controls the contractility of fibroblastic reticular cells and lymph node microarchitecture. *Nat Immunol*. 2015;16(1):75–84.

CHAPTER THREE

Understanding the Roles of the NF-κB Pathway in Regulatory T Cell Development, Differentiation and Function

Thomas S. Fulford, Darcy Ellis, Steve Gerondakis[1]
Infection and Immunity Program, Monash Biomedicine Discovery Institute and Department of Biochemistry and Molecular Biology, Monash University, Clayton, Victoria, Australia
[1]Corresponding author: e-mail address: steven.gerondakis@monash.edu

Contents

Abstract

The nuclear factor of kappaB (NF-κB) pathway has emerged as an important regulator of gene expression in $CD4^+$ regulatory T cells. Here, we review various aspects of the roles NF-κB signaling plays in the development and function of Tregs.

1. INTRODUCTION

$Foxp3^+CD4^+$ regulatory T cells (Tregs) serve an essential role controlling peripheral tolerance as well as lymphocyte homeostasis during the course of normal immune responses. Most Tregs develop in the thymus as a distinct T cell lineage (referred to as thymic or tTregs), but also emerge in the periphery (pTregs), particularly at locations of inflammation, when conventional $CD4^+$ T cells upregulate Foxp3 expression. The phenotypic and functional heterogeneity of Tregs is determined by the combined impact of the different immune signals these cells encounter during development and differentiation. Among the signal transduction pathways that coordinate gene expression in Tregs is the NF-κB pathway, a signaling network that has a prominent role in the development, differentiation, and

Progress in Molecular Biology and Translational Science, Volume 136
ISSN 1877-1173
http://dx.doi.org/10.1016/bs.pmbts.2015.08.002

function of many types of immune cells.[1] An outline of the biochemical features of the NF-κB pathway and its regulation has been described in detail elsewhere.[1–5] In brief, the transcriptional mediators of the pathway comprise a family of homodimers and heterodimers composed of the five related proteins, p65 (RelA), c-Rel, RelB, p50 (NF-κB1), and p52 (NF-κB2). Each of these NF-κB proteins has a conserved N-terminal Rel homology domain that encompasses sequences required for dimer formation, binding to specific DNA sequences (κB elements), nuclear import, and interaction with the inhibitor of kappa B (IκB) family of regulatory molecules.[6] Different NF-κB dimers have distinct transcriptional regulatory properties. RelA, c-Rel, and RelB each possess C-terminal transcriptional transactivation domains (TAD) that confer dimers with an intrinsic capacity to promote transcription.[7] The mature forms of NF-κB1 (p50) and NF-κB2 (p52), both of which are derived from the N-terminal halves of larger precursor proteins (p105 and p100, respectively), lack a TAD and instead regulate transcription in a context-dependent manner. p50/p50 and p52/p52 homodimers can repress transcription either by preventing transcriptionally active NF-κB heterodimers from occupying κB sites or by recruiting histone deacetylases.[8] Alternatively, these homodimers can promote gene expression through interactions with transcriptional coactivators.[9–11]

In most cells, NF-κB dimers typically reside within the cytosol in a dormant state, retained in this configuration by IκB proteins that mask the nuclear localization motif and DNA-binding domains of NF-κB dimers. NF-κB transcription factors are mobilized to the nucleus by a diverse range of stimuli that engage either the classical (also known as canonical) or alternate (noncanonical) arms of the pathway. The canonical arm activates dimers of RelA, c-Rel, and p50, whereas the nuclear translocation of p52–RelB dimers requires processing of p100–RelB heterodimers by the noncanonical pathway. Different signal transduction networks engaged by extracellular signals all converge on an IκB kinase (IKK) complex comprising distinct kinase subunits, IKKα (IKK1) and IKKβ (IKK2) and the regulatory protein, NEMO. Antigens, Toll-like receptor ligands or inflammatory cytokines such as tumor necrosis factor (TNF) are among the many immune signals that activate the IKK complex by triggering phosphorylation of IKKβ, which in turn phosphorylates IκB proteins associated with NF-κB dimers. This initiates IκB polyubiquitination and degradation that culminates in NF-κB nuclear translocation.[3] The noncanonical pathway engaged by ligands for TNF receptor superfamily members such as lymphotoxinβ receptor and CD40 activates the IKK complex through

NF-κB-inducing kinase phosphorylation of IKKα. Activated IKKα in turn generates p52/p52 homodimers and p52/RelB heterodimers by triggering the proteolytic processing of p100 bound to p52 and RelB, respectively, in response to IKKα phosphorylation of p100. NF-κB dimers activated by both pathways regulate transcription by recognizing and binding to consensus κB elements located in the regulatory regions of target genes.

2. NF-κB AND TREG DEVELOPMENT

Both the canonical and noncanonical pathways contribute to tTreg development.[12] The impact of the canonical pathway is mediated primarily via Treg intrinsic mechanisms,[13] whereas the role of the noncanonical pathway appears to be indirect through its involvement in regulating the development and maintenance of thymic medullary architecture.[14,15] During tTreg development, T cell receptor (TCR) and CD28 costimulatory signals are the main activators of the canonical pathway.[13,16–18] In the case of TCR-dependent signals, an intracellular network very similar to that employed to activate canonical NF-κB in mature conventional T cells appears to be utilized for tTregs. This is in stark contrast to the seemingly dispensable role these signaling intermediates play in NF-κB activation during conventional T cell development.[19] The activation of NF-κB in thymocytes destined to become tTregs begins with TCR ligation by peptide–MHC complexes triggering a phosphorylation cascade involving PI3K and PKCθ. Accompanying CD80 or CD86-dependent CD28 costimulatory signals that engage the tyrosine kinases Zap70 and Fyn also contribute to the optimal phosphorylation-dependent activation of PKCθ. Activated PKCθ in turn phosphorylates CARMA1, initiating the assembly and recruitment to the membrane of the CARMA1–Bcl10–Malt1 (CMB) complex that promotes the TRAF2–TRAF6-dependent generation of Lys63-linked polyubiquitin chains needed for the activation of TAK1, the kinase that directly phosphorylates and activates IKKβ. This signaling network and how it engages the NF-κB pathway is depicted schematically in Fig. 1.

Among the different NF-κB proteins, it is c-Rel that has the most important role in tTreg development, with the tTreg population reduced to ~15% of normal levels in *c-rel*$^{-/-}$ mice.[20–22] Yet the impact an absence of c-Rel has on tTreg development is still less profound than a loss of IKKβ function, indicating that other canonical NF-κB transcription factors must also contribute to tTreg development. With *nfkb1*$^{-/-}$ mice displaying normal tTreg numbers,[20,22] RelA emerges as the most likely candidate, a conclusion

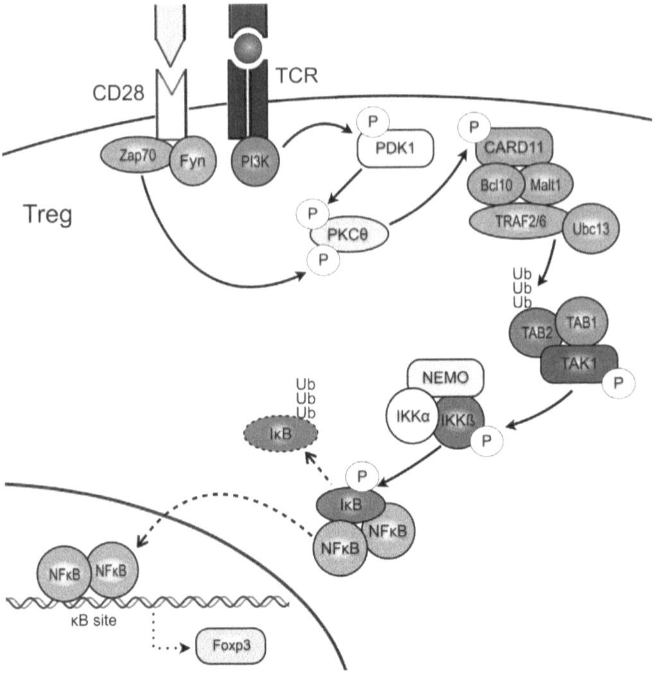

Figure 1 The NF-κB signaling pathway is activated by TCR engagement. Peptide–MHC binding to the TCR and CD80 or CD86 binding CD28 engage and activate a kinase signaling cascades involving Zap70, Fyn, PI(3)K, PDK1, and PKCθ. Phosphorylated PKCθ assembles a signaling network involving the CMB complex, TRAF (TRAF2 and TRAF6), Ubc13, TAB (TAB1 and TAB2), and TAK1 proteins, which activates the canonical NF-κB pathway via the IKKβ-dependent phosphorylation and consequential degradation of IκB. Subsequently, NF-κB dimers are released and translocate to the nucleus where they alter the expression of target genes, e.g. Foxp3, by recognizing and binding to consensus κB sites in the promoter. *Adapted from figure first published in Nature Immunology.*[12]

supported by reduced numbers of thymic Tregs (~30–50% of normal levels) in radiation chimeras reconstituted with *rela*$^{-/-}$ hemopoietic stem cells.[20,23] NF-κB involvement is required at two critical junctures in tTreg development. The first, which occurs during negative selection, involves $CD4^{+}CD8^{+}$ cells that express TCRs with an intermediate affinity for self-antigens differentiating into $CD25^{hi}GITR^{hi}Foxp3^{-}CD4^{+}$ thymocytes, a population highly enriched for tTreg precursors.[24] This is followed by the constitutive induction of Foxp3 expression, a second NF-κB regulated step dependent on the common γ-chain cytokines IL-2 and IL-15. The stages during which the different components in the NF-κB signaling pathway impact tTreg development are summarized in Fig. 2. With c-Rel serving

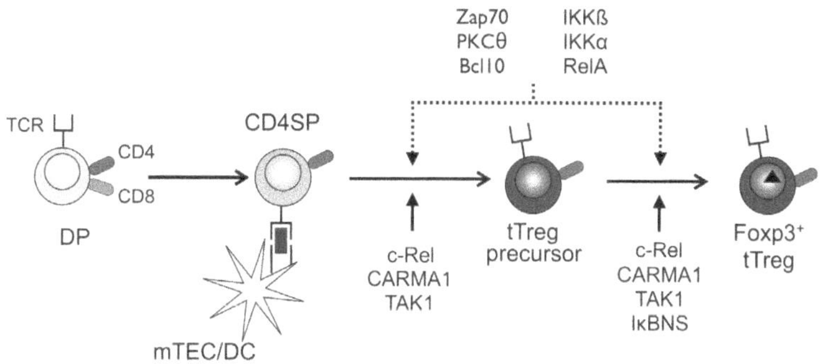

Figure 2 The two-step development of Tregs depends on NF-κB signaling. TCR signals of intermediate strength needed for tTreg cell development require CARMA1 and TAK1 signals and the nuclear localization of c-Rel, to promote tTreg precursor ($CD25^{hi}GITR^{hi}Foxp3^{-}CD4^{+}$) survival and differentiation. Subsequent tTreg cell maturation involving the cytokine (IL-2 or IL-15) induction of Foxp3 is also dependent on CARMA1 and TAK1 signaling, and c-Rel activation. The atypical nuclear IκB, IκBNS, also regulates NF-κB-dependent transcription by directly binding c-Rel dimers required for cytokine-mediated induction of Foxp3. Dotted line indicates NF-κB transcription factors and signaling proteins that play a role in the thymic development of Tregs, but have not been shown to act specifically at one step or the other of the developmental process.

as the key NF-κB protein during tTreg development, its expression pattern and activity not surprisingly is intimately linked with tTreg differentiation. c-Rel is absent in $CD4^{+}CD8^{+}$ thymocytes, but following positive selection is expressed in a subset of $CCR7^{+}CD4^{+}$ cells enriched for pre-tTregs, with constitutive c-Rel expression maintained thereafter in both thymic and peripheral tTregs. A reduction in *c-rel*$^{-/-}$ mice of both $CCR7^{+}$ thymocytes and pre-tTregs reinforces this association.[25,26] The induction of c-Rel expression in $CCR7^{+}CD4^{+}$ thymocytes subjected to Bim-dependent negative selection offers key insights into how c-Rel appears to control the commitment of thymocytes to the tTreg lineage. This $CCR7^{+}$ population, which has a pattern of gene expression reminiscent of a mature activated T cell,[26] includes a prominent NF-κB-regulated gene signature. With a dependence of tTreg development on TCR signals stronger than those compatible with the survival of conventional $CD4^{+}$ T cells during negative, it appears that the prosurvival role of c-Rel involving the induction of the Bcl2-like proteins A1 and Bcl-xL also serves to counter Bim-dependent negative selection in those $CCR7^{+}CD4^{+}$ cells earmarked for tTreg development. Whether the induction of c-Rel in $CCR7^{+}CD4^{+}$ thymocytes is employed exclusively to prevent the death of thymocytes destined to become tTreg precursors remains unclear, particularly given these

$CCR7^+$ cells also express NF-κB target genes that have no known prosurvival roles. Instead, a model whereby NF-κB signaling in $CCR7^+CD4^+$ thymocytes is geared toward helping promote a transcriptional program that mimics the phenotype of an activated $CD4^+$ T cell, a hallmark of Tregs, is better able to encompasses both survival and nonsurvival roles for c-Rel. With this "activated" phenotype established in thymocytes committed to the tTreg lineage prior to the induction of Foxp3,[24] the dilemma of how to stably maintain this transcriptional landscape created by NF-κB and other transcription factors in tTreg cell precursors is solved by the subsequent induction of Foxp3, the sustained expression of which is critical in stabilizing the epigenetic landscape of Tregs.

Following the development of tTreg precursors, NF-κB subsequently contributes to the cytokine-dependent (IL-2 or IL-15) induction of Foxp3 expression. c-Rel, most likely in the form of a p50/c-Rel heterodimer, bind to a conserved noncoding sequence (CNS3) within *foxp3* intron 1 that is required for the developmental induction of Foxp3 expression.[17,27,28] The importance of c-Rel interaction with CNS3 in promoting the induction of Foxp3 is supported by the observation that the reduction in tTreg numbers in mice with a targeted deletion of CNS3 is similar to that seen in *c-rel*$^{-/-}$ mice.[27] Binding of c-Rel to CNS3 is thought to generate a transcriptional landscape that allows additional transcription factors needed to establish and maintain stable Foxp3 expression to be recruited to the locus. Precisely, how this is achieved remains to be determined. It could involve c-Rel-promoting epigenetic changes within the *foxp3* locusor by c-Rel initiating the assembly of a transcription factor complex that precedes the cytokine-dependent induction of Foxp3 expression. Support for the idea that c-Rel has a primary role in the induction of Foxp3 expression by inducing epigenetic changes comes from the observation that c-Rel binds to the CpG island within CNS3 while it is still methylated,[17] a scenario that differs from other transcription factors involved in controlling Foxp3 expression such as ATF and CREB that only occupy binding sites in a demethylated CpG island.[29] Alternatively, c-Rel promotes Foxp3 transcription as part of a *foxp3*-specific enhancer complex, with c-Rel orchestrating the formation of this "enhancesome" by recruiting RelA, NFAT, Smad, and CREB.[21] Regardless of the precise mechanism by which c-Rel operates, with neither IL-2 nor IL-15 able to activate NF-κB, c-Rel appears to be active in tTreg precursors in order to orchestrate the cytokine-dependent induction of Foxp3. This notion is consistent with IκBNS, an atypical nuclear IκB that directly regulates NF-κB-dependent transcription by binding to these transcription factors being required for the cytokine-mediated induction of Foxp3.[28] Following the induction of Foxp3 expression in tTregs, a role for c-Rel in maintaining the stable transcription of *foxp3* appears to be redundant. Whether other NF-κB transcription factors such as RelA contribute to the postdevelopmental control of Foxp3 expression remains to be determined.

3. NF-κB INVOLVEMENT IN TREG FUNCTION

Notwithstanding the importance of NF-κB activity in tTreg development, surprisingly little is known about the postdevelopmental functions of the NF-κB pathway in Tregs. Certainly, the various manifestations of autoimmune disease that develop in mice with Treg-specific knockouts of different signaling molecules involved in IKK/NF-κB activation and regulation such as CARD11 and TRAF6,[30,31] or downstream NF-κB targets including IRF4, SOCS1, and Blimp,[32–34] strongly indicate an important role for NF-κB in mature Treg effector function or differentiation. However, the strongest evidence that NF-κB is indeed required by peripheral Tregs arises from the phenotype of mice with a Treg-specific inactivation of Ubc13, an E2 ubiquitin ligase responsible for the K63 polyubiquitin-dependent activation of IKKβ. The conditional inactivation of Ubc13 in Tregs ($Ube2n^{Treg\text{-}KO}$ mice) leads to the development of multifocal autoimmune disease characterized by markedly increased spleen and peripheral lymph node cellularity, plus excessive lymphocytic infiltrates in many nonlymphoid organs.[35] Ubc13 appears to play a pivotal role in maintaining the "*in vivo*" immunosuppressive function of Treg cells by preventing the conversion of Tregs into TH1 or TH17-like effector T cells. Although Ubc13 regulates the function of a number of proteins in addition to IKKβ, mice with a Treg-specific inactivation of IKKβ ($Ikk2^{Treg\text{-}KO}$ mice) were also found to develop an autoimmune phenotype that closely resembles their $Ube2n^{Treg\text{-}KO}$ counterparts. For example, IKKβ-deficient Tregs had an unstable phenotype and were prone to acquire TH17- and TH1-like effector functions "*in vivo*" that led to higher than normal numbers of memory and effector-like T cells in peripheral lymphoid organs that produced IL-17 and IFN-γ. Importantly, it was the capacity of a constitutively active (CA) IKKβ transgene to restore the suppressive function of Tregs in $Ube2n^{Treg\text{-}KO}$ mice that confirmed the phenotype in the $Ube2n^{Treg\text{-}KO}$ mice was largely, if not completely due to an IKKβ defect. Despite NF-κB being involved in controlling the developmental induction of Foxp3 expression, the Treg defects in Ube2n$^{Treg\text{-}KO}$ mice were not due reduced Foxp3 levels; instead, they coincided with the impaired expression of SOCS1 and IL-10, both of which are known NF-κB-regulated target genes.[35,36] Instability of the Treg phenotype in Ube2n$^{Treg\text{-}KO}$ mice could be overcome by enforced SOCS1 expression, which blocked Ubc13-deficient Treg cells being converted into Th17-like cells "*in vitro*" and into inflammatory effector T cells "*in vivo*." The IL-10 defect in Ubc13-deficient Treg cells was also largely overcome by expression of the IKKβ(CA) transgene, which was also able to restore SOCS1 expression in Ubc13-deficient Treg cells.

Precisely which NF-κB transcription factors are important for these and other Treg functions remains to be determined and is a task that is likely to be complicated by the unique and overlapping roles of these transcription factors. To date, most of the information about specific NF-κB protein function and Tregs has emerged from the analysis of mice with a non-conditional knockout of c-Rel. Despite *c-rel*$^{-/-}$ mice having only ~15% of normal Treg numbers and c-Rel being required for peripheral Treg homeostatic expansion, these mice do not develop lymphoproliferative or autoimmune disease.[20] While the remaining Tregs in *c-rel*$^{-/-}$ mice have a normal phenotype and are able to effectively inhibit T cell activity "*in vivo*,"[20] the most likely explanation for the absence of disease is the presence of additional immune defects that prevent the activation of autoreactive T cells that would normally occur when tTreg cell numbers are low. On the other hand, aged *nfkb1*$^{-/-}$ mice are prone to autoimmune disease, which appears to coincide with systemic low-grade chronic inflammation that afflicts this strain.[37] However, unlike the rapid autoimmune disease that develops in both Ube2n$^{Treg\text{-}KO}$ and Ikk2$^{Treg\text{-}KO}$ mice, the autoimmune disease seen in *nfkb1*$^{-/-}$ mice takes much longer to evolve. Whether Tregs play a role in the disease seen in *nfkb1*$^{-/-}$ mice remains unclear given *nfkb1*$^{-/-}$ Tregs are able to efficiently suppress T cell activity in culture (S. Gerondakis, unpublished results). The question of which NF-κB factors do what in Tregs is further complicated by the difficulty of determining the role of RelA in Tregs due to the embryonic lethality of *rela*$^{-/-}$ mice at E14.5.[38,39] While the generation of fetal liver-derived hemopoietic stem cell chimeras has shed some light on the role of RelA in Tregs, this technical limitation plus the confounding issues of overlapping NF-κB activities await the individual and combined conditional targeting of these transcription factors in Tregs.

4. CONCLUSION

Much remains to be learnt about the detailed role NF-κB signaling plays in the Treg lineage. Outstanding questions include determining what role the pathway plays in controlling the differentiation of naïve tTregs into effector Tregs, ascertaining if and how NF-κB signaling is important in pTreg differentiation or function, and understanding how particular NF-κB transcription factors control gene expression in response to specific immune signals through modification of the epigenetic landscape or regulating miRNA activity.

REFERENCES

1. Hayden MS, Ghosh S. Signaling to NF-κB. *Genes Dev.* 2004;18(18):2195–2224. http://dx.doi.org/10.1101/gad.1228704.
2. Ghosh S, Karin M. Missing pieces in the NF-κB puzzle. *Cell.* 2002;109(Suppl):S81–S96.
3. Gilmore TD. Introduction to NF-κB: players, pathways, perspectives. *Oncogene.* 2006;25(51):6680–6684. http://dx.doi.org/10.1038/sj.onc.1209954.
4. Hayden MS, Ghosh S. Shared principles in NF-κB signaling. *Cell.* 2008;132(3): 344–362. http://dx.doi.org/10.1016/j.cell.2008.01.020.
5. Paul S, Schaefer BC. A new look at T cell receptor signaling to nuclear factor-κB. *Trends Immunol.* 2013;34(6):269–281. http://dx.doi.org/10.1016/j.it.2013.02.002.
6. Ghosh S, May MJ, Kopp EB. NF-κB and Rel proteins: evolutionarily conserved mediators of immune responses. *Annu Rev Immunol.* 1998;16:225–260. http://dx.doi.org/10.1146/annurev.immunol.16.1.225.
7. Chen L-F, Greene WC. Shaping the nuclear action of NF-κB. *Nat Rev Mol Cell Biol.* 2004;5(5):392–401. http://dx.doi.org/10.1038/nrm1368.
8. Zhong H, May MJ, Jimi E, Ghosh S. The phosphorylation status of nuclear NF-κB determines its association with CBP/p300 or HDAC-1. *Mol Cell.* 2002;9(3): 625–636.
9. Kurland JF, Kodym R, Story MD, Spurgers KB, McDonnell TJ, Meyn RE. NF-κB1 (p50) homodimers contribute to transcription of the bcl-2 oncogene. *J Biol Chem.* 2001;276(48):45380–45386. http://dx.doi.org/10.1074/jbc.M108294200.
10. Cao S, Zhang X, Edwards JP, Mosser DM. NF-κB1 (p50) homodimers differentially regulate pro- and anti-inflammatory cytokines in macrophages. *J Biol Chem.* 2006;281(36):26041–26050. http://dx.doi.org/10.1074/jbc.M602222200.
11. Oeckinghaus A, Ghosh S. The NF-κB family of transcription factors and its regulation. *Cold Spring Harb Perspect Biol.* 2009;1(4):a000034. http://dx.doi.org/10.1101/cshperspect.a000034.
12. Gerondakis S, Fulford TS, Messina NL, Grumont RJ. NF-κB control of T cell development. *Nat Immunol.* 2014;15(1):15–25. http://dx.doi.org/10.1038/ni.2785.
13. Gückel E, Frey S, Zaiss MM, Schett G, Ghosh S, Voll RE. Cell-intrinsic NF-κB activation is critical for the development of natural regulatory T cells in mice. In: Kleinschnitz C, ed. *PLoS One.* 2011;6(5):e20003. http://dx.doi.org/10.1371/journal.pone.0020003.g004.

14. Zhu M, Fu Y. The complicated role of NF-κB. *Nature*. 2010;7(2):89–93. http://dx.doi.org/10.1038/cmi.2009.112.
15. Cowan JE, Parnell SM, Nakamura K, et al. The thymic medulla is required for Foxp3+ regulatory but not conventional CD4+ thymocyte development. *J Exp Med*. 2013;210(4):675–681.
16. Tai X, Cowan M, Feigenbaum L, Singer A. CD28 costimulation of developing thymocytes induces Foxp3 expression and regulatory T cell differentiation independently of interleukin 2. *Nat Immunol*. 2005;6(2):152–162. http://dx.doi.org/10.1038/ni1160.
17. Long M, Park S-G, Strickland I, Hayden MS, Ghosh S. Nuclear factor-κB modulates regulatory T cell development by directly regulating expression of Foxp3 transcription factor. *Immunity*. 2009;31(6):921–931. http://dx.doi.org/10.1016/j.immuni.2009.09.022.
18. Vang KB, Yang J, Pagán AJ, et al. Cutting edge: CD28 and c-Rel-dependent pathways initiate regulatory T cell development. *J Immunol*. 2010;184(8):4074–4077. http://dx.doi.org/10.4049/jimmunol.0903933.
19. Gerondakis S, Siebenlist U. Roles of the NF-κB pathway in lymphocyte development and function. *Cold Spring Harb Perspect Biol*. 2010;2(5):a000182. http://dx.doi.org/10.1101/cshperspect.a000182.
20. Isomura I, Palmer S, Grumont RJ, et al. c-Rel is required for the development of thymic Foxp3+ CD4 regulatory T cells. *J Exp Med*. 2009;206(13):3001–3014. http://dx.doi.org/10.1084/jem.20091411.
21. Ruan Q, Kameswaran V, Tone Y, et al. Development of Foxp3(+) regulatory t cells is driven by the c-Rel enhanceosome. *Immunity*. 2009;31(6):932–940. http://dx.doi.org/10.1016/j.immuni.2009.10.006.
22. Deenick EK, Elford AR, Pellegrini M, Hall H, Mak TW, Ohashi PS. c-Rel but not NF-κB1 is important for T regulatory cell development. *Eur J Immunol*. 2010;40(3):677–681. http://dx.doi.org/10.1002/eji.201040298.
23. Fukazawa T, Hiraiwa N, Umemura T, Mise-Omata S, Obata Y, Doi T. Egress of mature murine regulatory T cells from the thymus requires RelA. *J Immunol*. 2015;194(7):3020–3028. http://dx.doi.org/10.4049/jimmunol.1302756.
24. Lio C-WJ, Hsieh C-S. A two-step process for thymic regulatory T cell development. *Immunity*. 2008;28(1):100–111. http://dx.doi.org/10.1016/j.immuni.2007.11.021.
25. Grigoriadis G, Vasanthakumar A, Banerjee A, et al. c-Rel controls multiple discrete steps in the thymic development of Foxp3(+) CD4 regulatory T Cells. *PLoS One*. 2011;6(10):e26851. http://dx.doi.org/10.1371/journal.pone.0026851.
26. Daley SR, Hu DY, Goodnow CC. Helios marks strongly autoreactive CD4+ T cells in two major waves of thymic deletion distinguished by induction of PD-1 or NF-κB. *J Exp Med*. 2013;210(2):269–285. http://dx.doi.org/10.1084/jem.20121458.
27. Zheng Y, Josefowicz S, Chaudhry A, Peng XP, Forbush K, Rudensky AY. Role of conserved non-coding DNA elements in the Foxp3 gene in regulatory T-cell fate. *Nature*. 2010;463(7282):808–812. http://dx.doi.org/10.1038/nature08750.
28. Schuster M, Glauben R, Plaza-Sirvent C, et al. IκBNS protein mediates regulatory T cell development via induction of the Foxp3 transcription factor. *Immunity*. 2012;37(6):998–1008. http://dx.doi.org/10.1016/j.immuni.2012.08.023.
29. Kim HP, Leonard WJ. CREB/ATF-dependent T cell receptor-induced FoxP3 gene expression: a role for DNA methylation. *J Exp Med*. 2007;204(7):1543. http://dx.doi.org/10.1084/jem.20070109.
30. Altin JA, Tian L, Liston A, Bertram EM, Goodnow CC, Cook MC. Decreased T-cell receptor signaling through CARD11 differentially compromises forkhead box protein 3-positive regulatory versus T_H2 effector cells to cause allergy. *J Allergy Clin Immunol*. 2011;127(5). http://dx.doi.org/10.1016/j.jaci.2010.12.1081. 1277–1285.e5.

31. Muto G, Kotani H, Kondo T, et al. TRAF6 is essential for maintenance of regulatory T cells that suppress Th2 type autoimmunity. Unutmaz D, ed. *PLoS One*. 2013;8(9): e74639. http://dx.doi.org/10.1371/journal.pone.0074639.g006.
32. Zheng Y, Chaudhry A, Kas A, et al. Regulatory T-cell suppressor program co-opts transcription factor IRF4 to control T_H2 responses. *Nature*. 2009;458(7236):351–356. http://dx.doi.org/10.1038/nature07674.
33. Cretney E, Xin A, Shi W, et al. The transcription factors Blimp-1 and IRF4 jointly control the differentiation and function of effector regulatory T cells. *Nat Immunol*. 2011;12(4):304–311. http://dx.doi.org/10.1038/ni.2006.
34. Takahashi R, Nishimoto S, Muto G, et al. SOCS1 is essential for regulatory T cell functions by preventing loss of Foxp3 expression as well as IFN-γ and IL-17A production. *J Exp Med*. 2011;208(10):2055–2067.
35. Chang J-H, Xiao Y, Hu H, et al. Ubc13 maintains the suppressive function of regulatory T cells and prevents their conversion into effector-like T cells. *Nat Immunol*. 2012;13(5):481–490. http://dx.doi.org/10.1038/ni.2267.
36. Xu LG, Shu HB. TNFR-associated factor-3 is associated with BAFF-R and negatively regulates BAFF-R-mediated NF-κB activation and IL-10 production. *J Immunol*. 2002;169(12):6883–6889. http://dx.doi.org/10.4049/jimmunol.169.12.6883.
37. Jurk D, Wilson C, Passos JF, et al. Chronic inflammation induces telomere dysfunction and accelerates ageing in mice. *Nat Commun*. 2014;2:4172. http://dx.doi.org/10.1038/ncomms5172.
38. Beg AA, Sha WC, Bronson RT, Ghosh S, Baltimore D. Embryonic lethality and liver degeneration in mice lacking the RelA component of NF-κB. *Nature*. 1995;376(6536):167–170. http://dx.doi.org/10.1038/376167a0.
39. Doi TS, Takahashi T, Taguchi O, Azuma T, Obata Y. NF-κB RelA-deficient lymphocytes: normal development of T cells and B cells, impaired production of IgA and IgG1 and reduced proliferative responses. *J Exp Med*. 1997;185(5):953–961.

CHAPTER FOUR

The Molecular Control of Regulatory T Cell Induction

Annemarie van Nieuwenhuijze[*,†,1], **Adrian Liston**[*,†]
[*]Translational Immunology Laboratory, VIB, Leuven, Belgium
[†]Department of Microbiology and Immunology, University of Leuven, Leuven, Belgium
[1]Corresponding author: e-mail address: Annemarie.VanNieuwenhuijze@vib-kuleuven.be

Contents

Abstract

Regulatory T cells (Tregs) are characterized by the expression of the master transcription factor forkhead box P3 (Foxp3). Although Foxp3 expression is widely used as a marker of the Treg lineage, recent data show that the Treg fate is determined by a multifactorial signaling pathway, involving cytokines, nuclear factors, and epigenetic modifications. Foxp3 expression and the Treg phenotype can be acquired by T cells in the periphery, illustrating that the Treg fate is not necessarily conferred during thymic development. The two main Treg populations *in vivo*, thymic Tregs and peripheral Tregs, differ in the pathways followed for their maturation. This chapter discusses the molecular control of Treg induction, in the thymus as well as the periphery.

Progress in Molecular Biology and Translational Science, Volume 136
ISSN 1877-1173
http://dx.doi.org/10.1016/bs.pmbts.2015.09.001

1. INTRODUCTION

Regulatory T cells (Tregs) are a subset of T lymphocytes, characterized by the expression of the master transcription factor forkhead box P3 (Foxp3), the expression of the interleukin (IL)-2 receptor α chain (IL-2Rα, or CD25) and the ability to suppress other T cell subsets.[1] The expression of Foxp3 is essential for the suppressor function of Tregs, required to maintain self-tolerance.[2–5] Indeed, in the absence of Foxp3, such as in mice bearing the *Scurfy* mutation,[6] fatal autoimmunity arises. In humans, mutations in Foxp3 were found to cause the immune dysregulation, polyendocrinopathy, enteropathy, X-linked syndrome (IPEX).[7] In addition, forced expression of Foxp3 in conventional T cells (Tconv) induces suppressive capacities in these cells, illustrating that the expression of Foxp3 activates a specific differentiation program.[2,4]

In humans and rodents, two distinct populations of Tregs coexist *in vivo*: natural Tregs or thymic Tregs (tTregs), and peripheral Tregs (pTregs). These populations differ in the pathway followed for differentiation, and together make up for about 10% of the peripheral T cell pool.[8] While tTregs are defined as those Tregs that commit to the Treg fate in the thymus, pTregs exit the thymus as conventional single-positive (SP) CD4+ T cells and differentiate into Tregs in the periphery. As such, the difference between tTregs and pTregs includes the developmental timing, location, and causative stimuli of the Treg inductive event. This chapter discusses the molecular factors involved in the induction of the Treg fate *in vivo*.

2. TREG DIFFERENTIATION IN THE THYMUS

2.1 T Cell receptor Signaling and Downstream Mediators

The fate of T cell precursors in the thymus is determined by the strength and duration of T cell receptor (TCR) signaling. Where Tconv are negatively selected upon recognition of peptides in the context of major histocompatibility (MHC) molecules, a self-peptide-specific TCR combined with the appropriate costimulatory signals and peptide avidity favors tTreg differentiation.[9] The absolute requirement for self-reactivity was shown by taking advantage of TCR transgenic mice.[10–12] These studies showed that Tregs can only differentiate from TCR transgenic precursors in the thymus if the cognate antigen is expressed and presented as well. Despite the criticism that these studies received with regards to the nonphysiological expression of a TCR at the double-negative (DN) stage of thymic T cell development (the TCR is normally not expressed until the DP stage),[13] the basis was laid for the current understanding of Treg differentiation and selection in the thymus.

Two models have been proposed for the developmental pathway of a mature Foxp3+ tTreg: the instructive model and the selective model.[14] Both models find their basis in the fact that the TCR is able to distinguish between subtle differences in MHCII-peptide complex structure. As a result, the signaling cascade that is activated following TCR engagement reflects the binding strength of the MHCII-peptide complex to the TCR.[15]

The instructive model postulates that the strength of the TCR signal is the single determining factor for commitment to either SP CD4+ T cell or Treg. Upon binding to MHCII-peptide complexes on thymic antigen-presenting cells (APC), cells displaying low signal strength proceed to CD4 SP differentiation and will make up the Tconv pool. Cells that do not bind the MHCII-peptide complexes at all, or bind these with such low avidity that the TCR signal is too weak, will die by neglect. By contrast, cells that bind too strongly and as such are autoreactive and potentially harmful are negatively selected and die by activation-induced apoptosis, or agonist selection. Such strict selection criteria leave a narrow window for Treg differentiation, and indeed only an intermediate TCR signal leads to Foxp3 expression.[10,14] The strength of the TCR signal can be directly determined by the measurement of Nr4a1 (Nur77) expression. Nr4a1 is a member of an Nr4a family of nuclear receptors that are induced by TCR signaling.[16] Using Nr4a1-GFP reporter mice, it was shown that expression of the Nr4a receptors is increased in Tregs, confirming that these require stronger TCR signaling compared to Tconv for differentiation.[17] Mice deficient for all Nr4a receptors have a complete absence of Tregs and develop systemic autoimmune disease, which led to the hypothesis that Nr4a receptors are required to translate TCR signal strength to transcription of Foxp3.[18] In addition, increased expression of CD25, CD5, and cytotoxic T lymphocyte antigen 4 (CTLA4) by Tregs indicated that Tregs are indeed exposed to stronger TCR signals than Tconv. CD5 was shown to sense TCR signaling strength and adjust this by differentially recruiting SHP-1, a tyrosine phosphatase, leading to fine-tuning of the TCR response.[19–21] Mice deficient for CD5 or SHP-1 showed increased numbers of tTregs and perturbed negative feedback of TCR stimulation.[22] Another mechanism of fine-tuning TCR signaling is through the expression of OX40, GITR, and TNFR2. These molecules belong to the tumor necrosis factor receptor superfamily

(TNFRSF), and their expression on Treg precursors directly correlates with the TCR signal strength.[23] Interestingly, this expression is dependent on CD28 costimulation and the expression of the MAPK kinase TAK1 (MAP3k7), which plays a role in IL-2, IL-15 and transforming growth factor (TGF)β signaling.[24,25] Inhibition of TNFRSF expression led to reduced Treg differentiation, and conversely, stimulation/activation of TNFRSF led to increased Treg differentiation through STAT5 activation and increased sensitivity to IL-2.[23] These results led to the conclusion that TNFRSF expression plays an essential role in Treg precursors, by shaping the Treg repertoire through translation of TCR signaling in molecular parameters.

In the selective model of tTreg differentiation, TCR signaling and Foxp3 expression are two independent events. This two-step model of tTreg development was first suggested by Lio *et al.*[26] In this model, cells with a self-reactive TCR of intermediate affinity are selected to become Treg precursors. TCR engagement then leads to upregulation of CD25. At this point, TCR signaling is not further required, and thymic APC induce subsequent Foxp3 expression.[26] Not all cells that are selected for a self-reactive TCR of intermediate affinity upregulate CD25 and Foxp3, and these cells will undergo apoptosis. The cells that express Foxp3 have a survival advantage inferred by the increased sensitivity to IL-2 and will differentiate into mature Tregs.[14]

ZAP70 is a kinase important for signal transduction after TCR engagement. Foxp3 expression targets ZAP70, and the Foxp3-induced downregulation of ZAP70 allows modulation of the TCR signal strength, and adjustment to a level that allows Treg differentiation. Mice that have mutation in the gene encoding for ZAP70 have markedly reduced numbers of tTregs.[27] This is due to the reduced production of IL-2, as well as the defect in TCR signaling. Interestingly, Treg numbers in the periphery are normal, or even increased, illustrating that TCR modulation is important for the selection of Tregs in the thymus, but not the periphery.[27]

2.2 TGFβ

Early neonatal thymectomy experiments in mice showed that autoimmunity develops only if the thymectomy takes place 3 days after birth, but not 1 or 7 days.[28] These experiments suggested that the thymic development of Tregs is delayed compared to the development of Tconv.[29] It has long been unclear whether the reason for this delay is the requirement of additional developmental stages before Foxp3 expression is induced, whether it is due to the increased time required for TCR–MHCII interactions between Treg precursors and thymic APC, or whether other cell-intrinsic mechanisms are responsible. A recent report answers the question of delayed development that has eluded scientists for many years, by showing that the differentiation of tTregs is driven by the release of TGFβ from thymic DC, macrophages, and epithelial cells, induced by thymocyte apoptosis.[30] Previously, conflicting reports were published on the role of TGFβ in tTreg development, with one report claiming TGFβ signaling is critical,[31] and another report claiming that tTreg development is independent of TGFβ signaling.[32] In the latter report, TGFβ was suggested to function as a prosurvival factor, antagonizing negative selection and as such allowing for the survival of tTregs.[32] The current understanding is that TGFβ in fact induces Foxp3 expression in thymocytes and thereby drives tTreg differentiation. Support for this hypothesis was shown through specific depletion of the TGFβ-receptor I (TGFβRI) from Foxp3-expressing thymocytes by generating $Tgfbr1^{f/f}$Foxp3-cre+ mice.[30] These mice showed no difference in the tTreg compartment compared to wild-type mice, and a simple anti-apoptotic role for TGFβ was therefore excluded. On a molecular level, TGF-β activates Foxp3 via induction of Smad binding to a conserved Smad–NFAT response element (CNS1).[33] In neonatal mice, the level of TGFβ is low in the thymus but increases steadily over the following days and this increase coincided with an increase in thymocyte apoptosis (due to negative selection) and the emergence of tTregs. Apoptotic cells are potent inducers of TGFβ in phagocytic cells[34,35] and thymic DC, macrophages, and thymic epithelial cells were indeed found to be the source of the intrathymic increase in TGFβ.[30]

2.3 Costimulation

As discussed above, the relay of TCR signals of the appropriate strength is crucial for the development of functional Tregs in the thymus. Costimulation is a powerful pathway by which TCR signaling is regulated. A range of different costimulatory signals has been described in Tregs, many of which signal via the NF-κB pathway (described in more detail below). The most well-known costimulatory molecule is CD28, the receptor for CD80 and CD86, B7-proteins that are expressed on APC.[36]

2.3.1 CD28

CD28 signaling induces IL-2 production and cell proliferation, and importantly, stabilizes Bcl-xL, an antiapoptotic molecule required for T cell survival.[37,38] CD28 was one of the first molecules discovered to play a role in Treg-mediated control of autoimmunity, when it was found that diabetes was exacerbated in nonobese diabetic mice in the absence of CD28 or its ligands CD80 and CD86.[39,40] This exacerbation was due to the lack of Tregs and as such could be prevented by the transfer of Tregs from wild-type animals.[40] It was later shown that costimulation through CD28 is required for the induction of CD25 and CD122 expression, and c-Rel-mediated remodeling of the chromatin in the *Foxp3* locus.[41–43] This series of events forms the first step in the two-step model of tTreg development.[26] CD28 can signal through multiple pathways, but for Treg development, signaling through Lck is required.[44] Targeted deletion of CD28 in Treg revealed that after initial tTreg development, CD28 is merely required for Treg proliferation, survival, and subsequent suppression of Tconv expansion.[45,46] Mice deficient for CD28 in Tregs (CD28-ΔTreg mice) developed autoimmune lymphoproliferative disease, which could be prevented by transfer of CD28-sufficient Tregs.[45,46] Interestingly, CD28-ΔTreg mice had decreased numbers, but not a complete absence of tTregs.[45] This suggests that the role of CD28 in tTregs differentiation is indeed most important before precursors start expressing Foxp3, and the absence of CD25+Foxp3− Treg precursors in CD28 and CD80/86-deficient mice[47] strengthens the argument for the two-step model of tTreg development.

2.3.2 CD27–CD70

CD27 is another member of the aforementioned TNFRSF and is expressed on all thymocytes from the pro-T cell stage.[48,49]

CD27-deficient mice had reduced numbers of Tregs, but Tconv were not affected.[50] In Tconv differentiation, CD28 signaling induces apoptosis in autoreactive CD4+ T cells. It has long been unclear how self-reactive Tregs could escape apoptosis, in particular because CD28 signaling in Tregs precursors is required for the induction of Foxp3 expression. CD27 is not required for induction of Foxp3 or Treg function but rather allows Treg development by inhibition of mitochondrial apoptosis. Its ligand, CD70, is expressed by thymic APC (DC as well as thymic epithelial cells), and it is hypothesized that thymic APC form dedicated niches where CD27-expressing developing Tregs are rescued from apoptosis and can continue differentiation.[50]

2.3.3 CTLA4

CTLA4, or CD152, is highly activated on Tregs in a Foxp3-dependent manner and is a ligand for CD80 and CD86 expressed on APC.[51,52] CTLA4 has a higher affinity for CD80 and CD86 than CD28 and can therefore regulate the degree of costimulation by competing for binding.[53] In CTLA4-deficient mice, lethal lymphoproliferation is driven by excessive CD28 signaling,[53–55] which proves that CTLA4 is required as a negative regulator of T cell-mediated immune responses.[56] CTLA4 can be induced independently of Foxp3, but Foxp3 amplifies and stabilizes its expression, which is important for the suppressive function of Tregs, but less so for the differentiation of tTregs.[3,57] In contrast, differentiation of pTregs seems to depend on CTLA4, as CTLA4-deficient mice have impaired pTreg differentiation.[58]

2.4 CD25, IL-2, and Common γ Chain Signaling

The significance of IL-2 and IL-2 signaling for tTreg development has been reported in numerous studies, but controversy exists about the function IL-2 exerts during Treg differentiation. The development of autoimmunity and reduced Treg numbers in IL-2- or CD25-deficient mice[59] demonstrates the importance of IL-2 signaling for the differentiation, survival, and homeostasis of Tregs. It was suggested that IL-2 is redundant for Treg differentiation in the thymus,[60,61] but the upregulation of CD25 on tTreg precursors and subsequent IL-2 dependence suggests otherwise.[26,62–64] A partial redundancy is present, however, as mice deficient in CD25, CD122 (IL-2Rβ), or IL-2 only show a 50% reduction in the tTreg compartment, rather than a complete absence of tTregs.[65,66]

As mentioned above, Tregs constitutively express the high-affinity IL-2 receptor alpha (IL-2Rα or CD25), and, once in the periphery, are highly dependent on exogenous IL-2. However, the models for tTreg differentiation discussed above cannot fully explain the functional relationship between Foxp3 and CD25 expression, which remains controversial. The two-step model postulates that Treg precursors in the thymus express increased levels of CD25, induced by TCR signaling.[26] CD25 expression then allows signaling from IL-2, which in turn induces Foxp3. Alternatively, other groups have shown that Foxp3 + double-positive (DP) thymocytes are present in the thymic cortex. These cells express only low levels of CD25,[62,63] which suggests that Foxp3 expression preempts CD25 expression. In this model, CD25 is required *after* Foxp3 induction in order to provide a prosurvival buffer against the apoptotic effects of Foxp3 expression.[63] In another, more recent study, it was suggested that Foxp3 expression and CD25 expression are in fact regulated independently.[67] This study showed Foxp3−CD25+, as well as Foxp3+CD25−, SP cells could develop into Foxp3+CD25+ tTreg *in vitro* as well as *in vivo* but that both required either IL-2 or IL-15 to fully differentiate.[67] This finding was in agreement with an earlier study showing a prominent population of Foxp3+CD25− SP cells in the absence of CD122, which suggests that at least for a proportion of the developing tTregs, Foxp3 expression precedes CD25 expression.[61] In addition, Tregs recirculating to the thymus from the periphery were shown to inhibit *de novo* Treg differentiation by reduction of the available IL-2, but Foxp3−CD25+ precursor cells were not affected by the lack of IL-2.[68]

The requirement for signaling via the common γ chain (CD132, γc), which is common to the receptors for IL-2, IL-4, IL-7, IL-9, and IL-21, for the differentiation of Tregs was first shown in CD132-deficient mice.[61] These mice displayed a complete lack of Foxp3+ cells in the thymus and the periphery. In contrast, mice lacking one of the cytokine-specific receptor subunits IL-15Rα or CD127 (IL-7Rα) showed only a slight reduction in the percentage of tTregs.[69] Although IL-7Rα-deficient mice have a severe block in thymocyte differentiation and therefore severely reduced thymic cell numbers, normalized fractions were similar to wild-type mice.[69] However, the combined deficiency of CD127 and CD122 mimicked the phenotype of the CD132-deficient mice and caused deficiency in Foxp3+ cells in the thymus and the periphery, illustrating that both IL-2 and IL-7 signaling are important to generate and maintain functional Tregs.[69] Whether IL-7 acts mostly as a fitness factor, as has been suggested for IL-2, is an area of current investigation.

In the periphery, a decrease in IL-2 reduces Treg proliferation and survival and rapidly leads to T cell activation and compensatory production of IL-2 by effector T cells.[61,70] Interestingly, complete loss of CD25 leads to reduced production of tTregs but a relatively normal pool of Tregs in the periphery early in life.[71]

Although Tregs are highly dependent on IL-2, they are incapable of IL-2 production. This is due to the partial repression of the *Il2* locus in developing thymocytes destined for Treg fate.[57] In addition, Foxp3 binds to and inhibits NFAT, required for IL-2 transcription.[72] In differentiated Tregs, transcription factors Helios and Eos change the epigenetic modifications of the *Il2* promoter region and the Foxp3-binding efficiency, thereby completely suppressing transcription of the Il2 locus.[73] Whether the inability of Tregs to produce IL-2 is essential to prevent a chronic increase in Treg numbers is still unclear.

2.5 Nuclear Factors

TCR engagement in tTreg precursors leads to a number of downstream signaling events. The transcription factors nuclear factor κB (NF-κB), Akt, and mechanistic target of rapamycin (mTOR) have all been shown to be involved in the differentiation of tTregs.[74–77] In addition, calcium signaling is required, but it is unclear whether the downstream nuclear factor of activated T cells (NFAT) plays a role in tTreg differentiation, or whether (Foxp3-mediated inhibition of) NFAT merely plays a role after cells have entered into the Treg fate.[72,77,78] Another group of nuclear factors involved in Treg fate decisions are E-proteins, the role of which is discussed in detail below.

2.5.1 mTOR

mTOR is a highly conserved Ser/Thr protein kinase that regulates proliferation, translation, and metabolism.[79,80] The mTOR kinase is required in T cells for integrity of homeostasis, maintenance, and activation.[81] It functions as a sensor, translating environmental cues such as cell–cell contact, nutrients, cytokines, and growth factors into actions suited to the time and situation.[82] mTOR can bind to different transcription factors, and depending on its binding partners, two different complexes are formed: (i) -rapamycin-sensitive mTORC1, where mTOR is bound to Raptor, and (ii) rapamycin-insensitive mTORC2, where mTOR is bound to Rictor and mSin1.[83,84]

In the thymus, mTOR is activated by TCR signaling in a multistep process. TCR signaling activates PI3K, which catalyzes the conversion of PIP2 to PIP3. This leads to the recruitment of the enzymes Akt and PDK1 to the plasma membrane and activation of mTORC1 via a number of mechanisms.[85,86] Although Akt is required for mTORC1 activation in Th2 cells,[87] it may not be of importance to other T cell subsets such as Tregs. A pathway independent of Akt, via PI3K and PDK1, is active during CD8+ effector T cell activation. Interestingly, this pathway is activated upon IL-2 signaling[88]; however, IL-2 signaling has also been shown to activate the Jak3–STAT5 and the PI3K–Akt–mTORC1 pathway and enhances mTOR activation in Tregs.[87] Tregs have increased mTORC1 activity compared to Tconv,[87] but the exact role of mTOR signaling in the differentiation of Tregs remains unclear. Several studies have shown a negative role for mTORC1 and mTORC2. mTORC2 activates Akt and inhibits Treg differentiation and proliferation through Foxo1 signaling and mTORC1 synergism.[75,76,89–91] Indeed, Treg-specific deletion of mTORC1 led to increased numbers of Tregs, confirming the negative role of mTORC1 in Treg differentiation, but these Tregs had reduced suppressive function and these mice succumbed to autoinflammatory disorders resembling the *scurfy* phenotype early in life.[87] In summary, mTOR does play a role in Treg differentiation but appears more important for the conversion of Tconv to pTreg upon sensing environmental cues, than it is for tTreg fate commitment.

2.5.2 E-Proteins

E-proteins are a family of transcription factors important for thymocyte differentiation.[92–94] E-proteins (E12, E47, HEB, and E2-2[95,96]) play a role in thymocyte development taking place before lineage commitment but additionally inhibit the transition from DN to DP and DP to SP in later thymocyte differentiation.[92–94] This inhibition is counteracted by TCR engagement, which downregulates E-protein expression, and allows for progression to mature thymocytes. It is this latter function that led to the hypothesis that E-proteins could play a role in tTreg differentiation, by enabling fine-tuning of the TCR signal strength. Indeed, deletion of E-proteins led to increased tTreg differentiation, whereas deficiency for E-protein inhibitors Id2 and Id3 led to reduced tTreg differentiation.[17] E-proteins bind to a CD25 enhancer locus, suppressing CD25 transcription. In the absence of E-proteins, this suppression is removed leading to increased CD25 expression, increased STAT5 phosphorylation, and enhanced expression of c-Rel.[17,97] These results led to the hypothesis that the role of E-proteins in the differentiation of tTregs is to establish the threshold for STAT5 phosphorylation and activation of NF-κB family members, induced by TCR signaling.[17]

2.5.3 NF-κB

NF-κB family transcription factors are activated by TCR signaling, and although it is clear that many of these are required for tTreg development, specific gene targets are mostly still unknown. Targeted loss-of-function mutation in members of the NF-κB signaling pathway (such as PKCθ, CARMA1, Bcl10, and IκB kinase (IKK) 2) caused notable defects in the generation of tTregs.[98–101] The most direct effect on tTregs differentiation was shown for NF-κB family member c-Rel. c-Rel directly binds to the CNS2, CNS3, and promoter regions of the *Foxp3* locus,[43,74,102] possibly leading to chromatin remodeling of these regions and allowing for, or enhancing, transcription. In addition, c-Rel enables the formation of the c-Rel enhanceosome at the Foxp3 promoter region, driving Foxp3 expression.[43] In its turn, Foxp3 can interact with c-Rel to downregulate c-Rel-induced activation of NF-κB, which is required for to maintain a suppressive phenotype in mature Tregs.[103] Deletion of c-Rel or CNS3 indeed led to comparable defects in Foxp3 expression in developing thymocytes.[42] Constitutive expression of the IKKβ complex in wild-type mice, which enhances NF-κB activity, led to increased numbers of tTregs.[74] Interestingly, a complex consisting of CARMA1, Bcl10, and Malt-1 (CBM complex) is required for NF-κB activation, and mice deficient for any of these factors have reduced numbers of tTregs.[98–101] The CBM complex is proximal to the TCR, and therefore constitutive expression of the IKKβ complex in mice deficient for CARMA1 can restore Foxp3 expression in thymocytes.[74] Both classical cellular inhibitors (IκBs) and nonclassical inhibitors (Bcl3, IκBζ, and $I\kappa B_{NS}$) of NF-κB are essential for Foxp3 induction in tTreg precursors, and thus required for establishment of the tTreg compartment.[104,105]

2.6 Foxp3: At the Center of a Treg Transcriptional Complex

The expression of the transcription factor Foxp3 identifies the Treg lineage, plays a role in the differentiation of Tregs, and is required for the suppressive capacities of Tregs.[3] Foxp3 expression was shown to suppress effector genes and induce suppressor genes.[2,29,57] In addition, as discussed above, Foxp3 induces the expression of CD25, which is a hallmark of the Treg lineage and renders Tregs highly sensitive to changes in the level of IL-2.[57]

Although Foxp3 is still widely regarded as a master transcription factor of the Treg lineage, recent studies suggest that Foxp3 is merely one of the factors in an elaborate network of lineage-commitment signaling molecules. Due to this complex molecular control of Treg fate, Foxp3 can be viewed as a surrogate marker for the Treg lineage.[106] For instance, tTregs fail to develop in the absence of Foxo1 and Foxo3.[107–109] It was suggested that the Foxo transcription factors mediate Treg differentiation through modulation of the TGFβ pathway.[107]

Recent studies showed that Foxp3 expression alone is not sufficient to induce Treg phenotype—instead, Foxp3 is part of large heterogeneous transcriptional complexes, consisting of more than 200 binding partners.[110] Many of these proteins are transcriptional regulators, which act as facilitators for Foxp3 binding; this explains how many identified Foxp3-binding sites in Tregs are responsive even in the absence of a forkhead box-binding motif.[111–113] Interestingly, many of the transcription factors and cofactors that were identified as binding partners for Foxp3 are part of a feedback loop that regulates gene expression in Tregs. Foxp3 regulates expression of the binding partners, but the binding partners in their turn can bind to intronic enhancers and the Foxp3 promoter region, thereby controlling the expression of Foxp3 itself.[114–116] This mechanism allows for fine-tuning of gene expression and a large degree of flexibility depending on the required cellular function at any given moment.

2.7 Micro RNA

Micro RNAs (miRNAs) are small endogenous RNAs of approximately 23 nucleotides long, which are evolutionary conserved.[117] MiRNAs regulate gene expression by binding to the mRNA of protein-coding genes, thereby causing posttranscriptional repression.[117] Transcription of miRNAs takes place predominantly via RNA polymerase II, and the mature miRNA sequence including variable flanking regions is then processed by the nuclease Drosha into precursor miRNA.[118] This precursor miRNA is subsequently exported from the nucleus by exportin 5[119] and further processed in the cytoplasm by a second nuclease Dicer and loaded into the RNA-induced silencer complex RISC.[120] The RISC is then guided by a single strand of the miRNA and binds to 3′ untranslated regions resulting in sequence-specific targeting of the target gene.[121] Binding of miRNAs to the target mRNA generally leads to degradation of the complex, although other inhibitory mechanisms have been described.[122] Interestingly, miRNAs were shown to be critical for the differentiation of Tregs: mice specifically deficient for Dicer in CD4 T cells had decreased tTregs and reduced Foxp3 induction by TGFβ.[123] Phenotypically, these mice developed autoimmune pathology from 3 months of age. In contrast, conditional depletion of Drosha from CD4+ cells or Dicer from Foxp3+ cells led to fatal autoimmunity by 4 weeks of age.[124–126] Depletion of Drosha or Dicer also causes deficiency for other small RNA species such as shRNA and siRNA; therefore, these studies could not formally exclude a role for these in Treg differentiation. However, a recent study confirmed that miRNAs are in fact required for Treg development by showing that mice with Tregs deficient for the RNA-binding protein DCGR8, which is required for the processing of miRNAs, display a phenotype very similar to that of the Treg-specific Dicer knockouts.[127]

Due to the many RNAs that can be inhibited by a single miRNA, it is still unclear what the mechanism is behind miRNA-induced Treg development. However, a number of miRNAs was found to be differentially expressed in Tregs, such as miR21, miR31, miR155, and miR146a.[125,128,129] miR21 and miR31 have a direct, opposing effect on Foxp3 expression.[128] miR155 enhances Stat5 signaling via inhibition of suppressor of cytokine signaling 1 (SOCS1).[130] miR146a targets Stat1, and conditional deletion in Tregs leads to dysfunctional Tregs in the periphery.[129] Overall, miRNAs are required for the differentiation as well as the function of Tregs, but many of the targets are yet to be identified.

2.8 Epigenetic Control of Treg Fate

The function of Tregs relies heavily on the high, stable, and sustained expression of Foxp3.[2,57,131] Where *Scurfy* mice showed the absolute requirement for Foxp3 in preventing fatal autoimmunity,[6] FILIG mice demonstrated that also the level of Foxp3 expression is important.[132] FILIG mice developed a fatal autoimmune disorder that resembled the pathology seen in *Scurfy* mice, but in addition, this study showed that expression of low levels of Foxp3 facilitated differentiation into effector T cells, which contributed to disease development. In fact, T cells expressing reduced levels of Foxp3 secreted Th2 cytokines and expressed fewer Treg signature genes.[132] The requirement for sustained Foxp3 expression was demonstrated in conditional knockout mice, where Foxp3 was deleted in mature Tregs.[3] Loss of Foxp3 expression in these cells allowed for the production of IL-2 and Th1 cytokines such as IFNγ and TNF. In addition, it was shown that Foxp3 can be expressed transiently by Tconv cells, but without endowing these cells with suppressive capacities.[133–135] In humans, not all Foxp3+ T cells are suppressive, and it was suggested that there is functional plasticity in Tregs and that T cell regulation is directed by the dynamics of Foxp3 expression and antigenic history of the cells.[136–138] By contrast, Tregs may also lose Foxp3 expression in a lymphopenic or inflammatory environment, and thereby lose suppressive functions and gain the ability to produce inflammatory cytokines.[139,140] Forced expression of Foxp3 in T cells revealed that the gene signature of Tregs is not determined solely by Foxp3 but that regulation of Treg identity incorporates a combination of factors.[141] It has now become apparent that stability of Foxp3 is controlled by posttranslational modifications, but in contrast, the active transcription of Foxp3 is regulated by epigenetic mechanisms. Three conserved noncoding sequences (CNS1, 2, and 3) in the Foxp3 locus regulate Foxp3 transcription. Interestingly, induction of tTregs or pTregs requires the regulation of Foxp3 expression by separate CNS regions. Where binding of NFAT and AP1 to CNS1 is required for the induction of Foxp3 expression in pTregs, CNS3 is required for induction of Foxp3 expression in the thymus.[42] CNS2 is a binding site for CBF-β-Runx1 and is required to maintain a high and sustained Foxp3 expression level. Upon TCR stimulation, Rel is activated and binds to CNS3, which opens up the Foxp3 promoter region and allows active transcription.[115] Interestingly, a specific region exists in CNS2, which is one of multiple Treg-specific demethylated regions, or TSDRs. The TSDR in Foxp3 CNS2 is a hypomethylated CpG-island, and it is this specific methylation status of CNS2 that was shown to determine the level of Foxp3 expression.[142,143]

The importance of the TSDRs in the control of Treg fate determination *in vivo* is illustrated by the fact that only tTregs and pTregs generated *in vivo* display these methylation/demethylation patterns. By contrast, Tregs generated *in vitro* either by anti-CD3, TGFβ, and IL-2 stimulation or forced Foxp3 expression (by retroviral transduction), and other T cell subsets such as Th1, Th2, Th17 cells, do not possess these demethylated regions.[144,145] TSDRs are found in genes encoding for factors important for Treg function, such as Foxp3 (*Foxp3*), CD25 (*Il2ra*), Ctla4 (*Ctla4*), Helios and Eos (*Ikzf2* and *Ikzf4*), and GITR (*Tnfrsf18*).[144,146] DNA hypomethylation in Tregs provides a separate regulatory mechanism for gene expression from Foxp3 expression. Foxp3-binding sites are mostly found outside TSDRs, suggesting that Treg fate is regulated on different levels and requires both mechanisms to ensure lineage stability.[146]

Interestingly, both Foxp3 expression and demethylation of the TSDRs is dependent on TCR stimulation and IL-2. Foxp3–CD25hi CD4 SP Treg precursors in the thymus only acquire Foxp3 CNS2-specific demethylation upon TCR engagement in the presence of IL-2, leading to Foxp3 expression and conferring the epigenetic signature of Tregs.[147] Indeed, Foxp3+CD4 SP thymocytes contain TSDRs.[144] It is still unclear whether Foxp3+CD4 SP cells with TSDRs preferentially expand over Foxp3+CD4 SP cells that do not display Treg-specific demethylation patterns, or whether the latter do not survive due to the proapoptotic effects of Foxp3 expression alone.[144]

As discussed above, the molecular mechanism of TCR-induced Foxp3 expression is still unclear, as are the molecular events leading to TCR-induced DNA demethylation. However, recent data now suggest that TCR strength and duration are the deciding parameters that determine the likelihood of Treg precursors actually entering the Treg fate.[144] Treg precursors express CTLA4 during the conversion of Foxp3−CD25hi thymocytes to Foxp3+ thymocytes. CTLA4 is a regulator of thymic TCR signal strength, due to its ability to bind CD80 and CD86 on APC in the thymus, leading to reduced availability of these molecules to bind the costimulatory molecule CD28. In addition, CTLA4 can directly suppress CD28 signaling.[53] The demethylation of the gene coding for CTLA4 during formation of the Treg epigenome suggests an active role of this molecule in Treg fate decision. The attenuation of the TCR signal by CTLA4 could well lead to the escape of self-reactive Tregs from negative selection. In agreement with this hypothesis, Treg TCR were shown to have a higher affinity for the selecting ligand than the Tconv selected by the same ligand.[10,148,149] In addition, CTLA4-deficient Tregs do not display a skewing toward self-reactive TCR such as seen in wild-type mice, suggesting that CTLA4 is required to escape negative selection.[150]

In summary, epigenetic mechanisms are crucial for Treg fate decision. The Treg epigenome synergizes with, and complements, other Treg-specific strategies (such as Foxp3-dependent ZAP70 expression, discussed above) to influence TCR signal strength and the self-reactive TCR repertoire.

3. PERIPHERAL TREG

Besides differentiation of Tregs in the thymus, it has become clear that Foxp3− precursors in the periphery can differentiate into Foxp3+ Tregs upon antigenic stimulation with a relevant antigen.[151,152] Due to the fact that pTregs are derived from peripheral CD4+ T cells, the TCR repertoire is broader than that of the tTreg population.[153,154] A recent study found that in healthy mucosal tissues, Treg differentiation is efficiently supported but that recent thymic emigrants are the preferential pTreg precursors in inflammatory situations.[155] This also illustrates that pTreg differentiation takes place in steady state, but also in the context of inflammation, infection, tumor immunity, and mucosal immunity.[154,156,157] Interestingly, data obtained from mice deficient for CNS1 in the *Foxp3* locus showed that this region is not required for the generation of tTregs but is indispensable for the differentiation of Tregs in the periphery.[42] This suggests that a dedicated mechanism exists for the induction of pTregs. In healthy animals, the pool of Tregs generated in the thymus is sufficiently capable to keep self-reactive T cells in check and prevent tissue-specific autoimmunity. However, mice deficient for CNS1 and therefore deficient for pTregs, developed mucosal pathology and altered gut microbiota. This observation led to the hypothesis that the main purpose and function of pTregs is the prevention of allergic inflammation at mucosal sites.[154,158]

3.1 TCR Specificity and Signaling in pTreg Differentiation

A number of signaling pathways have been described for the induction of Tregs in the periphery. TCR specificity and TCR signaling, costimulation, and cytokine signaling are required but interact in a manner distinct from the tTreg differentiation for efficient pTreg generation. The TCR repertoire of tTregs and pTregs partially overlaps, which indicates that some self-reactive CD4 SP T cells that may have escaped negative selection are primed to become pTregs.[159] T cells that possess a TCR that is not specific to self-peptide might become pTregs if their cognate antigen is presented under tolerizing conditions.[156] It has been shown that Foxp3 induction in the periphery takes place upon weak TCR stimulation, in contrast with the optimal activating condition that requires strong TCR signaling.[75,160,161] More specifically, peripheral Foxp3 induction requires the correct balance of antigen availability and affinity: low doses of antigen with a high affinity and vice versa.[162] If signaling reaches the upper threshold, T cells with the appropriate TCRs will expand as effector cells, rather than undergo conversion to pTregs.[162] In addition to an appropriately balanced cumulative TCR signal, pTregs also require weak CD28 costimulation and functional CTLA4 for Foxp3 induction.[58,163] Strong CD28 costimulation leads to enhanced Lck signaling, which inhibits Foxp3 expression.[163] Other negative regulators of Foxp3 expression are the PI3K–Akt–mTOR axis, discussed in detail above. The activation of Akt inhibits Foxo proteins, which are important regulators of Foxp3 expression.[107,108] In agreement with this, inhibition of mTOR or PI3K, or overexpression of Akt enhances pTreg differentiation, whereas deletion of PTEN (an Akt inhibitor) or Cbl-b (which degrades a subunit of PI3K) inhibits pTreg differentiation.[75,91,161,164,165] It was recently shown that CD5 plays an important role in the inhibition of mTOR.[166] CD5 is expressed by Tregs and is upregulated on T cells in response to tolerizing antigens in periphery. This leads to modulation of TCR signaling and the promotion of pTreg differentiation.[166]

3.2 The Role of TGFβ, IL-2, Retinoic Acid, and Microbial Metabolites in pTreg Differentiation

The differentiation of pTregs is dependent on TGFβ and IL-2. Foxp3 expression is promoted by TGFβ-induced binding of Smad3 and NFAT to CNS1.[33] In addition, TGFβ promotes TCR-induced Foxp3 expression through inhibition of DNA methyltransferase I, preventing its recruitment to the *foxp3* locus.[167] IL-2 was shown to directly induce Foxp3 expression in "precommitted" precursor cells.[168] Perhaps, these precursors are those cells with low affinity self-reactive TCR that escaped negative selection and tTreg induction in the thymus. IL-2 also plays a role in inhibition of Th17 differentiation. The shared developmental pathway and precursor population for pTreg and Th17 cells require fine-tuning of the appropriate fate decision, which is mediated by TGFβ, IL-2, and retinoic acid.[169–171] Retinoic acid (RA), produced by CD103+ tolerogenic DC in the gut, induces expression and phosphorylation of Smad3, thereby enhancing TGFβ signal strength and subsequent Foxp3 expression.[171–174] RA signaling is partly dependent on IL-2, which induces STAT5 phosphorylation and allows Foxp3 expression while inhibiting Rorγt.[169,175,176] It was shown that short-chain fatty acids (SCFAs) could be the trigger for RA production by mucosal DC.[177] SCFAs, microbial metabolites produced upon fermentation of dietary fibers, drive the induction of Treg differentiation in the gut.[178–180] The importance of gut microbiota for Treg differentiation is illustrated by the dramatic reduction of Tregs in germ-free or antibiotic-treated mice[181]; however, to maintain intestinal immunological tolerance, both tTregs and pTregs are required.[154,182]

The molecular mechanism behind SCFA-induced pTreg differentiation and/or proliferation is still somewhat unclear. It was shown *in vitro* that SCFA butyrate induces acetylation of H3K27 in the Foxp3 promoter and the CNS1 and CNS3 enhancers in naïve CD4+ T cells, leading to Foxp3 expression.[178,180] While this might provide a possible mechanism for *de novo* pTreg differentiation in the gut, it has not been formally shown to occur *in vivo*. In addition, other groups found that SCFAs induce expansion of tTregs in the gut, rather than conversion of naïve T cells to pTregs.[179] A recent study showed that the DNA methylation adaptor molecule Uhrf1 is required for the maintenance of immunological homeostasis in the gut.[183]

Uhrf1-deficient CD4 T cells fail to differentiate into pTreg in the gut, leading to severe colitis in Uhrf1$^{fl/fl}$CD4-Cre mice.[183]

Tregs are required in the gut to maintain tolerance to beneficial microorganisms, whereas Th17 cells are required to counteract intestinal pathogens.[184] TGFβ regulates the differentiation of the Th17 lineage, as well as the Treg lineage. Whether naïve CD4+ T cells upregulate Rorγt (for Th17 differentiation) or Foxp3 is a decision made based on the local cytokine milieu, which is made possible by considerable plasticity between these two lineages and a shared progenitor cell.[170]

4. CONCLUDING REMARKS

The differentiation of Tregs, both in the thymus and the periphery, requires a unique and appropriately timed combination of TCR signaling, costimulation, cytokine milieu, Foxp3 expression, activation of Treg-specific transcriptional mechanisms, and epigenetic modeling. The use of conditional knockout mice and advanced (epi)genomic and proteomic techniques has advanced our knowledge of the molecular control of Treg induction significantly over the past years. Once these pivotal differentiation mechanisms are fully understood, extended opportunities may arise for *in vivo* and *in vitro* manipulation of the Treg pool, providing new avenues for the treatment of immune disorders with autoimmune manifestations.

ACKNOWLEDGMENTS

This review was supported by the T-TIME IAP Grant and the FWO.

REFERENCES

1. Sakaguchi S. Naturally arising CD4+ regulatory T cells for immunologic self-tolerance and negative control of immune responses. *Annu Rev Immunol.* 2004;22:531–562. http://dx.doi.org/10.1146/annurev.immunol.21.120601.141122.
2. Hori S, Nomura T, Sakaguchi S. Control of regulatory T cell development by the transcription factor Foxp3. *Science.* 2003;299(5609):1057–1061. http://dx.doi.org/10.1126/science.1079490.
3. Williams LM, Rudensky AY. Maintenance of the Foxp3-dependent developmental program in mature regulatory T cells requires continued expression of Foxp3. *Nat Immunol.* 2007;8(3):277–284. http://dx.doi.org/10.1038/ni1437.
4. Fontenot JD, Gavin MA, Rudensky AY. Foxp3 programs the development and function of CD4+CD25+ regulatory T cells. *Nat Immunol.* 2003;4(4):330–336. http://dx.doi.org/10.1038/ni904.
5. Khattri R, Cox T, Yasayko S-A, Ramsdell F. An essential role for Scurfin in CD4+ CD25+ T regulatory cells. *Nat Immunol.* 2003;4(4):337–342. http://dx.doi.org/10.1038/ni909.

6. Brunkow ME, Jeffery EW, Hjerrild KA, et al. Disruption of a new forkhead/winged-helix protein, scurfin, results in the fatal lymphoproliferative disorder of the scurfy mouse. *Nat Genet.* 2001;27(1):68–73. http://dx.doi.org/10.1038/83784.
7. Bennett CL, Christie J, Ramsdell F, et al. The immune dysregulation, polyendocrinopathy, enteropathy, X-linked syndrome (IPEX) is caused by mutations of FOXP3. *Nat Genet.* 2001;27(1):20–21. http://dx.doi.org/10.1038/83713.
8. Sakaguchi S, Sakaguchi N, Asano M, Itoh M, Toda M. Immunologic self-tolerance maintained by activated T cells expressing IL-2 receptor alpha-chains (CD25). Breakdown of a single mechanism of self-tolerance causes various autoimmune diseases. *J Immunol.* 1995;155(3):1151–1164.
9. Klein L, Hinterberger M, Wirnsberger G, Kyewski B. Antigen presentation in the thymus for positive selection and central tolerance induction. *Nat Rev Immunol.* 2009;9(12):833–844. http://dx.doi.org/10.1038/nri2669.
10. Jordan MS, Boesteanu A, Reed AJ, et al. Thymic selection of CD4+CD25+ regulatory T cells induced by an agonist self-peptide. *Nat Immunol.* 2001;2(4):301–306. http://dx.doi.org/10.1038/86302.
11. Apostolou I, Sarukhan A, Klein L, Boehmer von H. Origin of regulatory T cells with known specificity for antigen. *Nat Immunol.* 2002;3(8):756–763. http://dx.doi.org/10.1038/ni816.
12. Knoechel B, Lohr J, Kahn E, Bluestone JA, Abbas AK. Sequential development of interleukin 2-dependent effector and regulatory T cells in response to endogenous systemic antigen. *J Exp Med.* 2005;202(10):1375–1386. http://dx.doi.org/10.1084/jem.20050855.
13. Baldwin TA, Sandau MM, Jameson SC, Hogquist KA. The timing of TCR alpha expression critically influences T cell development and selection. *J Exp Med.* 2005;202(1):111–121. http://dx.doi.org/10.1084/jem.20050359.
14. Hsieh C-S, Lee H-M, Lio C-WJ. Selection of regulatory T cells in the thymus. *Nat Rev Immunol.* 2012;12(3):157–167. http://dx.doi.org/10.1038/nri3155.
15. Kersh GJ, Allen PM. Structural basis for T cell recognition of altered peptide ligands: a single T cell receptor can productively recognize a large continuum of related ligands. *J Exp Med.* 1996;184(4):1259–1268.
16. Fassett MS, Jiang W, D'Alise AM, Mathis D, Benoist C. Nuclear receptor Nr4a1 modulates both regulatory T-cell (Treg) differentiation and clonal deletion. *Proc Natl Acad Sci USA.* 2012;109(10):3891–3896. http://dx.doi.org/10.1073/pnas.1200090109.
17. Gao P, Han X, Zhang Q, et al. Dynamic changes in E-protein activity regulate T reg cell development. *J Exp Med.* 2014;211(13):2651–2668. http://dx.doi.org/10.1084/jem.20132681.
18. Sekiya T, Kashiwagi I, Yoshida R, et al. Nr4a receptors are essential for thymic regulatory T cell development and immune homeostasis. *Nat Immunol.* 2013;14(3):230–237. http://dx.doi.org/10.1038/ni.2520.
19. Azzam HS, Grinberg A, Lui K, Shen H, Shores EW, Love PE. CD5 expression is developmentally regulated by T cell receptor (TCR) signals and TCR avidity. *J Exp Med.* 1998;188(12):2301–2311.
20. Azzam HS, DeJarnette JB, Huang K, et al. Fine tuning of TCR signaling by CD5. *J Immunol.* 2001;166(9):5464–5472.
21. Wong P, Barton GM, Forbush KA, Rudensky AY. Dynamic tuning of T cell reactivity by self-peptide-major histocompatibility complex ligands. *J Exp Med.* 2001;193(10):1179–1187.
22. Carter JD, Calabrese GM, Naganuma M, Lorenz U. Deficiency of the Src homology region 2 domain-containing phosphatase 1 (SHP-1) causes enrichment of CD4+CD25+ regulatory T cells. *J Immunol.* 2005;174(11):6627–6638.

23. Mahmud SA, Manlove LS, Schmitz HM, et al. Costimulation via the tumor-necrosis factor receptor superfamily couples TCR signal strength to the thymic differentiation of regulatory T cells. *Nat Immunol.* 2014;15(5):473–481. http://dx.doi.org/10.1038/ni.2849.
24. Sato S, Sanjo H, Tsujimura T, et al. TAK1 is indispensable for development of T cells and prevention of colitis by the generation of regulatory T cells. *Int Immunol.* 2006;18(10):1405–1411. http://dx.doi.org/10.1093/intimm/dxl082.
25. Wan YY, Flavell RA. The roles for cytokines in the generation and maintenance of regulatory T cells. *Immunol Rev.* 2006;212:114–130. http://dx.doi.org/10.1111/j.0105-2896.2006.00407.x.
26. Lio C-WJ, Hsieh C-S. A two-step process for thymic regulatory T cell development. *Immunity.* 2008;28(1):100–111. http://dx.doi.org/10.1016/j.immuni.2007.11.021.
27. Hsu LY, Tan YX, Xiao Z, Malissen M, Weiss A. A hypomorphic allele of ZAP-70 reveals a distinct thymic threshold for autoimmune disease versus autoimmune reactivity. *J Exp Med.* 2009;206(11):2527–2541. http://dx.doi.org/10.1084/jem.20082902.
28. Nishizuka Y, Sakakura T. Thymus and reproduction: sex-linked dysgenesia of the gonad after neonatal thymectomy in mice. *Science.* 1969;166(3906):753–755. http://dx.doi.org/10.1126/science.166.3906.753.
29. Fontenot JD, Dooley JL, Farr AG, Rudensky AY. Developmental regulation of Foxp3 expression during ontogeny. *J Exp Med.* 2005;202(7):901–906. http://dx.doi.org/10.1084/jem.20050784.
30. Konkel JE, Jin W, Abbatiello B, Grainger JR, Chen W. Thymocyte apoptosis drives the intrathymic generation of regulatory T cells. *Proc Natl Acad Sci USA.* 2014;111(4):E465–E473. http://dx.doi.org/10.1073/pnas.1320319111.
31. Liu Y, Zhang P, Li J, Kulkarni AB, Perruche S, Chen W. A critical function for TGF-beta signaling in the development of natural CD4+CD25+Foxp3+ regulatory T cells. *Nat Immunol.* 2008;9(6):632–640. http://dx.doi.org/10.1038/ni.1607.
32. Ouyang W, Beckett O, Ma Q, Li MO. Transforming growth factor-beta signaling curbs thymic negative selection promoting regulatory T cell development. *Immunity.* 2010;32(5):642–653. http://dx.doi.org/10.1016/j.immuni.2010.04.012.
33. Tone Y, Furuuchi K, Kojima Y, Tykocinski ML, Greene MI, Tone M. Smad3 and NFAT cooperate to induce Foxp3 expression through its enhancer. *Nat Immunol.* 2008;9(2):194–202. http://dx.doi.org/10.1038/ni1549.
34. Fadok VA, Bratton DL, Konowal A, Freed PW, Westcott JY, Henson PM. Macrophages that have ingested apoptotic cells in vitro inhibit proinflammatory cytokine production through autocrine/paracrine mechanisms involving TGF-beta, PGE2, and PAF. *J Clin Invest.* 1998;101(4):890–898. http://dx.doi.org/10.1172/JCI1112.
35. Xiao YQ, Freire-de-Lima CG, Schiemann WP, Bratton DL, Vandivier RW, Henson PM. Transcriptional and translational regulation of TGF-beta production in response to apoptotic cells. *J Immunol.* 2008;181(5):3575–3585.
36. Carreno BM, Collins M. The B7 family of ligands and its receptors: new pathways for costimulation and inhibition of immune responses. *Annu Rev Immunol.* 2002;20:29–53. http://dx.doi.org/10.1146/annurev.immunol.20.091101.091806.
37. Jenkins MK, Taylor PS, Norton SD, Urdahl KB. CD28 delivers a costimulatory signal involved in antigen-specific IL-2 production by human T cells. *J Immunol.* 1991;147(8):2461–2466.
38. Boise LH, Minn AJ, Noel PJ, et al. CD28 costimulation can promote T cell survival by enhancing the expression of Bcl-XL. *Immunity.* 1995;3(1):87–98. http://dx.doi.org/10.1016/1074-7613(95)90161-2.
39. Lenschow DJ, Herold KC, Rhee L, et al. CD28/B7 regulation of Th1 and Th2 subsets in the development of autoimmune diabetes. *Immunity.* 1996;5(3):285–293.

40. Salomon B, Lenschow DJ, Rhee L, et al. B7/CD28 costimulation is essential for the homeostasis of the CD4+CD25+ immunoregulatory T cells that control autoimmune diabetes. *Immunity*. 2000;12(4):431–440.
41. Vang KB, Yang J, Pagán AJ, et al. Cutting edge: CD28 and c-Rel-dependent pathways initiate regulatory T cell development. *J Immunol*. 2010;184(8):4074–4077. http://dx.doi.org/10.4049/jimmunol.0903933.
42. Zheng Y, Josefowicz S, Chaudhry A, Peng XP, Forbush K, Rudensky AY. Role of conserved non-coding DNA elements in the Foxp3 gene in regulatory T-cell fate. *Nature*. 2010;463(7282):808–812. http://dx.doi.org/10.1038/nature08750.
43. Ruan Q, Kameswaran V, Tone Y, et al. Development of Foxp3(+) regulatory t cells is driven by the c-Rel enhanceosome. *Immunity*. 2009;31(6):932–940. http://dx.doi.org/10.1016/j.immuni.2009.10.006.
44. Tai X, Cowan M, Feigenbaum L, Singer A. CD28 costimulation of developing thymocytes induces Foxp3 expression and regulatory T cell differentiation independently of interleukin 2. *Nat Immunol*. 2005;6(2):152–162. http://dx.doi.org/10.1038/ni1160.
45. Zhang R, Huynh A, Whitcher G, Chang J, Maltzman JS, Turka LA. An obligate cell-intrinsic function for CD28 in Tregs. *J Clin Invest*. 2013;123(2):580–593. http://dx.doi.org/10.1172/JCI65013.
46. Franckaert D, Dooley J, Roos E, et al. Promiscuous Foxp3-cre activity reveals a differential requirement for CD28 in Foxp3(+) and Foxp3(-) T cells. *Immunol Cell Biol*. 2015;93(4):417–423. http://dx.doi.org/10.1038/icb.2014.108.
47. Hinterberger M, Wirnsberger G, Klein L. B7/CD28 in central tolerance: costimulation promotes maturation of regulatory T cell precursors and prevents their clonal deletion. *Front Immun*. 2011;2:30. http://dx.doi.org/10.3389/fimmu.2011.00030.
48. Gravestein LA, van Ewijk W, Ossendorp F, Borst J. CD27 cooperates with the pre-T cell receptor in the regulation of murine T cell development. *J Exp Med*. 1996;184(2):675–685.
49. Igarashi H, Gregory SC, Yokota T, Sakaguchi N, Kincade PW. Transcription from the RAG1 locus marks the earliest lymphocyte progenitors in bone marrow. *Immunity*. 2002;17(2):117–130.
50. Coquet JM, Ribot JC, Bąbała N, et al. Epithelial and dendritic cells in the thymic medulla promote CD4+Foxp3+ regulatory T cell development via the CD27-CD70 pathway. *J Exp Med*. 2013;210(4):715–728. http://dx.doi.org/10.1084/jem.20112061.
51. Ono M, Yaguchi H, Ohkura N, et al. Foxp3 controls regulatory T-cell function by interacting with AML1/Runx1. *Nature*. 2007;446(7136):685–689. http://dx.doi.org/10.1038/nature05673.
52. Wu Y, Borde M, Heissmeyer V, et al. FOXP3 controls regulatory T cell function through cooperation with NFAT. *Cell*. 2006;126(2):375–387. http://dx.doi.org/10.1016/j.cell.2006.05.042.
53. Wing K, Onishi Y, Prieto-Martin P, et al. CTLA-4 control over Foxp3+ regulatory T cell function. *Science*. 2008;322(5899):271–275. http://dx.doi.org/10.1126/science.1160062.
54. Tai X, Van Laethem F, Sharpe AH, Singer A. Induction of autoimmune disease in CTLA-4-/- mice depends on a specific CD28 motif that is required for in vivo costimulation. *Proc Natl Acad Sci USA*. 2007;104(34):13756–13761. http://dx.doi.org/10.1073/pnas.0706509104.
55. Schmidt EM, Wang CJ, Ryan GA, et al. CTLA-4 controls regulatory T cell peripheral homeostasis and is required for suppression of pancreatic islet autoimmunity. *J Immunol*. 2008;182(1):274–282. http://dx.doi.org/10.4049/jimmunol.182.1.274.

56. Walker LSK, Sansom DM. The emerging role of CTLA4 as a cell-extrinsic regulator of T cell responses. *Nat Rev Immunol.* 2011;11(12):852–863. http://dx.doi.org/10.1038/nri3108.
57. Gavin MA, Rasmussen JP, Fontenot JD, et al. Foxp3-dependent programme of regulatory T-cell differentiation. *Nature.* 2007;445(7129):771–775. http://dx.doi.org/10.1038/nature05543.
58. Zheng SG, Wang JH, Stohl W, Kim KS, Gray JD, Horwitz DA. TGF-beta requires CTLA-4 early after T cell activation to induce FoxP3 and generate adaptive CD4+CD25+ regulatory cells. *J Immunol.* 2006;176(6):3321–3329.
59. Sakaguchi S, Ono M, Setoguchi R, et al. Foxp3+ CD25+ CD4+ natural regulatory T cells in dominant self-tolerance and autoimmune disease. *Immunol Rev.* 2006;212:8–27. http://dx.doi.org/10.1111/j.0105-2896.2006.00427.x.
60. D'Cruz LM, Klein L. Development and function of agonist-induced CD25+Foxp3+ regulatory T cells in the absence of interleukin 2 signaling. *Nat Immunol.* 2005;6(11):1152–1159. http://dx.doi.org/10.1038/ni1264.
61. Fontenot JD, Rasmussen JP, Gavin MA, Rudensky AY. A function for interleukin 2 in Foxp3-expressing regulatory T cells. *Nat Immunol.* 2005;6(11):1142–1151. http://dx.doi.org/10.1038/ni1263.
62. Liston AA, Nutsch KMK, Farr AGA, et al. Differentiation of regulatory Foxp3+ T cells in the thymic cortex. *Proc Natl Acad Sci USA.* 2008;105(33):11903–11908. http://dx.doi.org/10.1073/pnas.0801506105.
63. Tai X, Erman B, Alag A, et al. Foxp3 transcription factor is proapoptotic and lethal to developing regulatory T cells unless counterbalanced by cytokine survival signals. *Immunity.* 2013;38(6):1116–1128. http://dx.doi.org/10.1016/j.immuni.2013.02.022.
64. Bayer AL, Yu A, Malek TR. Function of the IL-2R for thymic and peripheral CD4+CD25+ Foxp3+ T regulatory cells. *J Immunol.* 2007;178(7):4062–4071.
65. Burchill MA, Yang J, Vang KB, Farrar MA. Interleukin-2 receptor signaling in regulatory T cell development and homeostasis. *Immunol Lett.* 2007;114(1):1–8. http://dx.doi.org/10.1016/j.imlet.2007.08.005.
66. Malek TR, Yu A, Vincek V, Scibelli P, Kong L. CD4 regulatory T cells prevent lethal autoimmunity in IL-2Rbeta-deficient mice. Implications for the nonredundant function of IL-2. *Immunity.* 2002;17(2):167–178. http://dx.doi.org/10.1016/S1074-7613(02)00367-9.
67. Marshall D, Sinclair C, Tung S, Seddon B. Differential requirement for IL-2 and IL-15 during bifurcated development of thymic regulatory T cells. *J Immunol.* 2014;193(11):5525–5533. http://dx.doi.org/10.4049/jimmunol.1402144.
68. Thiault N, Darrigues J, Adoue V, et al. Peripheral regulatory T lymphocytes recirculating to the thymus suppress the development of their precursors. *Nat Immunol.* 2015;16(6):628–634. http://dx.doi.org/10.1038/ni.3150.
69. Bayer AL, Lee JY, de la Barrera A, Surh CD, Malek TR. A function for IL-7R for CD4+CD25+Foxp3+ T regulatory cells. *J Immunol.* 2008;181(1):225–234.
70. Pierson W, Cauwe B, Policheni A, et al. Antiapoptotic Mcl-1 is critical for the survival and niche-filling capacity of Foxp3^{+} regulatory T cells. *Nat Immunol.* 2013;14(9):959–965. http://dx.doi.org/10.1038/ni.2649.
71. Almeida ARM, Legrand N, Papiernik M, Freitas AA. Homeostasis of peripheral CD4+ T cells: IL-2R alpha and IL-2 shape a population of regulatory cells that controls CD4+ T cell numbers. *J Immunol.* 2002;169(9):4850–4860.
72. Bettelli E, Dastrange M, Oukka M. Foxp3 interacts with nuclear factor of activated T cells and NF-kappa B to repress cytokine gene expression and effector functions of T helper cells. *Proc Natl Acad Sci USA.* 2005;102(14):5138–5143. http://dx.doi.org/10.1073/pnas.0501675102.

73. Baine I, Basu S, Ames R, Sellers RS, Macian F. Helios induces epigenetic silencing of IL2 gene expression in regulatory T cells. *J Immunol.* 2013;190(3):1008–1016. http://dx.doi.org/10.4049/jimmunol.1200792.
74. Long M, Park S-G, Strickland I, Hayden MS, Ghosh S. Nuclear factor-kappaB modulates regulatory T cell development by directly regulating expression of Foxp3 transcription factor. *Immunity.* 2009;31(6):921–931. http://dx.doi.org/10.1016/j.immuni.2009.09.022.
75. Haxhinasto S, Mathis D, Benoist C. The AKT-mTOR axis regulates de novo differentiation of CD4+Foxp3+ cells. *J Exp Med.* 2008;205(3):565–574. http://dx.doi.org/10.1084/jem.20071477.
76. Delgoffe GM, Pollizzi KN, Waickman AT, et al. The kinase mTOR regulates the differentiation of helper T cells through the selective activation of signaling by mTORC1 and mTORC2. *Nat Immunol.* 2011;12(4):295–303. http://dx.doi.org/10.1038/ni.2005.
77. Bopp T, Palmetshofer A, Serfling E, et al. NFATc2 and NFATc3 transcription factors play a crucial role in suppression of CD4+ T lymphocytes by CD4+ CD25+ regulatory T cells. *J Exp Med.* 2005;201(2):181–187. http://dx.doi.org/10.1084/jem.20041538.
78. Oh-hora M, Yamashita M, Hogan PG, et al. Dual functions for the endoplasmic reticulum calcium sensors STIM1 and STIM2 in T cell activation and tolerance. *Nat Immunol.* 2008;9(4):432–443. http://dx.doi.org/10.1038/ni1574.
79. Guertin DA, Sabatini DM. Defining the role of mTOR in cancer. *Cancer Cell.* 2007;12(1):9–22. http://dx.doi.org/10.1016/j.ccr.2007.05.008.
80. Hay N, Sonenberg N. Upstream and downstream of mTOR. *Genes Dev.* 2004;18(16):1926–1945. http://dx.doi.org/10.1101/gad.1212704.
81. Powell JD, Delgoffe GM. The mammalian target of rapamycin: linking T cell differentiation, function, and metabolism. *Immunity.* 2010;33(3):301–311. http://dx.doi.org/10.1016/j.immuni.2010.09.002.
82. Chapman NM, Chi H. mTOR links environmental signals to T cell fate decisions. *Front Immun.* 2014;5:686. http://dx.doi.org/10.3389/fimmu.2014.00686.
83. Laplante M, Sabatini DM. mTOR signaling in growth control and disease. *Cell.* 2012;149(2):274–293. http://dx.doi.org/10.1016/j.cell.2012.03.017.
84. Chi H. Regulation and function of mTOR signalling in T cell fate decisions. *Nat Rev Immunol.* 2012;12(5):325–338. http://dx.doi.org/10.1038/nri3198.
85. Brownlie RJ, Zamoyska R. T cell receptor signalling networks: branched, diversified and bounded. *Nat Rev Immunol.* 2013;13(4):257–269. http://dx.doi.org/10.1038/nri3403.
86. Vander Haar E, Lee S-I, Bandhakavi S, Griffin TJ, Kim D-H. Insulin signalling to mTOR mediated by the Akt/PKB substrate PRAS40. *Nat Cell Biol.* 2007;9(3):316–323. http://dx.doi.org/10.1038/ncb1547.
87. Zeng H, Yang K, Cloer C, Neale G, Vogel P, Chi H. mTORC1 couples immune signals and metabolic programming to establish T(reg)-cell function. *Nature.* 2013;499(7459):485–490. http://dx.doi.org/10.1038/nature12297.
88. Macintyre AN, Finlay D, Preston G, et al. Protein kinase B controls transcriptional programs that direct cytotoxic T cell fate but is dispensable for T cell metabolism. *Immunity.* 2011;34(2):224–236. http://dx.doi.org/10.1016/j.immuni.2011.01.012.
89. Delgoffe GM, Powell JD. mTOR: taking cues from the immune microenvironment. *Immunology.* 2009;127(4):459–465. http://dx.doi.org/10.1111/j.1365-2567.2009.03125.x.
90. Liu G, Yang K, Burns S, Shrestha S, Chi H. The S1P1-mTOR axis directs the reciprocal differentiation of TH1 and Treg cells. *Nat Immunol.* 2010;11(11):1047–1056. http://dx.doi.org/10.1038/ni.1939.

91. Battaglia M. Rapamycin selectively expands CD4+CD25+FoxP3+ regulatory T cells. *Blood*. 2005;105(12):4743–4748. http://dx.doi.org/10.1182/blood-2004-10-3932.
92. Engel I, Murre C. The function of E- and Id proteins in lymphocyte development. *Nat Rev Immunol*. 2001;1(3):193–199. http://dx.doi.org/10.1038/35105060.
93. Jones ME, Zhuang Y. Acquisition of a functional T cell receptor during T lymphocyte development is enforced by HEB and E2A transcription factors. *Immunity*. 2007;27(6):860–870. http://dx.doi.org/10.1016/j.immuni.2007.10.014.
94. Jones-Mason ME, Zhao X, Kappes D, Lasorella A, Iavarone A, Zhuang Y. E protein transcription factors are required for the development of CD4(+) lineage T cells. *Immunity*. 2012;36(3):348–361. http://dx.doi.org/10.1016/j.immuni.2012.02.010.
95. Murre C. Helix-loop-helix proteins and lymphocyte development. *Nat Immunol*. 2005;6(11):1079–1086. http://dx.doi.org/10.1038/ni1260.
96. Kee BL. E and ID proteins branch out. *Nat Rev Immunol*. 2009;9(3):175–184. http://dx.doi.org/10.1038/nri2507.
97. Schwartz R, Engel I, Fallahi-Sichani M, Petrie HT, Murre C. Gene expression patterns define novel roles for E47 in cell cycle progression, cytokine-mediated signaling, and T lineage development. *Proc Natl Acad Sci USA*. 2006;103(26):9976–9981. http://dx.doi.org/10.1073/pnas.0603728103.
98. Schmidt-Supprian M, Tian J, Grant EP, et al. Differential dependence of CD4+CD25+ regulatory and natural killer-like T cells on signals leading to NF-kappaB activation. *Proc Natl Acad Sci USA*. 2004;101(13):4566–4571. http://dx.doi.org/10.1073/pnas.0400885101.
99. Gupta S, Manicassamy S, Vasu C, Kumar A, Shang W, Sun Z. Differential requirement of PKC-theta in the development and function of natural regulatory T cells. *Mol Immunol*. 2008;46(2):213–224. http://dx.doi.org/10.1016/j.molimm.2008.08.275.
100. Medoff BD, Sandall BP, Landry A, et al. Differential requirement for CARMA1 in agonist-selected T-cell development. *Eur J Immunol*. 2009;39(1):78–84. http://dx.doi.org/10.1002/eji.200838734.
101. Barnes MJ, Krebs P, Harris N, et al. Commitment to the regulatory T cell lineage requires CARMA1 in the thymus but not in the periphery. *PLoS Biol*. 2009;7(3):e51. http://dx.doi.org/10.1371/journal.pbio.1000051.
102. Isomura I, Palmer S, Grumont RJ, et al. c-Rel is required for the development of thymic Foxp3+ CD4 regulatory T cells. *J Exp Med*. 2009;206(13):3001–3014. http://dx.doi.org/10.1084/jem.20091411.
103. Loizou L, Andersen KG, Betz AG. Foxp3 interacts with c-Rel to mediate NF-κB repression. *PLoS One*. 2011;6(4):e18670. http://dx.doi.org/10.1371/journal.pone.0018670.
104. Schuster M, Glauben R, Plaza-Sirvent C, et al. IκB(NS) protein mediates regulatory T cell development via induction of the Foxp3 transcription factor. *Immunity*. 2012;37(6):998–1008. http://dx.doi.org/10.1016/j.immuni.2012.08.023.
105. Ghosh S, Hayden MS. New regulators of NF-κB in inflammation. *Nat Rev Immunol*. 2008;8(11):837–848. http://dx.doi.org/10.1038/nri2423.
106. Liston A. Is foxp3 the master regulator of regulatory T cells? *Prog Mol Biol Transl Sci*. 2010;92:315–317. http://dx.doi.org/10.1016/S1877-1173(10)92017-6.
107. Kerdiles YM, Stone EL, Beisner DR, et al. Foxo transcription factors control regulatory T cell development and function. *Immunity*. 2010;33(6):890–904. http://dx.doi.org/10.1016/j.immuni.2010.12.002.
108. Ouyang W, Beckett O, Ma Q, Paik J-H, DePinho RA, Li MO. Foxo proteins cooperatively control the differentiation of Foxp3+ regulatory T cells. *Nat Immunol*. 2010;11(7):618–627. http://dx.doi.org/10.1038/ni.1884.

109. Ouyang W, Liao W, Luo CT, et al. Novel Foxo1-dependent transcriptional programs control T(reg) cell function. *Nature*. 2012;491(7425):554–559. http://dx.doi.org/10.1038/nature11581.
110. Rudra D, deRoos P, Chaudhry A, et al. Transcription factor Foxp3 and its protein partners form a complex regulatory network. *Nat Immunol*. 2012;13(10):1010–1019. http://dx.doi.org/10.1038/ni.2402.
111. Marson A, Kretschmer K, Frampton GM, et al. Foxp3 occupancy and regulation of key target genes during T-cell stimulation. *Nature*. 2007;445(7130):931–935. http://dx.doi.org/10.1038/nature05478.
112. Zheng Y, Rudensky AY. Foxp3 in control of the regulatory T cell lineage. *Nat Immunol* 2007;8(5):457–462. http://dx.doi.org/10.1038/ni1455.
113. Samstein RM, Arvey A, Josefowicz SZ, et al. Foxp3 exploits a pre-existent enhancer landscape for regulatory T cell lineage specification. *Cell*. 2012;151(1):153–166. http://dx.doi.org/10.1016/j.cell.2012.06.053.
114. Rudra D, Egawa T, Chong MMW, Treuting P, Littman DR, Rudensky AY. Runx-CBFbeta complexes control expression of the transcription factor Foxp3 in regulatory T cells. *Nat Immunol*. 2009;10(11):1170–1177. http://dx.doi.org/10.1038/ni.1795.
115. Kitoh A, Ono M, Naoe Y, et al. Indispensable role of the Runx1-Cbfbeta transcription complex for in vivo-suppressive function of FoxP3+ regulatory T cells. *Immunity*. 2009;31(4):609–620. http://dx.doi.org/10.1016/j.immuni.2009.09.003.
116. Klunker S, Chong MMW, Mantel P-Y, et al. Transcription factors RUNX1 and RUNX3 in the induction and suppressive function of Foxp3+ inducible regulatory T cells. *J Exp Med*. 2009;206(12):2701–2715. http://dx.doi.org/10.1084/jem.20090596.
117. Bartel DP. MicroRNAs: genomics, biogenesis, mechanism, and function. *Cell*. 2004;116(2):281–297.
118. Han J, Lee Y, Yeom K-H, et al. Molecular basis for the recognition of primary microRNAs by the Drosha-DGCR8 complex. *Cell*. 2006;125(5):887–901. http://dx.doi.org/10.1016/j.cell.2006.03.043.
119. Yi R, Qin Y, Macara IG, Cullen BR. Exportin-5 mediates the nuclear export of pre-microRNAs and short hairpin RNAs. *Genes Dev*. 2003;17(24):3011–3016. http://dx.doi.org/10.1101/gad.1158803.
120. Chendrimada TP, Finn KJ, Ji X, et al. MicroRNA silencing through RISC recruitment of eIF6. *Nature*. 2007;447(7146):823–828. http://dx.doi.org/10.1038/nature05841.
121. Bartel DP. MicroRNAs: target recognition and regulatory functions. *Cell*. 2009;136(2):215–233. http://dx.doi.org/10.1016/j.cell.2009.01.002.
122. Lodish HF, Zhou B, Liu G, Chen C-Z. Micromanagement of the immune system by microRNAs. *Nat Rev Immunol*. 2008;8(2):120–130. http://dx.doi.org/10.1038/nri2252.
123. Cobb BS, Hertweck A, Smith J, et al. A role for Dicer in immune regulation. *J Exp Med*. 2006;203(11):2519–2527. http://dx.doi.org/10.1084/jem.20061692.
124. Chong MMW, Rasmussen JP, Rudensky AY, Rundensky AY, Littman DR. The RNAseIII enzyme Drosha is critical in T cells for preventing lethal inflammatory disease. *J Exp Med*. 2008;205(9):2005–2017. http://dx.doi.org/10.1084/jem.20081219.
125. Liston AA, Lu L-FL, O'Carroll DD, Tarakhovsky AA, Rudensky AYA. Dicer-dependent microRNA pathway safeguards regulatory T cell function. *J Exp Med*. 2008;205(9):1993–2004. http://dx.doi.org/10.1084/jem.20081062.
126. Zhou X, Jeker LT, Fife BT, et al. Selective miRNA disruption in T reg cells leads to uncontrolled autoimmunity. *J Exp Med*. 2008;205(9):1983–1991. http://dx.doi.org/10.1084/jem.20080707.

127. Jeker LT, Zhou X, Blelloch R, Bluestone JA. DGCR8-mediated production of canonical microRNAs is critical for regulatory T cell function and stability. *PLoS One*. 2013;8(5):e66282. http://dx.doi.org/10.1371/journal.pone.0066282.
128. Rouas R, Fayyad-Kazan H, Zein El N, et al. Human natural Treg microRNA signature: role of microRNA-31 and microRNA-21 in FOXP3 expression. *Eur J Immunol*. 2009;39(6):1608–1618. http://dx.doi.org/10.1002/eji.200838509.
129. Lu L-F, Boldin MP, Chaudhry A, et al. Function of miR-146a in controlling Treg cell-mediated regulation of Th1 responses. *Cell*. 2010;142(6):914–929. http://dx.doi.org/10.1016/j.cell.2010.08.012.
130. Lu L-F, Thai T-H, Calado DP, et al. Foxp3-dependent microRNA155 confers competitive fitness to regulatory T cells by targeting SOCS1 protein. *Immunity*. 2009;30(1):80–91. http://dx.doi.org/10.1016/j.immuni.2008.11.010.
131. Fontenot JDJ, Rasmussen JPJ, Williams LML, Dooley JLJ, Farr AGA, Rudensky AYA. Regulatory T cell lineage specification by the forkhead transcription factor foxp3. *Immunity*. 2005;22(3):329–341. http://dx.doi.org/10.1016/j.immuni.2005.01.016.
132. Wan YY, Flavell RA. Regulatory T-cell functions are subverted and converted owing to attenuated Foxp3 expression. *Nature*. 2007;445(7129):766–770. http://dx.doi.org/10.1038/nature05479.
133. Gavin MA, Torgerson TR, Houston E, et al. Single-cell analysis of normal and FOXP3-mutant human T cells: FOXP3 expression without regulatory T cell development. *Proc Natl Acad Sci USA*. 2006;103(17):6659–6664. http://dx.doi.org/10.1073/pnas.0509484103.
134. Tran DQ, Ramsey H, Shevach EM. Induction of FOXP3 expression in naive human CD4+FOXP3 T cells by T-cell receptor stimulation is transforming growth factor-beta dependent but does not confer a regulatory phenotype. *Blood*. 2007;110(8):2983–2990. http://dx.doi.org/10.1182/blood-2007-06-094656.
135. Miyao T, Floess S, Setoguchi R, et al. Plasticity of Foxp3(+) T cells reflects promiscuous Foxp3 expression in conventional T cells but not reprogramming of regulatory T cells. *Immunity*. 2012;36(2):262–275. http://dx.doi.org/10.1016/j.immuni.2011.12.012.
136. d'Hennezel E, Yurchenko E, Sgouroudis E, Hay V, Piccirillo CA. Single-cell analysis of the human T regulatory population uncovers functional heterogeneity and instability within FOXP3+ cells. *J Immunol*. 2011;186(12):6788–6797. http://dx.doi.org/10.4049/jimmunol.1100269.
137. Bin Dhuban K, Kornete M, Mason ES, Piccirillo CA. Functional dynamics of Foxp3$^+$ regulatory T cells in mice and humans. *Immunol Rev*. 2014;259(1):140–158. http://dx.doi.org/10.1111/imr.12168.
138. Da Silva Martins M, Piccirillo CA. Functional stability of Foxp3+ regulatory T cells. *Trends Mol Med*. 2012;18(8):454–462. http://dx.doi.org/10.1016/j.molmed.2012.06.001.
139. Zhou X, Bailey-Bucktrout SL, Jeker LT, et al. Instability of the transcription factor Foxp3 leads to the generation of pathogenic memory T cells in vivo. *Nat Immunol*. 2009;10(9):1000–1007. http://dx.doi.org/10.1038/ni.1774.
140. Komatsu N, Mariotti-Ferrandiz ME, Wang Y, Malissen B, Waldmann H, Hori S. Heterogeneity of natural Foxp3+ T cells: a committed regulatory T-cell lineage and an uncommitted minor population retaining plasticity. *Proc Natl Acad Sci USA*. 2009;106(6):1903–1908. http://dx.doi.org/10.1073/pnas.0811556106.
141. Hill JA, Feuerer M, Tash K, et al. Foxp3 transcription-factor-dependent and -independent regulation of the regulatory T cell transcriptional signature. *Immunity*. 2007;27(5):786–800. http://dx.doi.org/10.1016/j.immuni.2007.09.010.
142. Baron U, Floess S, Wieczorek G, et al. DNA demethylation in the human FOXP3 locus discriminates regulatory T cells from activated FOXP3(+) conventional

T cells. *Eur J Immunol.* 2007;37(9):2378–2389. http://dx.doi.org/10.1002/eji.200737594.

143. Huehn J, Polansky JK, Hamann A. Epigenetic control of FOXP3 expression: the key to a stable regulatory T-cell lineage? *Nat Rev Immunol.* 2009;9(2):83–89. http://dx.doi.org/10.1038/nri2474.
144. Ohkura NN, Hamaguchi MM, Morikawa HH, et al. T cell receptor stimulation-induced epigenetic changes and foxp3 expression are independent and complementary events required for treg cell development. *Immunity.* 2012;37(5):785–799. http://dx.doi.org/10.1016/j.immuni.2012.09.010.
145. Schmidl C, Klug M, Boeld TJ, et al. Lineage-specific DNA methylation in T cells correlates with histone methylation and enhancer activity. *Genome Res.* 2009;19(7):1165–1174. http://dx.doi.org/10.1101/gr.091470.109.
146. Morikawa H, Ohkura N, Vandenbon A, et al. Differential roles of epigenetic changes and Foxp3 expression in regulatory T cell-specific transcriptional regulation. *Proc Natl Acad Sci USA.* 2014;111(14):5289–5294. http://dx.doi.org/10.1073/pnas.1312717110.
147. Toker A, Engelbert D, Garg G, et al. Active demethylation of the Foxp3 locus leads to the generation of stable regulatory T cells within the thymus. *J Immunol.* 2013;190(7):3180–3188. http://dx.doi.org/10.4049/jimmunol.1203473.
148. Itoh M, Takahashi T, Sakaguchi N, et al. Thymus and autoimmunity: production of CD25+CD4+ naturally anergic and suppressive T cells as a key function of the thymus in maintaining immunologic self-tolerance. *J Immunol.* 1999;162(9):5317–5326.
149. Kawahata K, Misaki Y, Yamauchi M, et al. Generation of CD4(+)CD25(+) regulatory T cells from autoreactive T cells simultaneously with their negative selection in the thymus and from nonautoreactive T cells by endogenous TCR expression. *J Immunol.* 2002;168(9):4399–4405.
150. Yamaguchi T, Kishi A, Osaki M, et al. Construction of self-recognizing regulatory T cells from conventional T cells by controlling CTLA-4 and IL-2 expression. *Proc Natl Acad Sci USA.* 2013;110(23):E2116–E2125. http://dx.doi.org/10.1073/pnas.1307185110.
151. Furtado GC, Curotto de Lafaille MA, Kutchukhidze N, Lafaille JJ. Interleukin 2 signaling is required for CD4(+) regulatory T cell function. *J Exp Med.* 2002; 196(6):851–857.
152. Thornton AM, Korty PE, Tran DQ, et al. Expression of Helios, an Ikaros transcription factor family member, differentiates thymic-derived from peripherally induced Foxp3+ T regulatory cells. *J Immunol.* 2010;184(7):3433–3441. http://dx.doi.org/10.4049/jimmunol.0904028.
153. Haribhai D, Williams JB, Jia S, et al. A requisite role for induced regulatory T cells in tolerance based on expanding antigen receptor diversity. *Immunity.* 2011;35(1):109–122. http://dx.doi.org/10.1016/j.immuni.2011.03.029.
154. Josefowicz SZ, Niec RE, Kim HY, et al. Extrathymically generated regulatory T cells control mucosal TH2 inflammation. *Nature.* 2012;482(7385):395–399. http://dx.doi.org/10.1038/nature10772.
155. Paiva RS, Lino AC, Bergman M-L, et al. Recent thymic emigrants are the preferential precursors of regulatory T cells differentiated in the periphery. *Proc Natl Acad Sci USA.* 2013;110(16):6494–6499. http://dx.doi.org/10.1073/pnas.1221955110.
156. Curotto de Lafaille MA, Lafaille JJ. Natural and adaptive foxp3+ regulatory T cells: more of the same or a division of labor? *Immunity.* 2009;30(5):626–635. http://dx.doi.org/10.1016/j.immuni.2009.05.002.
157. Liu VC, Wong LY, Jang T, et al. Tumor evasion of the immune system by converting CD4+CD25- T cells into CD4+CD25+ T regulatory cells: role of tumor-derived TGF-beta. *J Immunol.* 2007;178(5):2883–2892.

158. Zheng SG, Wang J, Horwitz DA. Cutting edge: Foxp3+CD4+CD25+ regulatory T cells induced by IL-2 and TGF-beta are resistant to Th17 conversion by IL-6. *J Immunol*. 2008;180(11):7112–7116.
159. Lathrop SK, Santacruz NA, Pham D, Luo J, Hsieh CS. Antigen-specific peripheral shaping of the natural regulatory T cell population. *J Exp Med*. 2008;205(13):3105–3117. http://dx.doi.org/10.1073/pnas.0610289104.
160. Kretschmer K, Apostolou I, Hawiger D, Khazaie K, Nussenzweig MC, Boehmer von H. Inducing and expanding regulatory T cell populations by foreign antigen. *Nat Immunol*. 2005;6(12):1219–1227. http://dx.doi.org/10.1038/ni1265.
161. Sauer S, Bruno L, Hertweck A, et al. T cell receptor signaling controls Foxp3 expression via PI3K, Akt, and mTOR. *Proc Natl Acad Sci USA*. 2008;105(22):7797–7802. http://dx.doi.org/10.1073/pnas.0800928105.
162. Gottschalk RA, Corse E, Allison JP. TCR ligand density and affinity determine peripheral induction of Foxp3 in vivo. *J Exp Med*. 2010;207(8):1701–1711. http://dx.doi.org/10.1084/jem.20091999.
163. Semple K, Nguyen A, Yu Y, Wang H, Anasetti C, Yu X-Z. Strong CD28 costimulation suppresses induction of regulatory T cells from naive precursors through Lck signaling. *Blood*. 2011;117(11):3096–3103. http://dx.doi.org/10.1182/blood-2010-08-301275.
164. Wohlfert EA, Gorelik L, Mittler R, Flavell RA, Clark RB. Cutting edge: deficiency in the E3 ubiquitin ligase Cbl-b results in a multifunctional defect in T cell TGF-beta sensitivity in vitro and in vivo. *J Immunol*. 2006;176(3):1316–1320.
165. Harada Y, Harada Y, Elly C, et al. Transcription factors Foxo3a and Foxo1 couple the E3 ligase Cbl-b to the induction of Foxp3 expression in induced regulatory T cells. *J Exp Med*. 2010;207(7):1381–1391. http://dx.doi.org/10.1084/jem.20100004.
166. Henderson JG, Opejin A, Jones A, Gross C, Hawiger D. CD5 instructs extrathymic regulatory T cell development in response to self and tolerizing antigens. *Immunity*. 2015;42(3):471–483. http://dx.doi.org/10.1016/j.immuni.2015.02.010.
167. Josefowicz SZ, Wilson CB, Rudensky AY. Cutting edge: TCR stimulation is sufficient for induction of Foxp3 expression in the absence of DNA methyltransferase 1. *J Immunol*. 2009;182(11):6648–6652. http://dx.doi.org/10.4049/jimmunol.0803320.
168. Schallenberg S, Tsai P-Y, Riewaldt J, Kretschmer K. Identification of an immediate Foxp3(-) precursor to Foxp3(+) regulatory T cells in peripheral lymphoid organs of nonmanipulated mice. *J Exp Med*. 2010;207(7):1393–1407. http://dx.doi.org/10.1084/jem.20100045.
169. Laurence A, Tato CM, Davidson TS, et al. Interleukin-2 signaling via STAT5 constrains T helper 17 cell generation. *Immunity*. 2007;26(3):371–381. http://dx.doi.org/10.1016/j.immuni.2007.02.009.
170. Zhou LL, Lopes JEJ, Chong MMWM, et al. TGF-beta-induced Foxp3 inhibits T(H)17 cell differentiation by antagonizing RORgammat function. *Nature*. 2008;453(7192):236–240. http://dx.doi.org/10.1038/nature06878.
171. Xiao S, Jin H, Korn T, et al. Retinoic acid increases Foxp3+ regulatory T cells and inhibits development of Th17 cells by enhancing TGF-beta-driven Smad3 signaling and inhibiting IL-6 and IL-23 receptor expression. *J Immunol*. 2008;181(4):2277–2284.
172. Coombes JL, Siddiqui KRR, Arancibia-Carcamo CV, et al. A functionally specialized population of mucosal CD103+ DCs induces Foxp3+ regulatory T cells via a TGF- and retinoic acid dependent mechanism. *J Exp Med*. 2007;204(8):1757–1764. http://dx.doi.org/10.1093/nar/29.9.e45.
173. Sun CM, Hall JA, Blank RB, et al. Small intestine lamina propria dendritic cells promote de novo generation of Foxp3 T reg cells via retinoic acid. *J Exp Med*. 2007;204(8):1775–1785. http://dx.doi.org/10.1084/jem.188.2.287.

174. Benson MJ, Pino-Lagos K, Rosemblatt M, Noelle RJ. All-trans retinoic acid mediates enhanced T reg cell growth, differentiation, and gut homing in the face of high levels of co-stimulation. *J Exp Med.* 2007;204(8):1765–1774. http://dx.doi.org/10.1084/jem.20070719.
175. Mucida D, Park Y, Kim G, et al. Reciprocal TH17 and regulatory T cell differentiation mediated by retinoic acid. *Science.* 2007;317(5835):256–260. http://dx.doi.org/10.1126/science.1145697.
176. Davidson TS, DiPaolo RJ, Andersson J, Shevach EM. Cutting Edge: IL-2 is essential for TGF-beta-mediated induction of Foxp3+ T regulatory cells. *J Immunol.* 2007;178(7):4022–4026.
177. Singh N, Gurav A, Sivaprakasam S, et al. Activation of Gpr109a, receptor for niacin and the commensal metabolite butyrate, suppresses colonic inflammation and carcinogenesis. *Immunity.* 2014;40(1):128–139. http://dx.doi.org/10.1016/j.immuni.2013.12.007.
178. Arpaia N, Campbell C, Fan X, et al. Metabolites produced by commensal bacteria promote peripheral regulatory T-cell generation. *Nature.* 2013;504(7480):451–455. http://dx.doi.org/10.1038/nature12726.
179. Smith PM, Howitt MR, Panikov N, et al. The microbial metabolites, short-chain fatty acids, regulate colonic Treg cell homeostasis. *Science.* 2013;341(6145):569–573. http://dx.doi.org/10.1126/science.1241165.
180. Furusawa Y, Obata Y, Fukuda S, et al. Commensal microbe-derived butyrate induces the differentiation of colonic regulatory T cells. *Nature.* 2013;504(7480):446–450. http://dx.doi.org/10.1038/nature12721.
181. Atarashi K, Tanoue T, Shima T, et al. Induction of colonic regulatory T cells by indigenous Clostridium species. *Science.* 2011;331(6015):337–341. http://dx.doi.org/10.1126/science.1198469.
182. Cebula A, Seweryn M, Rempala GA, et al. Thymus-derived regulatory T cells contribute to tolerance to commensal microbiota. *Nature.* 2013;497(7448):258–262. http://dx.doi.org/10.1038/nature12079.
183. Obata Y, Furusawa Y, Endo TA, et al. The epigenetic regulator Uhrf1 facilitates the proliferation and maturation of colonic regulatory T cells. *Nat Immunol.* 2014;15(6):571–579. http://dx.doi.org/10.1038/ni.2886.
184. Basu R, Hatton RD, Weaver CT. The Th17 family: flexibility follows function. *Immunol Rev.* 2013;252(1):89–103. http://dx.doi.org/10.1111/imr.12035.

CHAPTER FIVE

The Special Relationship in the Development and Function of T Helper 17 and Regulatory T Cells

Matthias Lochner*,[1], Zuobai Wang*,[1], Tim Sparwasser*,[2]

*Institute of Infection Immunology, TWINCORE, Centre for Experimental and Clinical Infection Research: A Joint Venture Between the Medical School Hannover (MHH) and the Helmholtz Centre for Infection Research (HZI), Hannover, Germany

[2]Corresponding author: e-mail address: Sparwasser.Tim@mh-hannover.de

Contents

Abstract

T helper 17 (Th17) cells play an essential role in the clearance of extracellular pathogenic bacteria and fungi. However, this subset is critically involved in the pathology of many autoimmune diseases, e.g., psoriasis, multiple sclerosis, allergy, rheumatoid arthritis, and inflammatory bowel diseases in humans. Therefore, Th17 responses need to be tightly regulated *in vivo* to mediate effective host defenses against pathogens without causing excessive host tissue damage. $Foxp3^{+}$ regulatory T (Treg) cells play an important role in maintaining peripheral tolerance to self-antigens and in counteracting the inflamma-tory activity of effector T helper cell subsets. Although Th17 and Treg cells represent two $CD4^{+}$ T cell subsets with opposing principal functions, these cell types are function-ally connected. In this review, we will first give an overview on the biology of Th17 cells

[1] Contributed equally.

Progress in Molecular Biology and Translational Science, Volume 136
ISSN 1877-1173
http://dx.doi.org/10.1016/bs.pmbts.2015.07.013

and describe their development and *in vivo* function, followed by an account on the special developmental relationship between Th17 and Treg cells. We will describe the identification of Treg/Th17 intermediates and consider their lineage stability and function *in vivo*. Finally, we will discuss how Treg cells may regulate the Th17 cell response in the context of infection and inflammation, and elude on findings demonstrating that Treg cells can also have a prominent function in promoting the differentiation of Th17 cells.

1. T HELPER 17 CELLS

Th17 cells were recognized as an independent T helper cell lineage and the cellular source of IL-17 in the year 2005[1–3], although the production of this cytokine by T cells had already been described earlier.[4,5] Since then, a plethora of studies and publications have shed light into the development and function of this important T cell lineage. It has become clear that, besides the "signature" cytokine IL-17A, Th17 cells can express an array of factors that include IL-17F, IL-6, IL-10, IL-22, GM-CSF, TNFα, CCL20, CCL22, IL1Rα, or the IL-23R. In addition to the $CD4^+$ $TCR\alpha\beta^+$ Th17 cells, other IL-17-producing cells have been identified, including subsets of $TCR\gamma\delta^+$ T cells,[6] invariant NKT cells[7] or subpopulations of group 3 innate lymphoid cells (ILC).[8]

1.1 The Development of Th17 Cells

The path for the discovery of Th17 cells was opened in the year 2003 by the seminal finding that the autoimmune pathology observed in experimental autoimmune encephalomyelitis (EAE), the murine model of multiple sclerosis, was not dependent on IL-12 and thus Th1 cells, but rather on the cytokine IL-23.[9] This was followed by the discovery that IL-23 promotes the development and expansion of an IL-17-producing $CD4^+$ T helper cell lineage that was eventually termed Th17 cells.[1–3] Nevertheless, IL-23 alone cannot drive the differentiation of naïve T cells into Th17 cells, and it was shown simultaneously by several groups, that the addition of TGF-β1 and IL-6 during initial TCR activation promotes the differentiation of Th17 cells.[10–12] IL-6 plays a crucial role during the initial phase by activating signal transducer and activator of transcription (STAT)3, which directly promotes the transcription of Th17 specific genes such as *Rorc* (encoding for retinoic acid receptor-related orphan receptor (ROR)γt in T cells), *Il17* and *Il23r*.[13] RORγt was first cloned as a shorter isoform of the hormone receptor RORγ and described to be expressed in the thymus and in

T cell lines.[14] Besides its function in the context of thymocyte development,[15–18] further studies demonstrated that RORγt is expressed in lymphoid tissue inducer cells,[19] and thus plays an important role for lymphoid tissue development as well as for the function of group 3 innate lymphoid cells (ILC).[20,21] However, work by the Littman group in 2006 demonstrated that RORγt was necessary for the development and function of the newly discovered $CD4^+$ Th17 cell lineage, clearly defining RORγt as the "master transcription factor" for Th17 cells.[22] RORγt directly regulates the transcription of Th17 specific genes such as *IL17a*, *Il17f*, and *Il23r* and is a critical component of a larger network of transcription factors like STAT3, IRF4, or BATF. This transcriptional network collectively initiates the complete Th17 differentiation program and is important to maintain Th17 lineage stability.[13,23] Besides IL-6, which is crucial for initiating Th17 lineage specification, IL-21 has been demonstrated to serve as an autocrine factor important for growth and expansion of Th17 cells.[24,25]

Another important proinflammatory cytokine for the induction of Th17 cells is IL-1β. The expression of IL1R during the initiation phase of Th17 cells is induced by IL-6, and signaling through the IL1R leads to enhanced IRF4 expression, which in turn further promotes the expression of RORγt.[26] Interestingly, it has been demonstrated that mechanistically the role of IL-1β in enforcing Th17 differentiation is mediated via its capacity to activate the mTOR pathway.[27] mTOR is an important regulator of the cellular metabolism, and its activation is crucial for the TCR-mediated development of naïve T cells into effector T cells. One important function of mTOR is the activation of catabolic pathways, especially the induction of aerobic glycolysis and *de novo* fatty acid synthesis, which play an essential role in Th17 differentiation.[28–30] In line with these findings, activation of the transcription factor HIF1α, a downstream target of mTOR and an important positive regulator of glycolysis, has been shown to critically contribute to Th17 lineage development.[31,32] These latter results nicely illustrate how signaling pathways influence changes in basic cellular metabolic processes that are necessary to allow the development of Th17 cells.

1.2 Evidence for a High Functional Diversity and Lineage Plasticity of Th17 Cells

In contrast to IL-6 and IL-1β, the role of TGF-β in Th17 differentiation is less well understood. There is ample evidence that TGF-β is of significance for Th17 development *in vivo*[33–35] and, as mentioned above, the combination of TGF-β and IL-6 efficiently generates Th17 cells *in vitro*.

Nevertheless, it is possible that a prominent function of TGF-β during Th17 differentiation is to suppress skewing into Th1 or Th2 direction.[36] In line with this notion, it was shown that Th17 cells could be efficiently generated in response to IL-1β, IL-6, and IL-23 in a TGF-β-independent manner.[37] However, while Th17 cells that are generated in the presence of TGF-β and IL-6 are only weakly pathogenic and express significant levels of IL-10,[38] it has become clear that the presence of IL-23 is critical for the differentiation of pathogenic Th17 cells.[37,39,40] In fact, it is very likely that the *in vivo* environment and the presence of specific factors and cytokines govern a high degree of functional plasticity within the Th17 lineage. In that respect, it has recently been shown in a model of anti-CD3-induced intestinal inflammation that Th17 cells can be redirected into a regulatory phenotype. This was shown to be dependent on the CCL20–CCR6-dependent recruitment of Th17 cells into the intestine and the induction of immune-suppressive factors such as IL-10 in these cells.[41] Along the same line, it was demonstrated that in the absence of sufficient amounts of IL-23, the cytokines IL-12 and IL-27 can drive precommitted Th17 cells into an IL-10-secreting, Tr1-like phenotype.[42] In contrast, presence of IL-23 induces autocrine production of TGF-β3 from developing Th17 cells, which, in concert with IL-6, induced the development of very pathogenic Th17 cells, further demonstrating the importance of IL-23 for the development of proinflammatory Th17 cells.[43] In addition, this study by the Kuchroo group demonstrated that TGF-β3-induced pathogenic Th17 cells displayed a molecular distinct pattern compared to the classic TGF-β1 induced Th17 cells, including the expression of several Th1-associated molecules such as T-bet. Indeed, several studies have shown that IL-23 induces the upregulation of the Th1 transcription factor T-bet with concomitant IFN-γ expression in Th17 cells,[37,44–46] and such IL-17^+IFN-γ^+ cells have been associated with an enhanced reactivity and pathological potential, e.g. in the intestine.[47–50] This functional diversity of Th17 cells is accompanied by a considerable tendency for lineage plasticity. In other words, the expression of IL-17A/F and RORγt (the minimal requirements for $CD4^+$ T cell to be considered as a Th17 cell) may not reflect the end stage of Th17 effector/memory development, but represent a more or less stable intermediate within the transition into other functional T helper cell lineages. Consistent with this idea, IL-17 production in *in vitro* polarized Th17 cells is unstable and lost after several rounds of restimulation, leading eventually to an $IFN\gamma^+$ Th1 phenotype.[51–53] In addition, studies with purified *in vitro*-generated Th17 cells transferred to NOD mice revealed tissue infiltrating cells changing their phenotype to become Th1 cells.[54,55] Using a fate mapping strategy *in vivo*, it was demonstrated that the majority of T cells that expressed IL-17A/F at some stage in their life cycle had lost IL-17 expression at the time point of analysis.[44,56] Fate mapping of IL-17A-expressing reporter T cells in the context of EAE revealed that Th17 cells downregulate their IL-17 and RORγt expression and switch to a Th1-like phenotype.[44] Importantly, the results of this study suggest that IFN-γ and other proinflammatory cytokines in the spinal cord were produced almost exclusively by cells that had produced IL-17 before their conversion, indicating that Th17 cells indeed converted into more pathogenic "exTh17" cells, at least under the chronic inflammatory conditions of this model. Nevertheless, recent evidence suggests that Th17 cells can also go along a different path and switch into IL-17-negative Tr1-like cells with a regulatory function. Using a triple transgenic IL-17 fate mapping/IL-10 reporter approach, Gagliani and colleagues showed that Th17 cells transdifferentiate in the anti-CD3-induced model of intestinal inflammation as well as during *S. aureus*-mediated sepsis into suppressive IL-10$^+$ "exTh17" cells with low expression of RORγt, Foxp3 and genes associated with Th1 or Th2 cells.[57] Interestingly, the authors found that this conversion was dependent on TGF-β1 signaling through Smad3 and could be enhanced by activation of the aryl hydrocarbon receptor (AhR) expressed on TGF-β1-induced Th17 cells. Another example of Th17 lineage conversion was described by the Stockinger group again using IL-17A fate mapping mice.[58] Under non-inflammatory conditions, transferred Th17 cells preferentially homed to the intestine and were found to convert into functional follicular T(fh) helper cells in the Peyer's patches, as illustrated by loss of IL-17 and RORγt expression and the concomitant upregulation of Tfh-specific factors such as CXCR5, IL-21, and Bcl6.

Although these studies indicated that Th17 cells can readily lose their phenotype, it should be noted that converted Th17 cells may still carry specific traits that differentiates them from other Th cell lineages. The study by Muranski *et al.* suggests that IFN-γ$^+$T-bet$^+$ "exTh17" cells differ significantly in their molecular expression profile compared to classical derived Th1 cells.[59] In addition, it has been demonstrated that Th17 cells can be long lived and display a high potential for self-renewal in mice as well as in humans.[59,60]

1.3 Induction of Th17 Cells under Physiological Conditions *in vivo*

Under normal physiological conditions, the greatest number of RORγt$^+$Th17 cells is found in barrier tissues like the skin, the mucosa of the respiratory tract and especially in the lamina propria of the intestine.[22,61] Interestingly, the presence of microbiota is absolutely required for the induction of Th17 cells in the intestine, since germ-free mice do not contain IL-17/IL-22 producing Th17 cells.[62–64] A particular potent bacterial inducer of Th17 cells is *Candidatus arthromitus* (also known as segmented filamentous bacteria, SFB),[63] and it was demonstrated that Th17 cells within the small intestine possess a distinct TCR repertoire that recognizes and specifically responds to SFB-antigens.[65,66] However, even though SFB is a very efficient inducer of Th17 cells, it also induces other types of T helper cells such as Th1, Th2, and Treg as well as the production of IgA.[62,67] Furthermore, other groups of bacteria can also efficiently induce Th17 cells in the intestine.[62,64] Despite the strict requirement for the presence of microbes, intestinal Th17-induction was initially shown to be independent of MyD88/TRIF- and NOD-transmitted signals, and thus independent of classic microbe-associated molecular patterns.[63,68,69] Instead, colonization with SFB results in a strong upregulation of Serum Amyloid A (SAA), a member of the family of acute-phase response proteins induced during infection, tissue damage or inflammatory disease. SAA can induce Th17 differentiation, at least *in vitro*, by stimulating IL-6 and IL-23 expression by dendritic cells (DC).[63] This finding is in line with several reports demonstrating the important role of DC for the induction of intestinal Th17 cells. In the intestine, loss or reduction of a specific population of DC characterized by the expression of CD103 and CD11b, as observed in mice with a DC-specific deletion in IRF4, Notch2 or in human Langerin-DTA mice, results in significantly reduced numbers of Th17 cells.[70–74] $CD103^+CD11b^+$ DC can express high levels of IL-6 and/or IL-23,[71–73] suggesting that $CD103^+CD11b^+$ DC contribute to the development of intestinal Th17 mainly by the production of those cytokines. Interestingly, the addition of ligands for several TLRs, including TLR2, TLR5 and TLR9 to *in vitro* cultures strongly induce the production of IL-6 by $CD103^+CD11b^+$ DC, indicating that MyD88-mediated inflammatory signals are required for their ability to induce Th17 cells.[71,75,76] Previous data highlighted the specific expression of IL-23 by a subset of intestinal $TLR5^+CD11b^+$ DC in promoting Th17 T helper cells.[77] In addition, a recent study further emphasized the crucial role of $CD103^+CD11b^+$ DC for IL-23 production after stimulation by the TLR5 ligand flagellin, although the direct impact on Th17 induction was not analyzed in this study.[78] A direct role for IL-1β in the microbiota-mediated induction of intestinal Th17 cells was also shown. $CD11b^+CD11c^{-/low}$ macrophages were identified as the main source of IL-1β and, importantly, the induction of IL-1 production was shown to be dependent on MyD88-signaling.[79] Thus, while there is now good evidence for an essential role of DC, the exact contribution of MyD88-mediated signals for Th17 development is less clear. It is possible that redundant layers of Th17-inducing factors exist *in vivo* that depend on the actual presence of specific environmental factors or the status and composition of the intestinal flora. In this regard, it has been shown that microbiota-derived ATP can increase Th17 cell numbers in the colon, probably by binding to purinergic receptors on gut residing $CD70^+$ DC/macrophage subpopulations, leading to the induction of MyD88-independent IL-6, TGF-β, and IL-23 production.[68] In addition, signals through Ahr, which recognizes ligands such as xenobiotic or natural arising toxins as well as ligands typically found in fruits and vegetables, may influence either directly or indirectly the development and function of Th17 cells. Initial studies have shown that Ahr is expressed on Th17 cells and that ligation with specific agonistic ligands can enhance Th17 development.[80,81] Furthermore, T cells derived from Ahr-deficient animals displayed reduced IL-17 and IL-22 expression and less potential to induce a Th17-mediated disease.[81–83] Nevertheless, the exact role of Ahr in Th17 cell development and function *in vivo* is not completely understood. Recent studies indicate that Ahr deficiency mainly affects the function and maintenance of innate cells such as $TCR\gamma\delta^+$ intra epithelial cells and group 3 ILCs.[84,85]

1.4 Role of Th17 Cells in Infection and Inflammation

IL-17A, IL-17F, and IL-22, produced by Th17 cells, bind to the receptors for IL-17A/F and IL-22 on the basal surface of intestinal epithelial cells, to produce antimicrobial peptides/proteins, e.g., lipocalin-2, β-defensins, Calprotectin, RegIIIβ, RegIIIγ or S100A. These antimicrobial molecules in turn limit bacterial penetration through the single cell-epithelial barrier and further mediate the clearance of invading microorganisms.[86–88] In addition, IL-22 also induces epithelial proliferation via STAT3 signaling, which facilitates wound healing and recovery from pathological damaged caused by inflammation.[89] In response to self-produced IL-21, Th17 cells in turn expand during pathogen infection, which will boost the pathogen-specific Th17 cell response. Moreover, IL-21 together with IL-17 support the formation of lymphoid tissues in the gut and facilitates the production of antipathogen antibodies to kill pathogens.[90] In addition to these mechanisms mediated by Th17 cells, this cell type is regarded as a master regulator of neutrophil-meditated clearance of invading microbes. Th17-derived cytokines induce granulopoietic factors like G-CSF or GM-CSF and several CXC chemokines indirectly facilitate neutrophil development, proliferation, recruitment and function.[90] In addition, a recent study illustrates that human Th17 cells also directly secrete CXCL8 (also known as IL-8), IFN-γ, TNF and GM-CSF to recruit, activate and sustain the survival of neutrophils at the site of infection or inflammation.[91] As a result of their production of the inflammatory cytokines and molecules, Th17 cells mediate effective protective immune defenses against invading pathogens, especially to extracellular bacteria and fungi at mucosal sites.[86]

Initial reports have indicated that Th17 cells play a pivotal role in the host response against pulmonary infections. IL-17RA knockout (KO) mice showed a delay in neutrophil recruitment and displayed greater dissemination of *Klebsiella pneumonia*. This effect was mediated mainly through decreased production of IL-17 downstream molecules such as G-CSF and CXCL2.[92] The importance of Th17 cells in host defense against pulmonary infections is further emphasized in patients suffering from the autosomal dominant hyper IgE syndrome, also known as Job's syndrome. This disease is characterized by recurrent staphylococcal pneumonia, mucocutaneous candidiasis, elevated serum IgE and abnormalities of bone and connective tissues. Interestingly, it has recently been demonstrated that the underlying cause for this disease is a dominant negative mutation in STAT3.[93,94] The involvement of STAT3 and the consequent deficiency in Th17 induction seems to play an important role in the increased susceptibility toward the pulmonary infection, especially in mucocutaneous candidiasis.[95,96] Data derived from mouse models of candidiasis underline the significance of Th17 responses for the protection of the host. Mice deficient for molecules involved in the Th17 response such as IL-17RA, IL-17RC, IL-23, or Act display enhanced susceptibility towards oropharyngeal candidiasis.[95,97,98] Similar, IL-23- as well as IL-17A-deficient mice were shown to be susceptible toward dermal candidiasis.[99] *Candida albicans* can also cause disseminated infections leading to high mortality rates, and mice deficient for IL-17RA or IL-17A show enhanced susceptibility to this type of infection.[100–102] Although there is evidence that IL-17 produced by innate cells such as ILC or TCRγδ$^+$ T cells may be essential for the host defense against *C. albicans*,[103] a recent report indicated that Th17 cells can confer long-term adaptive immunity to oral mucosal *C. albicans* infections.[104]

Th17 cells are likely to be involved also in the protection of the host from intestinal infections. It has been shown that infection with the murine gut-specific extracellular pathogen *C. rodentium* induces a robust Th17 cell response in the colon of the hosts.[11,105] Mice deficient for IL-17A and/or IL-17F showed impaired clearance of *C. rodentium* and greater morbidity after *C. rodentium* infection at the late time point of infection,[106] illustrating the essential role of IL-17 producing cells in this model. In line with this study, the SFB-induced Th17 cell response protects the host from *C. rodentium*-mediated morbidity and mortality, further confirming the importance of Th17 cell response against *C. rodentium*.[63] Initial studies have shown the pivotal role of IL-23 for protection against *C. rodentium* infection.[11] Although this cytokine plays a critical role for the pathogenicity and maintenance of Th17 cells, as mentioned above, it is more likely that its major function during infection with *C. rodentium* is the activation of the ILC/IL-22 axis. In fact, $ROR\gamma t^{+}$ group 3 ILCs are major producers of IL-22,[107–109] and mice that lack ILC3 or IL-22 are highly susceptible to intestinal infection with *C. rodentium*.[109–111] IL-22 production by ILC3 can be induced by IL-23, which is secreted by intestinal DC after *C. rodentium* infection or intestinal TLR5 ligation.[72,78,110,112] Nevertheless, the actual contribution of ILC3 and Th17 cells for host protection may depend on the colonization status of the mice. In that respect, Th17 cells that are imprinted by signals derived from the steady-state microbiota (designated as "innate" Th17 cells) have been assigned a crucial role for the protection against *C. rodentium* as well as *S. typhimurium* at the very early phase of infection in the caecum of mice.[113] Together, these studies suggest that Th17 cells are crucial to mediate effective host protection from the pathogenic infections, indicating that the concomitant low grade Th17-mediated tissue inflammation can be beneficial for the host. However, there are circumstances in which Th17 mediated immunity, including the recruitment of neutrophils to sites of infection, does not result in improved pathogen clearance, but instead leads to dysregulated inflammation without benefit to the host. Such examples, in which the Th17 response results in exacerbation of disease which enhances inflammation and pathology, have been described for several infections including *Aspergillus fumigatus*-induced pneumonia as well as intestinal infection with *C. albicans*,[114] gut infection with *Helicobacter pylori*[115,116] and after viral infections including respiratory syncytial virus, influenza virus,[117] Herpes virus and human rhinovirus.[118]

The critical contribution of Th17 cells to inflammation and autoimmunity in several disorders such as multiple sclerosis, psoriasis, rheumatoid arthritis, asthma or type 1 diabetes has been well documented.[119] The cytokines and chemokines that are expressed by Th17 cells are found to be elevated in many of the described autoimmune diseases. Preclinical models have demonstrated that neutralization (or loss) of cytokines within the IL-17–IL-23 axis can have impressive beneficial effects on the inflammatory response and the disease-associated pathology.[119] Moreover, targeting of different Th17 factors by antibodies (e.g. against IL17A, IL-17RA, IL17A/F, IL17A/TNF or IL-23) has given encouraging results in human clinical trials.[120] It should be noted, however, that clinical trials for the treatment of Crohn's disease, a form of inflammatory bowel disease, with IL-17-specific antibodies were terminated owing to a lack of efficacy or even disease exacerbation.[121,122] These findings suggest that it is not IL-17 production, but rather the IL-23-mediated induction of a pathogenic Th17 phenotype that drives inflammation. In line with this, a large study found that a single-nucleotide polymorphism in the IL23R was significantly associated with chronic inflammatory bowel disease in humans.[123] Indeed there is evidence that in the intestine some levels of IL-17 (or Th17 cells) are beneficial and protect from inflammation. Transfer of IL17A-deficient T cells in the transfer colitis model of intestinal inflammation induced a stronger inflammatory response, associated with increased levels of IFN-γ producing T cells.[124] Likewise, antibody-mediated neutralization of IL-17 in the DSS colitis model aggravated disease, suggesting a protective role for IL-17 in intestinal inflammation.[125] In line with these studies, a protective role for IL-17 has been proposed in Graft versus host disease (GvHD). Although different studies have shown partially contradicting results regarding the role of IL-17 cytokines in GvHD, there is growing evidence that deficiency in IL-17 leads to aggravated GvHD in lung and colon, presumably due to upregulated Th1-responses.[126] Thus, there is now clear evidence that Th17 cells, especially in concert with the IL-23 axis, have a pathogenic function in chronic inflammation and autoimmunity. Nevertheless, Th17-derived cytokines, including IL-17 itself as well as IL-22, can contribute to tissue protection especially in the intestine, which emphasizes the importance of this cells type for normal intestinal homeostasis.

2. TH17 CELLS AND TREG CELLS: A SPECIAL RELATIONSHIP

T cells with suppressive functions were initially characterized as CD25 (α subunit of IL-2 receptor) expressing $CD4^+$ T cells.[127] It was shown that transfer of such "suppressor" $CD4^+CD25^+$ T cells confer protection from effector T cell-induced autoimmune disease. Scurfy mice and patients with immunodysregulation, polyendocrinopathy, enteropathy X linked syndrome share similar symptoms of massive aggressive autoimmunity and early death,[128–130] and it was found that mutations in the *Foxp3* locus on the X chromosome are associated with the severe autoimmunity both in humans and mice.[130–132] Further studies illustrated that Foxp3 is exclusively expressed in $CD4^+CD25^+$ Treg and is critical for the suppressive function of Treg.[133,134] By using $Foxp3^{GFP}$ knockin reporter mice, Fontenot and colleagues confirmed that Foxp3 is the critical lineage marker for $CD4^+TCR\alpha\beta^+$ Treg.[133] Since their discovery, $Foxp3^+$ Treg cells have been shown by many studies to be essential for suppression of effector T cells and maintaining tolerance against self and foreign antigens. Genetic approaches that allow for specific depletion of Foxp3-expressing cells in mice[135–137] have been instrumental to unravel the function of $Foxp3^+$ Treg in inflammation, infection and cancer.[135–140] According to the distinct origins and developmental programs, Treg are further divided into two subgroups—thymus-derived Treg (tTreg) and peripherally derived Treg (pTreg). Recent findings indicate that under homeostatic conditions, pTreg are of functional importance mainly at feto-maternal interfaces and for maintaining tolerance to food- and microbiota-derived antigens at mucosal sites.[141–144]

2.1 Th17 and Treg Cells develop via a common $Foxp3^+ROR\gamma t^+$ Phenotype

An interesting aspect of Th17 development is the fact that it is functionally linked to the development of $Foxp3^+$ Treg cells. Despite their different functions in tolerance and inflammation, Treg and Th17 cells share the requirement for TGF-β to develop from naïve T cells. When activated in the presence of TGF-β or TGF-β + IL-6, naïve T cells start to simultaneously upregulate both Foxp3 and RORγt, and it has been shown that these transcription factors can directly interact with each other.[61,145,146] From this intermediate $Foxp3^+ROR\gamma t^+$ stage, the cells can either differentiate into $Foxp3^+$ Treg or $ROR\gamma t^+$ Th17 cells. Importantly, Foxp3, in complex with RORγt, can inhibit the transcriptional activity of RORγt.[146,147] Thus, under prevailing Treg-inducing conditions (e.g. high TGF-β concentrations) the cells cannot differentiate into Th17 cells and eventually develop into $Foxp3^+$ Treg cells.[146] However, when inflammatory signals dominate (low TGF-β, presence of IL-6, IL-21, or IL-1β), Foxp3 activity is dampened and the cells develop into Th17 cells.[145,146] Several mechanisms have been described that might account for the loss of Foxp3 under these conditions. HIF1α, which is induced and stabilized by inflammatory cytokines during

Th17 development, can form a complex with Foxp3 and target it for proteasomal degradation.[31] In addition, inflammatory signals like IL-1β or TNF-α can induce enhanced recruitment of E3 ubiquitin-protein ligase STUB1, which mediates Foxp3 ubiquitylation and marks it for subsequent degradation.[148] In conclusion, Treg and Th17 cells develop from naïve T cells via a common $Foxp3^{+}ROR\gamma t^{+}$ intermediate, and it is the specific environmental context of pro- or anti-inflammatory signals that induce the fundamental different outcomes from the same cellular target. This close developmental relationship between Treg cells and Th17 cells has gained considerable attention, since it may allow for a direct *in vivo* modulation of the Th17/Treg balance in inflammatory disorders.[149]

2.2 Function and Stability of $Foxp3^{+}$ T Cells that express the Th17 molecules RORγt and IL-17

Reminiscent of the special relationship between Th17 and Treg cells described above, a high proportion of steady-state intestinal $ROR\gamma t^{+}$ T cells have been found to simultaneously express Foxp3.[61] Such $Foxp3^{+}ROR\gamma t^{+}$ T cells have the ability to express IL-10 as well as IL-17 and their expansion in the intestine depends at least in part on the presence of a complex microflora.[61,64] Nevertheless, it is so far not clear whether these $Foxp3^{+}ROR\gamma t^{+}$ T cells represent a stable phenotype with suppressive *in vivo* function. Moreover, their ability to induce IL-17 secretion indicates that they can participate in inflammatory reactions and may even convert into pathogenic $ROR\gamma t^{+}$ Th17 cells.

Using murine cells, it has been demonstrated that purified $Foxp3^{+}$ Treg cells can be converted into IL-17 producing cells by DC activated with the dectin-1 ligand curdlan.[150] IL-17-production from $Foxp3^{+}$ Treg can also be induced by co-culture of sorted Treg with APCs in the presence of IL-2 and IL-1β.[151,152] As expected, IL-17 expression in $Foxp3^{+}$ cells is accompanied by expression of RORγt in those cells. *In vivo*, $Foxp3^{+}ROR\gamma t^{+}$ cells have also been identified in mice under inflammatory conditions, such as type I autoimmune diabetes (TID) or relapsing-remitting experimental autoimmune encephalomyelitis (RR-EAE).[153,154] When analyzed for their function, $Foxp3^{+}(ROR\gamma t^{+})IL\text{-}17^{+}$ cells showed a clear suppressive function and a stable expression of Foxp3, at least in the *in vitro* converted Treg cells.[150–152,154] The *in vivo* situation may still be more complicated and it has been argued that the $Foxp3^{+}ROR\gamma t^{+}$ cells observed in TID represent intermediates that can differentiate either toward $Foxp3^{+}ROR\gamma t^{-}$ Treg or $ROR\gamma t^{+}Foxp3^{-}$ Th17 cells.[154] Nevertheless, the analysis of the phenotype of $Foxp3^{+}IL\text{-}17^{+}$ cells in the RR-EAE model revealed that most of the $Foxp3^{+}$ cells expressing IL-17 display a $CD25^{-}$ phenotype and hence might not represent a stable suppressive Treg population.[153] In line with these considerations, a recent report demonstrated that $Foxp3^{+}$ T cells can lose Foxp3 expression and convert into pathogenic Th17 cells in autoimmune arthritis.[155] Importantly, such "exFoxp3^{+}" cells were derived from $Foxp3^{+}$ cells with a $CD25^{lo}$ phenotype, whereas $CD25^{hi}Foxp3^{+}$ cells were stable and did not convert into Th17 cells in this model.

In fact, it is likely that a significant fraction of the $Foxp3^{+}$ cells that convert into Th17 cells under inflammatory conditions are derived from the developmental intermediate described in Section 2.1, and not from stable, suppressive $Foxp3^{+}$ Treg cells. In accordance with this notion, it has become clear that Foxp3-expression alone is not a reliable marker for Treg function. It has been shown that Treg lineage stability and suppressive function require a specific epigenetic modification: the hypomethylation of CpG motifs within a number of Treg-specific signature loci in genes encoding for Foxp3, Ctla4, GITR, Eos and IL2RA (CD25).[156–159] Thus, T cells possessing Foxp3 expression but not the Treg-cell-type DNA hypomethylation pattern fail to exhibit sufficient Treg cell suppressive activity and instead secrete effector cytokines upon loss of Foxp3 expression.[157,158,160] Consequently, the Treg-cell-specific DNA hypomethylation pattern, together with Foxp3 expression, can be taken as a reliable marker for defining functional Treg cells. Data from our lab indicated that *ex vivo*-purified $Foxp3^{+}ROR\gamma t^{+}$ T cells from lymphoid organs and peripheral tissues such as the colonic lamina propria exhibit a hypomethylation pattern that is indicative for lineage stable and functional suppressive Treg cells (unpublished data).

There is growing evidence that Treg having undergone antigen-specific stimulation may differ significantly from their naive recirculating counterparts in terms of activation status, migratory potential and regulatory function.[161,162] Together with the notion that Treg having different origins or anatomical locations display distinct but specific gene expression profiles,[163] this has led to the discovery of different functional suppressive Treg subpopulations, termed "effector" Treg lineages.[164,165] Recent reports indicate that $Foxp3^{+}$ Treg can upregulate the expression of transcription factors that are associated with the differentiation and function of effector $CD4^{+}$ T cell lineages such as Th1 or Th2. In the context of Th1-mediated inflammation, for example, $Foxp3^{+}$ Treg can upregulate the Th1-specific transcription factor T-bet, leading to the expression of CXCR3 and the accumulation of

Foxp3$^+$T-bet$^+$ Treg at sites of inflammation. This T-bet expression in Treg was shown to be essential for their homeostasis and function during type-1 inflammation.[165] Likewise, the Th2-associated transcription factor GATA3 was found to play an important role for Treg function, as it was shown that GATA3-deficient Treg display profound defects in peripheral homeostasis and suppressive function.[166,167] Effector Treg cells can express factors endowing them with specific tissue migration potential or enhanced suppressive capacity.[168] Interestingly, our data show that *ex vivo*-isolated Foxp3$^+$RORγt$^+$ T cells express high levels of genes encoding for IRF4, Blimp-1, ICOS, GITR, Grzmb or CCR9 that were found to be upregulated in effector Treg and small intestine lamina propria Treg (Refs. 163,169 and unpublished data). Together these findings suggest that the majority of Foxp3$^+$RORγt$^+$ T cells observed *in vivo* represent a stable, suppressive effector Treg lineage, at least under homeostatic conditions. Although such cells may start to express certain levels of IL-17, it is unlikely that they will lose their suppressive function and convert into Th17 cells.

Similar to mice, IL-17-expressing Foxp3$^+$ T cells have also been identified in humans. It was shown that Foxp3$^+$CD25$^+$ (CD45RA$^-$CCR6$^+$ memory) T cells isolated from peripheral blood of healthy donors can start to express IL-17 when activated *ex vivo* in the presence of proinflammatory cytokines like IL-6, IL-1β as well as IL-21 or IL-23, and expression of IL-17 came along with the expression of RORγt.[170–172] Moreover, high levels of IL-17A-producing Foxp3$^+$RORγt$^+$ CD4$^+$ T cells were also isolated from human tonsils.[172] Importantly, these cells exhibited suppressive capacity *in vitro*, indicating that they have Treg phenotype.[170–172] In addition, enhanced frequencies of Foxp3$^+$IL-17$^+$ cells have been described in the colonic microenvironments of patients suffering from ulcerative colitis and colon carcinoma.[173] Although these colonic Foxp3$^+$IL-17$^+$ cells were shown to be suppressive when tested *ex vivo*, it was speculated that they may contribute to inflammation and pathology due to their expression of IL-17. A recent report demonstrated that Foxp3$^+$CD25$^+$ Treg cells expressing TLR2 can be converted into IL-17$^+$ T cells by stimulation with TLR2 ligands in concert with IL-6. These cells downregulated Foxp3-expression and displayed decreased suppressive capacity.[174] In a follow-up study, it was shown that Treg cells derived from patients with multiple sclerosis express enhanced levels of TLR2. It was therefore suggested that TLR2-stimulation may contribute to reduced Treg function and a shift toward Th17 cells in this disease.[175] Conversion of Foxp3$^+$CD25$^+$ cells into IL17$^+$Foxp3$^-$ cells was reported when cells were stimulated by allogeneic APC, especially in the presence of proinflammatory cytokines such as IL-1β, IL-23, and IL-21.[176] Nevertheless, none of these studies assessed the epigenetic status of Treg-specific hypomethylated regions. It can therefore not be ruled out, that such converted Treg cells were derived from unstable Foxp3-expressing subpopulations. Of note, it was shown that human Th17 cells differentiate from a Foxp3-expressing precursor,[177] suggesting that similar to what has been described for mouse cells above, human Th17 development is also marked by a $Foxp3^+$ intermediate stage.

2.3 $Foxp3^+$ Treg can inhibit Th17 cells

As described above, Treg and Th17 cells arise in the periphery via a common $Foxp3^+ROR\gamma t^+$ intermediate. Thus, the competition of Treg with Th17 cells for their reciprocal development from this common precursor can already be seen as a way to control Th17 or Treg development, respectively. In the intestine, for example, nutrition-derived factors like retinoic acid (RA)[178–180] or short-chain fatty acids produced by the commensal microbiota will shift the balance toward the induction of Treg cells under normal homeostatic conditions.[181–183] This balance, however, can be modulated by inflammatory factors. In a recent report, it was shown that IL-1β can override the RA-mediated Treg induction by increasing the amplitude and duration of STAT3 phosphorylation. This led to an altered balance in the binding of STAT3 and STAT5 to shared consensus sequences in developing T cells and consequently shifted the balance toward Th17 induction.[184]

Nevertheless, beside this developmental competition, Treg cells can also directly inhibit the function of Th17 cells (Fig. 1A). Recently, it has been described that Th17 cells express the IL-10 receptor on their surface.[185,186] IL-10 secreted by $Foxp3^+$ as well as $Foxp3^-$ Tr1 cells selectively inhibited Th17 cell proliferation *in vitro* and suppressed Th17 cell-related T cell transfer colitis *in vivo*.[186] In addition, it was reported that IL-10 itself endows $Foxp3^+$ Treg with the ability to suppress pathogenic Th17 responses.[185] Interestingly, STAT3-activation downstream of IL10R signaling in Treg was shown to be crucial for their ability to suppress Th17 cell responses *in vivo*, probably by the ability of STAT3 to amplify the IL-10 expression by Treg themselves.[185] Since effector Treg cells can express high amounts of IL-10, these cells should exhibit a strong suppressive effect on Th17 cells. Yet, effector Treg cells may restrain the function of Th17 also via other mechanisms. A recent publication described Treg cells expressing the coinhibitory molecule TIGIT as potent inhibitors of Th17 and Th1, but

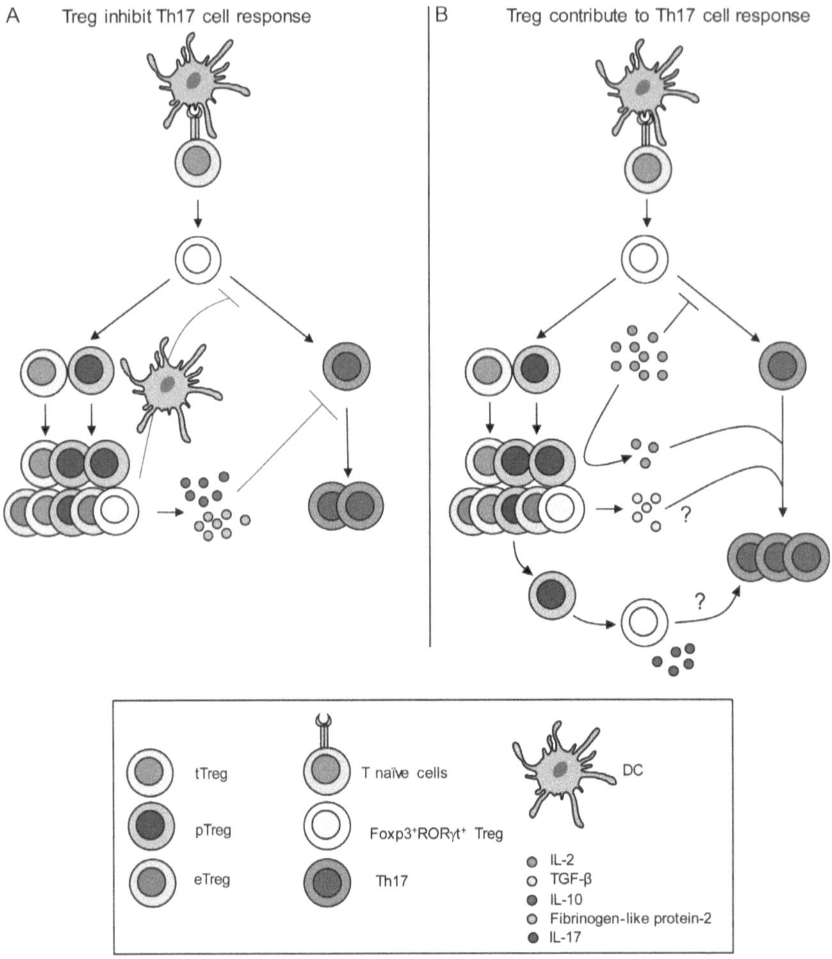

Figure 1 Dual roles of Treg in regulating Th17 cell response. (A) Treg can inhibit the Th17 cell response. Under homeostatic condition, naïve T cells differentiate into pTreg rather than Th17 cells in the periphery. pTreg together with tTreg may also tolerate dendritic cells that trigger Th17 cell induction. Established effector (e)Treg such as TIGIT$^+$ or stable Foxp3$^+$RORγt$^+$ T cells can produce factors like IL-10 or fibrinogen-like protein-2 that inhibits Th17 cells. (B) Treg can contribute to Th17 cell response. By consumption of local IL-2, Treg release the inhibitory effect of high-level of IL-2 on Th17 cell induction and promote Th17 cell differentiation. In addition, TGF-β production by Treg may also lead to enhanced Th17 induction. Under inflammatory condition, Treg may start to produce IL-17 and may even lose Foxp3 expression and convert into pathogenic Th17 cells.

not Th2 responses.[187] It was demonstrated that ligation of TIGIT induces this population to express fibrinogen-like protein 2 (Fgl2), which suppressed the expansion of effector T cells in a manner independent of IL-10. In addition, Treg may also indirectly influence Th17 cell response via tolerizing dendritic cells. In that respect, it has been shown that TIGIT can interact with its ligand on DC to induce IL-10 and reduce IL-12 production from these cells.[188] TIGIT-induced IL-10 was shown to suppress expression of both IL-12p35 and IL-12p40, the subunit shared with IL-23. Hence, Treg may directly suppress both IL-12 and IL-23 production from DC and thereby inhibit the development of both Th1 and Th17 responses.

It has been speculated that the expression of lineage-specific transcription factors like T-bet or Gata3 drives the generation of effector Treg that are specifically suited to regulate immune responses mediated by their corresponding conventional effector $CD4^+$ T cell lineages (see Section 2.2). It is therefore tempting to speculate that $Foxp3^+ROR\gamma t^+$ Treg, due to similar tissue location or migratory potential, may exhibit specific suppressive function on Th17 cells. However, the potential of the $Foxp3^+ROR\gamma t^+$ Treg subpopulation to specifically suppress Th17 or other T helper cell responses remains to be investigated.

2.4 Treg can promote Th17 induction

Although it is clear that Treg cells and Th17 cells have opposing functions, there is now ample evidence that Treg cells can also promote the differentiation of Th17 cells (Fig. 1B). It was already shown in one of the first reports on Th17 cells that the presence of Treg can stimulate Th17 development *in vitro*. Veldhoen *et al.* demonstrated that the production of TGF-β1 by Treg contributed to the Th17 development in the *in vitro* culture system.[12] However, whether Treg-produced TGF-β1 promotes Th17 cell induction *in vivo*, especially under conditions of pathogen infection, remains unclear. Given that Th17 cell development is induced upon low levels of TGF-β1, but is rather impaired under higher concentrations of TGF-β1 *in vitro*,[146] the maintenance of a proper low level of TGF-β1 by Treg in the local niche *in vivo* would be a critical step for Th17 cell development. Thus, although Treg may contribute to Th17 development *in vitro* by secreting TGF-β1, recent results from *in vivo* studies suggested that Treg-produced TGF-β1 does not influence *in vivo* Th17 cell development.[189,190]

In vivo, Treg cells can also favor the accumulation of Th17 cells by inhibiting the development of Th1 or Th1-like "exTh17" cells. This feature of Treg has been observed in context of co-transfer of Treg cells together with naïve T cells into immunodeficient RAG-deficient animals, the classic transfer colitis model.[191] Intriguingly, the transfer of Treg not only inhibited the development of classic Th1 T cells, but also increased the amount of IL-17 producing effector T cells in the colon of the mice by inhibiting the transition of Th17 cells into Th1-like "exTh17 cells."[192] It is thus possible that besides inhibiting the accumulation of pathogenic Th1 and "exTh17 cell" in this model, Treg protect the mice from intestinal immunopathology by favoring the presence of IL-17 producing cells with beneficial effects on intestinal barrier function and epithelial repair mechanisms (see Section 1.4).

In addition to TGF-β, IL-2 plays an important role for the differentiation of pTreg.[193] Although tTreg do not require IL-2 for their development, both tTreg and pTreg rely on IL-2 for the survival expansion.[194] Due to the expression of high levels of CD25 on their surface, Treg possess the trimeric high-affinity IL-2 receptor containing CD25, CD122, and CD132, whereas other effector T cells have dimeric low-affinity IL-2 receptor comprising CD122 and CD132.[195] As a result, Treg have an advantage over effector T cells in competing for the common T cell survival factor IL-2, and thus inhibiting effector T cell responses by depriving them of IL-2. Importantly, IL-2 induces phosphorylation of STAT5 in developing Th17 cells, which in turn blocks phosphorylated-STAT3 mediated induction of *Il17a* and *Rorc* transcription. Thus, presence of IL-2 can directly impair Th17 cell development.[196] Several reports have recently shown that Treg can indeed promote a Th17 cell response via consumption of IL-2.[189,197,198] Using the Foxp3-luci-DTR model,[137] Chen *et al.* showed that depletion of $Foxp3^+$ Treg by diphtheria toxin (DT) decreased the frequency of antigen-specific Th17 cells in draining lymph nodes and blood.[189] They further showed that the Th17-inducing ability of Treg is not attributed to their secretion of TGF-β1, but due to their IL-2 consumption. Pandiyan and coworkers also found that Treg promoted the differentiation of $CD4^+$ naïve T cells into Th17 cells under Th17-polarizing condition *in vitro* and upon *C. albicans* infection *in vivo*.[198] Using IL-2-deficient TCR transgenic mice, they further illustrated that Treg-mediated Th17 cell development depended on the consumption of IL-2 by Treg. Moreover, they confirmed the protective effect of Treg co-transfer with $CD4^+$ naïve T cells into $Rag2^{-/-}$ mice on the subsequent *C. albicans* infection, whereas recipients receiving only $CD4^+$ naïve T cells succumbed to this infection as a result of the insufficient induction of a Th17 cell response. In agreement with these previous studies, a study from our lab showed that Treg promote Th17 cell response against *C. rodentium* infection in the intestine.[199] Using the Foxp3-DTR DEREG mouse model,[136] we could demonstrate that depletion of $Foxp3^+$ Treg by administration of DT resulted in enhanced susceptibility towards *C. rodentium* infection due to an impaired local colonic Th17 cell response.[199] Interestingly, the failure to develop a sufficient Th17 cell response after Treg depletion resulted in decreased inflammation and immunopathology in the colon of the mice. Nevertheless, reduced Th17 induction and the consequent defect in neutrophil recruitment also led to enhanced systemic dissemination of the pathogen and eventually to higher mortality. This demonstrates that Treg can help to establish an effective Th17 response in the gut environment, although this comes at the cost of immunopathology at the site of infection. Blockade of IL-2 in Treg-depleted mice using neutralizing anti-IL-2 antibody restored colonic Th17 cell induction after *C. rodentium* infection. However, neutralization of IL-2 did not restore the Th17 response in terms of absolute numbers of IL-17-producing T cells in the colon. This suggests that even if IL-2 does inhibit initial Th17 induction, some levels of this cytokine are required for later expansion of these cells. In line with this, it has also been described that IL-2 facilitates the expansion of committed Th17 cells.[200] Together, these data indicate that by consuming IL-2, Treg can support the initial induction of Th17 cells. However, this mechanism may not play a role for the subsequent expansion and maintenance of Th17 cells.[189]

By their ability to modulate IL-2 levels, Treg have been shown to also be involved in negatively regulating natural killer (NK) cell expansion, activation, maturation, and function *in vivo*.[201,202] Gasteiger and coworkers have shown that Treg compete with NK cells for IL-2. In their system, depletion of Treg lead to maturation of NK cells and IL-2 treatment boosted NK cell mediated cytotoxicity. Moreover, a recent study also illustrated that Treg, by consuming IL-2, promote germinal center (GC) formation during influenza virus infection.[190] As a result of Treg depletion, accumulation of IL-2 impaired the differentiation of follicular T helper cells (Tfh) and impaired the protective GC B cell response against influenza viral infection.[190] Therefore, it seems that Treg-mediated consumption of IL-2 represents a general mechanism by which Treg can regulate multiple immune cell responses *in vivo*.

3. CONCLUDING REMARKS

As two important immune cell lineages, Treg and Th17 cells are involved in multiple physiological events. Treg are regarded as immune regulators that prevent an excessive immune response, while Th17 cells are normally involved in the defense against extracellular bacterial and fungal infections. However, both excessive Treg or Th17 cell responses can threaten the health of the host. Enhanced Treg suppression may facilitate tumorigenesis and can be used by pathogens to escape the immune surveillance system, whereas uncontrolled Th17 cell responses are involved in inflammation and autoimmune disease. According to the model of the immunological balance, a fine-tuned balance between Treg and Th17 cells would not only be a critical prerequisite for healthy homeostasis, but also direct the outcome of immune responses. Regarding the close interconnection of Treg/Th17 developmental and functional cues, direct modulation of the immunological balance between Treg and Th17 cells represents a promising option for immune therapy.

Factors that influence the regulation of the Treg/Th17 balance thus represent potential targets for therapeutic intervention. Recently, the field of immune metabolism has gained increasing attention in immunological research. Manipulation of the intermediate cellular metabolism such as glycolysis, fatty acid, or the mitochondrial metabolism can be achieved using small molecule inhibitors of key regulatory enzymes of these processes. Direct manipulation of the cellular fatty acid metabolism for example can shift the balance between Th17 and Treg development,[28] but many other metabolic checkpoints may be exploited as targets to control the Treg/Th17 balance.[203]

Yet, our increasing knowledge on the relationship between Treg and Th17 cells that we have reviewed here suggests multiple layers of complexity. As we have shown, Treg cells not only suppress the function of Th17 cells, but can also support their development by the uptake of IL-2. In addition, the inhibition of Th17 cells may not always be beneficial, as has been demonstrated in the context of intestinal inflammation and epithelial repair. Finally, features of Th17 cells such as the expression of RORγt in specific populations of effector Treg cells may be important for their optimal suppressive function. Although the exact role of RORγt in $Foxp3^+$ $ROR\gamma t^+$ remains to be assessed, it should be taken into account that approaches aiming at the inhibition of RORγt function will not only impact on Th17 cells, but may also affect such effector Treg cells.

ACKNOWLEDGMENTS

This work was supported by a grant from the Deutsche Forschungsgemeinschaft (DFG KFO250). A part of this work is content of the "habilitation" thesis of M. L. The authors would like to thank Peyman Ghorbani for critical reading of the chapter.

REFERENCES

1. Harrington LE, Hatton RD, Mangan PR, et al. Interleukin 17-producing CD4+ effector T cells develop via a lineage distinct from the T helper type 1 and 2 lineages. *Nat Immunol.* 2005;6(11):1123–1132.
2. Langrish CL, Chen Y, Blumenschein WM, et al. IL-23 drives a pathogenic T cell population that induces autoimmune inflammation. *J Exp Med.* 2005;201(2):233–240.
3. Park H, Li Z, Yang XO, et al. A distinct lineage of CD4 T cells regulates tissue inflammation by producing interleukin 17. *Nat Immunol.* 2005;6(11):1133–1141.
4. Aarvak T, Chabaud M, Miossec P, Natvig JB. IL-17 is produced by some proinflammatory Th1/Th0 cells but not by Th2 cells. *J Immunol.* 1999;162(3): 1246–1251.
5. Infante-Duarte C, Horton HF, Byrne MC, Kamradt T. Microbial lipopeptides induce the production of IL-17 in Th cells. *J Immunol.* 2000;165(11):6107–6115.
6. Cua DJ, Tato CM. Innate IL-17-producing cells: the sentinels of the immune system. *Nat Rev Immunol.* 2010;10(7):479–489.
7. Michel ML, Mendes-da-Cruz D, Keller AC, et al. Critical role of ROR-gammat in a new thymic pathway leading to IL-17-producing invariant NKT cell differentiation. *Proc Natl Acad Sci USA.* 2008;105(50):19845–19850.
8. Eberl G, Colonna M, Di Santo JP, McKenzie AN. Innate lymphoid cells. Innate lymphoid cells: a new paradigm in immunology. *Science.* 2015;348(6237):aaa6566.
9. Cua DJ, Sherlock J, Chen Y, et al. Interleukin-23 rather than interleukin-12 is the critical cytokine for autoimmune inflammation of the brain. *Nature.* 2003;421(6924): 744–748.
10. Bettelli E, Carrier Y, Gao W, et al. Reciprocal developmental pathways for the generation of pathogenic effector TH17 and regulatory T cells. *Nature.* 2006;441(7090): 235–238.
11. Mangan PR, Harrington LE, O'Quinn DB, et al. Transforming growth factor-beta induces development of the T(H)17 lineage. *Nature.* 2006;441(7090):231–234.
12. Veldhoen M, Hocking RJ, Atkins CJ, Locksley RM, Stockinger B. TGFbeta in the context of an inflammatory cytokine milieu supports de novo differentiation of IL-17-producing T cells. *Immunity.* 2006;24(2):179–189.
13. Durant L, Watford WT, Ramos HL, et al. Diverse targets of the transcription factor STAT3 contribute to T cell pathogenicity and homeostasis. *Immunity.* 2010;32(5): 605–615.
14. Ortiz MA, Piedrafita FJ, Pfahl M, Maki R. TOR: a new orphan receptor expressed in the thymus that can modulate retinoid and thyroid hormone signals. *Mol Endocrinol.* 1995;9(12):1679–1691.
15. Guo J, Hawwari A, Li H, et al. Regulation of the TCRalpha repertoire by the survival window of CD4(+)CD8(+) thymocytes. *Nat Immunol.* 2002;3(5):469–476.
16. He YW, Beers C, Deftos ML, Ojala EW, Forbush KA, Bevan MJ. Down-regulation of the orphan nuclear receptor ROR gamma t is essential for T lymphocyte maturation. *J Immunol.* 2000;164(11):5668–5674.
17. Sun Z, Unutmaz D, Zou YR, et al. Requirement for RORgamma in thymocyte survival and lymphoid organ development. *Science.* 2000;288(5475):2369–2373.
18. Villey I, de Chasseval R, de Villartay JP. RORgammaT, a thymus-specific isoform of the orphan nuclear receptor RORgamma/TOR, is up-regulated by signaling through the pre-T cell receptor and binds to the TEA promoter. *Eur J Immunol.* 1999;29(12): 4072–4080.
19. Eberl G, Marmon S, Sunshine MJ, Rennert PD, Choi Y, Littman DR. An essential function for the nuclear receptor RORgamma(t) in the generation of fetal lymphoid tissue inducer cells. *Nat Immunol.* 2004;5(1):64–73.

20. Cording S, Medvedovic J, Cherrier M, Eberl G. Development and regulation of RORgammat(+) innate lymphoid cells. *FEBS Lett*. 2014;588(22):4176–4181.
21. Diefenbach A, Colonna M, Koyasu S. Development, differentiation, and diversity of innate lymphoid cells. *Immunity*. 2014;41(3):354–365.
22. Ivanov II, McKenzie BS, Zhou L, et al. The orphan nuclear receptor RORgammat directs the differentiation program of proinflammatory IL-17+ T helper cells. *Cell*. 2006;126(6):1121–1133.
23. Ciofani M, Madar A, Galan C, et al. A validated regulatory network for Th17 cell specification. *Cell*. 2012;151(2):289–303.
24. Korn T, Oukka M, Kuchroo V, Bettelli E. Th17 cells: effector T cells with inflammatory properties. *Semin Immunol*. 2007;19(6):362–371.
25. Nurieva R, Yang XO, Martinez G, et al. Essential autocrine regulation by IL-21 in the generation of inflammatory T cells. *Nature*. 2007;448(7152):480–483.
26. Chung Y, Chang SH, Martinez GJ, et al. Critical regulation of early Th17 cell differentiation by interleukin-1 signaling. *Immunity*. 2009;30(4):576–587.
27. Gulen MF, Kang Z, Bulek K, et al. The receptor SIGIRR suppresses Th17 cell proliferation via inhibition of the interleukin-1 receptor pathway and mTOR kinase activation. *Immunity*. 2010;32(1):54–66.
28. Berod L, Friedrich C, Nandan A, et al. De novo fatty acid synthesis controls the fate between regulatory T and T helper 17 cells. *Nat Med*. 2014;20(11):1327–1333.
29. Michalek RD, Gerriets VA, Jacobs SR, et al. Cutting edge: distinct glycolytic and lipid oxidative metabolic programs are essential for effector and regulatory CD4+ T cell subsets. *J Immunol*. 2011;186(6):3299–3303.
30. Powell JD, Pollizzi KN, Heikamp EB, Horton MR. Regulation of immune responses by mTOR. *Annu Rev Immunol*. 2012;30:39–68.
31. Dang EV, Barbi J, Yang HY, et al. Control of T(H)17/T(reg) balance by hypoxia-inducible factor 1. *Cell*. 2011;146(5):772–784.
32. Shi LZ, Wang R, Huang G, et al. HIF1alpha-dependent glycolytic pathway orchestrates a metabolic checkpoint for the differentiation of TH17 and Treg cells. *J Exp Med*. 2011;208(7):1367–1376.
33. Gutcher I, Donkor MK, Ma Q, Rudensky AY, Flavell RA, Li MO. Autocrine transforming growth factor-beta1 promotes in vivo Th17 cell differentiation. *Immunity*. 2011;34(3):396–408.
34. Li MO, Wan YY, Flavell RA. T cell-produced transforming growth factor-beta1 controls T cell tolerance and regulates Th1- and Th17-cell differentiation. *Immunity*. 2007;26(5):579–591.
35. Veldhoen M, Hocking RJ, Flavell RA, Stockinger B. Signals mediated by transforming growth factor-beta initiate autoimmune encephalomyelitis, but chronic inflammation is needed to sustain disease. *Nat Immunol*. 2006;7(11):1151–1156.
36. Das J, Ren G, Zhang L, et al. Transforming growth factor beta is dispensable for the molecular orchestration of Th17 cell differentiation. *J Exp Med*. 2009;206(11): 2407–2416.
37. Ghoreschi K, Laurence A, Yang XP, et al. Generation of pathogenic T(H)17 cells in the absence of TGF-beta signalling. *Nature*. 2010;467(7318):967–971.
38. McGeachy MJ, Bak-Jensen KS, Chen Y, et al. TGF-beta and IL-6 drive the production of IL-17 and IL-10 by T cells and restrain T(H)-17 cell-mediated pathology. *Nat Immunol*. 2007;8(12):1390–1397.
39. Haines CJ, Chen Y, Blumenschein WM, et al. Autoimmune memory T helper 17 cell function and expansion are dependent on interleukin-23. *Cell Rep*. 2013;3(5): 1378–1388.

40. McGeachy MJ, Chen Y, Tato CM, et al. The interleukin 23 receptor is essential for the terminal differentiation of interleukin 17-producing effector T helper cells in vivo. *Nat Immunol.* 2009;10(3):314–324.
41. Esplugues E, Huber S, Gagliani N, et al. Control of TH17 cells occurs in the small intestine. *Nature.* 2011;475(7357):514–518.
42. Heinemann C, Heink S, Petermann F, et al. IL-27 and IL-12 oppose pro-inflammatory IL-23 in CD4+ T cells by inducing Blimp1. *Nat Commun.* 2014;5:3770.
43. Lee Y, Awasthi A, Yosef N, et al. Induction and molecular signature of pathogenic TH17 cells. *Nat Immunol.* 2012;13(10):991–999.
44. Hirota K, Duarte JH, Veldhoen M, et al. Fate mapping of IL-17-producing T cells in inflammatory responses. *Nat Immunol.* 2011;12(3):255–263.
45. Lee YK, Turner H, Maynard CL, et al. Late developmental plasticity in the T helper 17 lineage. *Immunity.* 2009;30(1):92–107.
46. Yang Y, Weiner J, Liu Y, et al. T-bet is essential for encephalitogenicity of both Th1 and Th17 cells. *J Exp Med.* 2009;206(7):1549–1564.
47. Ahern PP, Schiering C, Buonocore S, et al. Interleukin-23 drives intestinal inflammation through direct activity on T cells. *Immunity.* 2010;33(2):279–288.
48. Globig AM, Hennecke N, Martin B, et al. Comprehensive intestinal T helper cell profiling reveals specific accumulation of IFN-gamma+IL-17+coproducing CD4+ T cells in active inflammatory bowel disease. *Inflamm Bowel Dis.* 2014;20(12):2321–2329.
49. Hue S, Ahern P, Buonocore S, et al. Interleukin-23 drives innate and T cell-mediated intestinal inflammation. *J Exp Med.* 2006;203(11):2473–2483.
50. Kullberg MC, Jankovic D, Feng CG, et al. IL-23 plays a key role in Helicobacter hepaticus-induced T cell-dependent colitis. *J Exp Med.* 2006;203(11):2485–2494.
51. Bending D, Newland S, Krejci A, Phillips JM, Bray S, Cooke A. Epigenetic changes at Il12rb2 and Tbx21 in relation to plasticity behavior of Th17 cells. *J Immunol.* 2011;186(6):3373–3382.
52. Mukasa R, Balasubramani A, Lee YK, et al. Epigenetic instability of cytokine and transcription factor gene loci underlies plasticity of the T helper 17 cell lineage. *Immunity.* 2010;32(5):616–627.
53. Wei G, Wei L, Zhu J, et al. Global mapping of H3K4me3 and H3K27me3 reveals specificity and plasticity in lineage fate determination of differentiating CD4+ T cells. *Immunity.* 2009;30(1):155–167.
54. Bending D, De la Pena H, Veldhoen M, et al. Highly purified Th17 cells from BDC2.5NOD mice convert into Th1-like cells in NOD/SCID recipient mice. *J Clin Invest.* 2009;119(3):565–572.
55. Martin-Orozco N, Chung Y, Chang SH, Wang YH, Dong C. Th17 cells promote pancreatic inflammation but only induce diabetes efficiently in lymphopenic hosts after conversion into Th1 cells. *Eur J Immunol.* 2009;39(1):216–224.
56. Croxford AL, Kurschus FC, Waisman A. Cutting edge: an IL-17F-CreEYFP reporter mouse allows fate mapping of Th17 cells. *J Immunol.* 2009;182(3):1237–1241.
57. Gagliani N, Vesely MC, Iseppon A, et al. Th17 cells transdifferentiate into regulatory T cells during resolution of inflammation. *Nature.* 2015;523(7559):221–225.
58. Hirota K, Turner JE, Villa M, et al. Plasticity of Th17 cells in Peyer's patches is responsible for the induction of T cell-dependent IgA responses. *Nat Immunol.* 2013;14(4):372–379.
59. Muranski P, Borman ZA, Kerkar SP, et al. Th17 cells are long lived and retain a stem cell-like molecular signature. *Immunity.* 2011;35(6):972–985.
60. Kryczek I, Zhao E, Liu Y, et al. Human TH17 cells are long-lived effector memory cells. *Sci Transl Med.* 2011;3(104):104ra100.

61. Lochner M, Peduto L, Cherrier M, et al. In vivo equilibrium of proinflammatory IL-17+ and regulatory IL-10+ Foxp3+ RORgamma t+ T cells. *J Exp Med*. 2008;205(6): 1381–1393.
62. Gaboriau-Routhiau V, Rakotobe S, Lecuyer E, et al. The key role of segmented filamentous bacteria in the coordinated maturation of gut helper T cell responses. *Immunity*. 2009;31(4):677–689.
63. Ivanov II, Atarashi K, Manel N, et al. Induction of intestinal Th17 cells by segmented filamentous bacteria. *Cell*. 2009;139(3):485–498.
64. Lochner M, Berard M, Sawa S, et al. Restricted microbiota and absence of cognate TCR antigen leads to an unbalanced generation of Th17 cells. *J Immunol*. 2011;186(3):1531–1537.
65. Goto Y, Panea C, Nakato G, et al. Segmented filamentous bacteria antigens presented by intestinal dendritic cells drive mucosal Th17 cell differentiation. *Immunity*. 2014;40(4):594–607.
66. Yang Y, Torchinsky MB, Gobert M, et al. Focused specificity of intestinal TH17 cells towards commensal bacterial antigens. *Nature*. 2014;510(7503):152–156.
67. Talham GL, Jiang HQ, Bos NA, Cebra JJ. Segmented filamentous bacteria are potent stimuli of a physiologically normal state of the murine gut mucosal immune system. *Infect Immun*. 1999;67(4):1992–2000.
68. Atarashi K, Nishimura J, Shima T, et al. ATP drives lamina propria T(H)17 cell differentiation. *Nature*. 2008;455(7214):808–812.
69. Ivanov II, Frutos Rde L, Manel N, et al. Specific microbiota direct the differentiation of IL-17-producing T-helper cells in the mucosa of the small intestine. *Cell Host Microbe*. 2008;4(4):337–349.
70. Lewis KL, Caton ML, Bogunovic M, et al. Notch2 receptor signaling controls functional differentiation of dendritic cells in the spleen and intestine. *Immunity*. 2011;35(5): 780–791.
71. Persson EK, Uronen-Hansson H, Semmrich M, et al. IRF4 transcription-factor-dependent CD103(+)CD11b(+) dendritic cells drive mucosal T helper 17 cell differentiation. *Immunity*. 2013;38(5):958–969.
72. Satpathy AT, Briseno CG, Lee JS, et al. Notch2-dependent classical dendritic cells orchestrate intestinal immunity to attaching-and-effacing bacterial pathogens. *Nat Immunol*. 2013;14(9):937–948.
73. Schlitzer A, McGovern N, Teo P, et al. IRF4 transcription factor-dependent CD11b+ dendritic cells in human and mouse control mucosal IL-17 cytokine responses. *Immunity*. 2013;38(5):970–983.
74. Welty NE, Staley C, Ghilardi N, Sadowsky MJ, Igyarto BZ, Kaplan DH. Intestinal lamina propria dendritic cells maintain T cell homeostasis but do not affect commensalism. *J Exp Med*. 2013;210(10):2011–2024.
75. Cerovic V, Houston SA, Scott CL, et al. Intestinal CD103(−) dendritic cells migrate in lymph and prime effector T cells. *Mucosal Immunol*. 2013;6(1):104–113.
76. Fujimoto K, Karuppuchamy T, Takemura N, et al. A new subset of CD103 +CD8alpha+ dendritic cells in the small intestine expresses TLR3, TLR7, and TLR9 and induces Th1 response and CTL activity. *J Immunol*. 2011;186(11): 6287–6295.
77. Uematsu S, Fujimoto K, Jang MH, et al. Regulation of humoral and cellular gut immunity by lamina propria dendritic cells expressing Toll-like receptor 5. *Nat Immunol*. 2008;9(7):769–776.
78. Kinnebrew MA, Buffie CG, Diehl GE, et al. Interleukin 23 production by intestinal CD103(+)CD11b(+) dendritic cells in response to bacterial flagellin enhances mucosal innate immune defense. *Immunity*. 2012;36(2):276–287.

79. Shaw MH, Kamada N, Kim YG, Nunez G. Microbiota-induced IL-1beta, but not IL-6, is critical for the development of steady-state TH17 cells in the intestine. *J Exp Med*. 2012;209(2):251–258.
80. Quintana FJ, Basso AS, Iglesias AH, et al. Control of T(reg) and T(H)17 cell differentiation by the aryl hydrocarbon receptor. *Nature*. 2008;453(7191):65–71.
81. Veldhoen M, Hirota K, Westendorf AM, et al. The aryl hydrocarbon receptor links TH17-cell-mediated autoimmunity to environmental toxins. *Nature*. 2008;453(7191):106–109.
82. Nakahama T, Kimura A, Nguyen NT, et al. Aryl hydrocarbon receptor deficiency in T cells suppresses the development of collagen-induced arthritis. *Proc Natl Acad Sci USA*. 2011;108(34):14222–14227.
83. Quintana FJ, Murugaiyan G, Farez MF, et al. An endogenous aryl hydrocarbon receptor ligand acts on dendritic cells and T cells to suppress experimental autoimmune encephalomyelitis. *Proc Natl Acad Sci USA*. 2010;107(48):20768–20773.
84. Li Y, Innocentin S, Withers DR, et al. Exogenous stimuli maintain intraepithelial lymphocytes via aryl hydrocarbon receptor activation. *Cell*. 2011;147(3):629–640.
85. Qiu J, Guo X, Chen ZM, et al. Group 3 innate lymphoid cells inhibit T-cell-mediated intestinal inflammation through aryl hydrocarbon receptor signaling and regulation of microflora. *Immunity*. 2013;39(2):386–399.
86. Khader SA, Gaffen SL, Kolls JK. Th17 cells at the crossroads of innate and adaptive immunity against infectious diseases at the mucosa. *Mucosal Immunol*. 2009;2(5):403–411.
87. Kolls JK, Khader SA. The role of Th17 cytokines in primary mucosal immunity. *Cytokine Growth Factor Rev*. 2010;21(6):443–448.
88. Pappu R, Rutz S, Ouyang W. Regulation of epithelial immunity by IL-17 family cytokines. *Trends Immunol*. 2012;33(7):343–349.
89. Pickert G, Neufert C, Leppkes M, et al. STAT3 links IL-22 signaling in intestinal epithelial cells to mucosal wound healing. *J Exp Med*. 2009;206(7):1465–1472.
90. Ouyang W, Kolls JK, Zheng Y. The biological functions of T helper 17 cell effector cytokines in inflammation. *Immunity*. 2008;28(4):454–467.
91. Pelletier M, Maggi L, Micheletti A, et al. Evidence for a cross-talk between human neutrophils and Th17 cells. *Blood*. 2009;115(2):335–343.
92. Ye P, Rodriguez FH, Kanaly S, et al. Requirement of interleukin 17 receptor signaling for lung CXC chemokine and granulocyte colony-stimulating factor expression, neutrophil recruitment, and host defense. *J Exp Med*. 2001;194(4):519–527.
93. Milner JD, Sandler NG, Douek DC. Th17 cells, Job's syndrome and HIV: opportunities for bacterial and fungal infections. *Curr Opin HIV AIDS*. 2010;5(2):179–183.
94. Paulson ML, Freeman AF, Holland SM. Hyper IgE syndrome: an update on clinical aspects and the role of signal transducer and activator of transcription 3. *Curr Opin Allergy Clin Immunol*. 2008;8(6):527–533.
95. Conti HR, Shen F, Nayyar N, et al. Th17 cells and IL-17 receptor signaling are essential for mucosal host defense against oral candidiasis. *J Exp Med*. 2009;206(2):299–311.
96. Puel A, Cypowyj S, Marodi L, Abel L, Picard C, Casanova JL. Inborn errors of human IL-17 immunity underlie chronic mucocutaneous candidiasis. *Curr Opin Allergy Clin Immunol*. 2012;12(6):616–622.
97. Ferreira MC, Whibley N, Mamo AJ, Siebenlist U, Chan YR, Gaffen SL. Interleukin-17-induced protein lipocalin 2 is dispensable for immunity to oral candidiasis. *Infect Immun*. 2014;82(3):1030–1035.
98. Ho AW, Shen F, Conti HR, et al. IL-17RC is required for immune signaling via an extended SEF/IL-17R signaling domain in the cytoplasmic tail. *J Immunol*. 2010;185(2):1063–1070.

99. Kagami S, Rizzo HL, Kurtz SE, Miller LS, Blauvelt A. IL-23 and IL-17A, but not IL-12 and IL-22, are required for optimal skin host defense against Candida albicans. *J Immunol*. 2010;185(9):5453–5462.
100. Huang W, Na L, Fidel PL, Schwarzenberger P. Requirement of interleukin-17A for systemic anti-Candida albicans host defense in mice. *J Infect Dis*. 2004;190(3):624–631.
101. Saijo S, Ikeda S, Yamabe K, et al. Dectin-2 recognition of alpha-mannans and induction of Th17 cell differentiation is essential for host defense against Candida albicans. *Immunity*. 2010;32(5):681–691.
102. van de Veerdonk FL, Kullberg BJ, Verschueren IC, et al. Differential effects of IL-17 pathway in disseminated candidiasis and zymosan-induced multiple organ failure. *Shock*. 2010;34(4):407–411.
103. Whibley N, Gaffen SL. Brothers in arms: Th17 and Treg responses in Candida albicans immunity. *PLoS Pathog*. 2014;10(12):e1004456.
104. Hernandez-Santos N, Huppler AR, Peterson AC, Khader SA, McKenna KC, Gaffen SL. Th17 cells confer long-term adaptive immunity to oral mucosal Candida albicans infections. *Mucosal Immunol*. 2013;6(5):900–910.
105. Symonds EL, Riedel CU, O'Mahony D, Lapthorne S, O'Mahony L, Shanahan F. Involvement of T helper type 17 and regulatory T cell activity in Citrobacter rodentium invasion and inflammatory damage. *Clin Exp Immunol*. 2009;157(1): 148–154.
106. Ishigame H, Kakuta S, Nagai T, et al. Differential roles of interleukin-17A and -17F in host defense against mucoepithelial bacterial infection and allergic responses. *Immunity*. 2009;30(1):108–119.
107. Cella M, Fuchs A, Vermi W, et al. A human natural killer cell subset provides an innate source of IL-22 for mucosal immunity. *Nature*. 2009;457(7230):722–725.
108. Sanos SL, Bui VL, Mortha A, et al. RORgammat and commensal microflora are required for the differentiation of mucosal interleukin 22-producing NKp46+ cells. *Nat Immunol*. 2009;10(1):83–91.
109. Satoh-Takayama N, Vosshenrich CA, Lesjean-Pottier S, et al. Microbial flora drives interleukin 22 production in intestinal NKp46+ cells that provide innate mucosal immune defense. *Immunity*. 2008;29(6):958–970.
110. Sonnenberg GF, Monticelli LA, Elloso MM, Fouser LA, Artis D. CD4(+) lymphoid tissue-inducer cells promote innate immunity in the gut. *Immunity*. 2011;34(1): 122–134.
111. Zheng Y, Valdez PA, Danilenko DM, et al. Interleukin-22 mediates early host defense against attaching and effacing bacterial pathogens. *Nat Med*. 2008;14(3):282–289.
112. Manta C, Heupel E, Radulovic K, et al. CX(3)CR1(+) macrophages support IL-22 production by innate lymphoid cells during infection with Citrobacter rodentium. *Mucosal Immunol*. 2013;6(1):177–188.
113. Geddes K, Rubino SJ, Magalhaes JG, et al. Identification of an innate T helper type 17 response to intestinal bacterial pathogens. *Nat Med*. 2011;17(7):837–844.
114. Zelante T, De Luca A, Bonifazi P, et al. IL-23 and the Th17 pathway promote inflammation and impair antifungal immune resistance. *Eur J Immunol*. 2007;37(10): 2695–2706.
115. Caruso R, Fina D, Paoluzi OA, et al. IL-23-mediated regulation of IL-17 production in Helicobacter pylori-infected gastric mucosa. *Eur J Immunol*. 2008;38(2):470–478.
116. Luzza F, Parrello T, Monteleone G, et al. Up-regulation of IL-17 is associated with bioactive IL-8 expression in Helicobacter pylori-infected human gastric mucosa. *J Immunol*. 2000;165(9):5332–5337.
117. Way EE, Chen K, Kolls JK. Dysregulation in lung immunity—the protective and pathologic Th17 response in infection. *Eur J Immunol*. 2013;43(12):3116–3124.

118. Wiehler S, Proud D. Interleukin-17A modulates human airway epithelial responses to human rhinovirus infection. *Am J Physiol Lung Cell Mol Physiol.* 2007;293(2): L505–L515.
119. Singh RP, Hasan S, Sharma S, et al. Th17 cells in inflammation and autoimmunity. *Autoimmun Rev.* 2014;13(12):1174–1181.
120. Gaffen SL, Jain R, Garg AV, Cua DJ. The IL-23-IL-17 immune axis: from mechanisms to therapeutic testing. *Nat Rev Immunol.* 2014;14(9):585–600.
121. Hueber W, Sands BE, Lewitzky S, et al. Secukinumab, a human anti-IL-17A monoclonal antibody, for moderate to severe Crohn's disease: unexpected results of a randomised, double-blind placebo-controlled trial. *Gut.* 2012;61(12):1693–1700.
122. Targan SR, Feagan BG, Vermeire S, et al. A randomized, double-blind, placebo-controlled study to evaluate the safety, tolerability, and efficacy of AMG 827 in subjects with moderate to severe Crohn's disease. *Gastroenterology.* 2012;143(3):E26.
123. Duerr RH, Taylor KD, Brant SR, et al. A genome-wide association study identifies IL23R as an inflammatory bowel disease gene. *Science.* 2006;314(5804):1461–1463.
124. O'Connor Jr W, Kamanaka M, Booth CJ, et al. A protective function for interleukin 17A in T cell-mediated intestinal inflammation. *Nat Immunol.* 2009;10(6):603–609.
125. Ogawa A, Andoh A, Araki Y, Bamba T, Fujiyama Y. Neutralization of interleukin-17 aggravates dextran sulfate sodium-induced colitis in mice. *Clin Immunol.* 2004;110(1): 55–62.
126. Yi T, Zhao D, Lin CL, et al. Absence of donor Th17 leads to augmented Th1 differentiation and exacerbated acute graft-versus-host disease. *Blood.* 2008;112(5): 2101–2110.
127. Sakaguchi S, Sakaguchi N, Asano M, Itoh M, Toda M. Immunologic self-tolerance maintained by activated T cells expressing IL-2 receptor alpha-chains (CD25). Breakdown of a single mechanism of self-tolerance causes various autoimmune diseases. *J Immunol.* 1995;155(3):1151–1164.
128. Clark LB, Appleby MW, Brunkow ME, Wilkinson JE, Ziegler SF, Ramsdell F. Cellular and molecular characterization of the scurfy mouse mutant. *J Immunol.* 1999;162(5):2546–2554.
129. Lyon MF, Peters J, Glenister PH, Ball S, Wright E. The scurfy mouse mutant has previously unrecognized hematological abnormalities and resembles Wiskott-Aldrich syndrome. *Proc Natl Acad Sci USA.* 1990;87(7):2433–2437.
130. Wildin RS, Ramsdell F, Peake J, et al. X-linked neonatal diabetes mellitus, enteropathy and endocrinopathy syndrome is the human equivalent of mouse scurfy. *Nat Genet.* 2001;27(1):18–20.
131. Bennett CL, Christie J, Ramsdell F, et al. The immune dysregulation, polyendocrinopathy, enteropathy, X-linked syndrome (IPEX) is caused by mutations of FOXP3. *Nat Genet.* 2001;27(1):20–21.
132. Brunkow ME, Jeffery EW, Hjerrild KA, et al. Disruption of a new forkhead/winged-helix protein, scurfin, results in the fatal lymphoproliferative disorder of the scurfy mouse. *Nat Genet.* 2001;27(1):68–73.
133. Fontenot JD, Rasmussen JP, Williams LM, Dooley JL, Farr AG, Rudensky AY. Regulatory T cell lineage specification by the forkhead transcription factor foxp3. *Immunity.* 2005;22(3):329–341.
134. Hori S, Nomura T, Sakaguchi S. Control of regulatory T cell development by the transcription factor Foxp3. *Science.* 2003;299(5609):1057–1061.
135. Kim JM, Rasmussen JP, Rudensky AY. Regulatory T cells prevent catastrophic autoimmunity throughout the lifespan of mice. *Nat Immunol.* 2007;8(2):191–197.
136. Lahl K, Loddenkemper C, Drouin C, et al. Selective depletion of Foxp3+ regulatory T cells induces a scurfy-like disease. *J Exp Med.* 2007;204(1):57–63.

137. Suffner J, Hochweller K, Kuhnle MC, et al. Dendritic cells support homeostatic expansion of Foxp3+ regulatory T cells in Foxp3.LuciDTR mice. *J Immunol.* 2010;184(4):1810–1820.
138. Berod L, Puttur F, Huehn J, Sparwasser T. Tregs in infection and vaccinology: heroes or traitors? *Microb Biotechnol.* 2012;5(2):260–269.
139. Kim J, Lahl K, Hori S, et al. Cutting edge: depletion of Foxp3+ cells leads to induction of autoimmunity by specific ablation of regulatory T cells in genetically targeted mice. *J Immunol.* 2009;183(12):7631–7634.
140. Klages K, Mayer CT, Lahl K, et al. Selective depletion of Foxp3+ regulatory T cells improves effective therapeutic vaccination against established melanoma. *Cancer Res.* 2010;70(20):7788–7799.
141. Haribhai D, Lin W, Edwards B, et al. A central role for induced regulatory T cells in tolerance induction in experimental colitis. *J Immunol.* 2009;182(6):3461–3468.
142. Josefowicz SZ, Niec RE, Kim HY, et al. Extrathymically generated regulatory T cells control mucosal TH2 inflammation. *Nature.* 2012;482(7385):395–399.
143. Lathrop SK, Bloom SM, Rao SM, et al. Peripheral education of the immune system by colonic commensal microbiota. *Nature.* 2011;478(7368):250–254.
144. Samstein RM, Josefowicz SZ, Arvey A, Treuting PM, Rudensky AY. Extrathymic generation of regulatory T cells in placental mammals mitigates maternal-fetal conflict. *Cell.* 2012;150(1):29–38.
145. Yang XO, Nurieva R, Martinez GJ, et al. Molecular antagonism and plasticity of regulatory and inflammatory T cell programs. *Immunity.* 2008;29(1):44–56.
146. Zhou L, Lopes JE, Chong MM, et al. TGF-beta-induced Foxp3 inhibits T(H)17 cell differentiation by antagonizing RORgammat function. *Nature.* 2008;453(7192):236–240.
147. Ichiyama K, Yoshida H, Wakabayashi Y, et al. Foxp3 inhibits RORgammat-mediated IL-17A mRNA transcription through direct interaction with RORgammat. *J Biol Chem.* 2008;283(25):17003–17008.
148. Chen Z, Barbi J, Bu S, et al. The ubiquitin ligase Stub1 negatively modulates regulatory T cell suppressive activity by promoting degradation of the transcription factor Foxp3. *Immunity.* 2013;39(2):272–285.
149. Noack M, Miossec P. Th17 and regulatory T cell balance in autoimmune and inflammatory diseases. *Autoimmun Rev.* 2014;13(6):668–677.
150. Osorio F, LeibundGut-Landmann S, Lochner M, et al. DC activated via dectin-1 convert Treg into IL-17 producers. *Eur J Immunol.* 2008;38(12):3274–3281.
151. Li L, Kim J, Boussiotis VA. IL-1beta-mediated signals preferentially drive conversion of regulatory T cells but not conventional T cells into IL-17-producing cells. *J Immunol.* 2010;185(7):4148–4153.
152. Li L, Patsoukis N, Petkova V, Boussiotis VA. Runx1 and Runx3 are involved in the generation and function of highly suppressive IL-17-producing T regulatory cells. *PLoS One.* 2012;7(9):e45115.
153. Esposito M, Ruffini F, Bergami A, et al. IL-17- and IFN-gamma-secreting Foxp3+ T cells infiltrate the target tissue in experimental autoimmunity. *J Immunol.* 2010;185(12):7467–7473.
154. Tartar DM, VanMorlan AM, Wan X, et al. FoxP3+RORgammat+ T helper intermediates display suppressive function against autoimmune diabetes. *J Immunol.* 2010;184(7):3377–3385.
155. Komatsu N, Okamoto K, Sawa S, et al. Pathogenic conversion of Foxp3+ T cells into TH17 cells in autoimmune arthritis. *Nat Med.* 2014;20(1):62–68.
156. Floess S, Freyer J, Siewert C, et al. Epigenetic control of the foxp3 locus in regulatory T cells. *PLoS Biol.* 2007;5(2):e38.

157. Miyao T, Floess S, Setoguchi R, et al. Plasticity of Foxp3(+) T cells reflects promiscuous Foxp3 expression in conventional T cells but not reprogramming of regulatory T cells. *Immunity*. 2012;36(2):262–275.
158. Ohkura N, Hamaguchi M, Morikawa H, et al. T cell receptor stimulation-induced epigenetic changes and Foxp3 expression are independent and complementary events required for Treg cell development. *Immunity*. 2012;37(5):785–799.
159. Polansky JK, Kretschmer K, Freyer J, et al. DNA methylation controls Foxp3 gene expression. *Eur J Immunol*. 2008;38(6):1654–1663.
160. Miyara M, Yoshioka Y, Kitoh A, et al. Functional delineation and differentiation dynamics of human CD4+ T cells expressing the FoxP3 transcription factor. *Immunity*. 2009;30(6):899–911.
161. Huehn J, Hamann A. Homing to suppress: address codes for Treg migration. *Trends Immunol*. 2005;26(12):632–636.
162. Liston A, Gray DH. Homeostatic control of regulatory T cell diversity. *Nat Rev Immunol*. 2014;14(3):154–165.
163. Feuerer M, Hill JA, Kretschmer K, von Boehmer H, Mathis D, Benoist C. Genomic definition of multiple ex vivo regulatory T cell subphenotypes. *Proc Natl Acad Sci USA*. 2010;107(13):5919–5924.
164. Delacher M, Schreiber L, Richards DM, Farah C, Feuerer M, Huehn J. Transcriptional control of regulatory T cells. *Curr Top Microbiol Immunol*. 2014;381:83–124.
165. Koch MA, Tucker-Heard G, Perdue NR, Killebrew JR, Urdahl KB, Campbell DJ. The transcription factor T-bet controls regulatory T cell homeostasis and function during type 1 inflammation. *Nat Immunol*. 2009;10(6):595–602.
166. Wang Y, Su MA, Wan YY. An essential role of the transcription factor GATA-3 for the function of regulatory T cells. *Immunity*. 2011;35(3):337–348.
167. Wohlfert EA, Grainger JR, Bouladoux N, et al. GATA3 controls Foxp3(+) regulatory T cell fate during inflammation in mice. *J Clin Invest*. 2011;121(11):4503–4515.
168. Cretney E, Kallies A, Nutt SL. Differentiation and function of Foxp3(+) effector regulatory T cells. *Trends Immunol*. 2013;34(2):74–80.
169. Cretney E, Xin A, Shi W, et al. The transcription factors Blimp-1 and IRF4 jointly control the differentiation and function of effector regulatory T cells. *Nat Immunol*. 2011;12(4):304–311.
170. Ayyoub M, Deknuydt F, Raimbaud I, et al. Human memory FOXP3+ Tregs secrete IL-17 ex vivo and constitutively express the T(H)17 lineage-specific transcription factor RORgamma t. *Proc Natl Acad Sci USA*. 2009;106(21):8635–8640.
171. Beriou G, Costantino CM, Ashley CW, et al. IL-17-producing human peripheral regulatory T cells retain suppressive function. *Blood*. 2009;113(18):4240–4249.
172. Voo KS, Wang YH, Santori FR, et al. Identification of IL-17-producing FOXP3+ regulatory T cells in humans. *Proc Natl Acad Sci U S A*. 2009;106(12):4793–4798.
173. Kryczek I, Wu K, Zhao E, et al. IL-17+ regulatory T cells in the microenvironments of chronic inflammation and cancer. *J Immunol*. 2011;186(7):4388–4395.
174. Nyirenda MH, Sanvito L, Darlington PJ, et al. TLR2 stimulation drives human naive and effector regulatory T cells into a Th17-like phenotype with reduced suppressive function. *J Immunol*. 2011;187(5):2278–2290.
175. Nyirenda MH, Morandi E, Vinkemeier U, et al. TLR2 stimulation regulates the balance between regulatory T cell and Th17 function: a novel mechanism of reduced regulatory T cell function in multiple sclerosis. *J Immunol*. 2015;194(12):5761–5774.
176. Koenen HJ, Smeets RL, Vink PM, van Rijssen E, Boots AM, Joosten I. Human CD25highFoxp3pos regulatory T cells differentiate into IL-17-producing cells. *Blood*. 2008;112(6):2340–2352.

177. Valmori D, Raffin C, Raimbaud I, Ayyoub M. Human RORgammat+ TH17 cells preferentially differentiate from naive FOXP3+Treg in the presence of lineage-specific polarizing factors. *Proc Natl Acad Sci USA*. 2010;107(45):19402–19407.
178. Benson MJ, Pino-Lagos K, Rosemblatt M, Noelle RJ. All-trans retinoic acid mediates enhanced T reg cell growth, differentiation, and gut homing in the face of high levels of co-stimulation. *J Exp Med*. 2007;204(8):1765–1774.
179. Coombes JL, Siddiqui KR, Arancibia-Carcamo CV, et al. A functionally specialized population of mucosal CD103+ DCs induces Foxp3+ regulatory T cells via a TGF-beta and retinoic acid-dependent mechanism. *J Exp Med*. 2007;204(8):1757–1764.
180. Sun CM, Hall JA, Blank RB, et al. Small intestine lamina propria dendritic cells promote de novo generation of Foxp3 T reg cells via retinoic acid. *J Exp Med*. 2007;204(8):1775–1785.
181. Arpaia N, Campbell C, Fan X, et al. Metabolites produced by commensal bacteria promote peripheral regulatory T-cell generation. *Nature*. 2013;504(7480):451–455.
182. Furusawa Y, Obata Y, Fukuda S, et al. Commensal microbe-derived butyrate induces the differentiation of colonic regulatory T cells. *Nature*. 2013;504(7480):446–450.
183. Smith PM, Howitt MR, Panikov N, et al. The microbial metabolites, short-chain fatty acids, regulate colonic Treg cell homeostasis. *Science*. 2013;341(6145):569–573.
184. Basu R, Whitley SK, Bhaumik S, et al. IL-1 signaling modulates activation of STAT transcription factors to antagonize retinoic acid signaling and control the TH17 cell-iTreg cell balance. *Nat Immunol*. 2015;16(3):286–295.
185. Chaudhry A, Samstein RM, Treuting P, et al. Interleukin-10 signaling in regulatory T cells is required for suppression of Th17 cell-mediated inflammation. *Immunity*. 2011;34(4):566–578.
186. Huber S, Gagliani N, Esplugues E, et al. Th17 cells express interleukin-10 receptor and are controlled by Foxp3(−) and Foxp3+ regulatory CD4+ T cells in an interleukin-10-dependent manner. *Immunity*. 2011;34(4):554–565.
187. Joller N, Lozano E, Burkett PR, et al. Treg cells expressing the coinhibitory molecule TIGIT selectively inhibit proinflammatory Th1 and Th17 cell responses. *Immunity*. 2014;40(4):569–581.
188. Yu X, Harden K, Gonzalez LC, et al. The surface protein TIGIT suppresses T cell activation by promoting the generation of mature immunoregulatory dendritic cells. *Nat Immunol*. 2009;10(1):48–57.
189. Chen Y, Haines CJ, Gutcher I, et al. Foxp3(+) regulatory T cells promote T helper 17 cell development in vivo through regulation of interleukin-2. *Immunity*. 2011;34(3):409–421.
190. Leon B, Bradley JE, Lund FE, Randall TD, Ballesteros-Tato A. FoxP3+ regulatory T cells promote influenza-specific Tfh responses by controlling IL-2 availability. *Nat Commun*. 2014;5:3495.
191. Izcue A, Coombes JL, Powrie F. Regulatory lymphocytes and intestinal inflammation. *Annu Rev Immunol*. 2009;27:313–338.
192. Sujino T, Kanai T, Ono Y, et al. Regulatory T cells suppress development of colitis, blocking differentiation of T-helper 17 into alternative T-helper 1 cells. *Gastroenterology*. 2011;141(3):1014–1023.
193. Davidson TS, DiPaolo RJ, Andersson J, Shevach EM. Cutting edge: IL-2 is essential for TGF-beta-mediated induction of Foxp3+ T regulatory cells. *J Immunol*. 2007;178(7):4022–4026.
194. Sakaguchi S, Yamaguchi T, Nomura T, Ono M. Regulatory T cells and immune tolerance. *Cell*. 2008;133(5):775–787.
195. Shevach EM. Application of IL-2 therapy to target T regulatory cell function. *Trends Immunol*. 2012;33(12):626–632.

196. Laurence A, Tato CM, Davidson TS, et al. Interleukin-2 signaling via STAT5 constrains T helper 17 cell generation. *Immunity*. 2007;26(3):371–381.
197. Moore-Connors JM, Fraser R, Halperin SA, Wang J. CD4(+)CD25(+)Foxp3(+) regulatory T cells promote Th17 responses and genital tract inflammation upon intracellular Chlamydia muridarum infection. *J Immunol*. 2013;191(6):3430–3439.
198. Pandiyan P, Conti HR, Zheng L, et al. CD4(+)CD25(+)Foxp3(+) regulatory T cells promote Th17 cells in vitro and enhance host resistance in mouse Candida albicans Th17 cell infection model. *Immunity*. 2011;34(3):422–434.
199. Wang Z, Friedrich C, Hagemann SC, et al. Regulatory T cells promote a protective Th17-associated immune response to intestinal bacterial infection with C. rodentium. *Mucosal Immunol*. 2014;7(6):1290–1301.
200. Amadi-Obi A, Yu CR, Liu X, et al. TH17 cells contribute to uveitis and scleritis and are expanded by IL-2 and inhibited by IL-27/STAT1. *Nat Med*. 2007;13(6):711–718.
201. Gasteiger G, Hemmers S, Bos PD, Sun JC, Rudensky AY. IL-2-dependent adaptive control of NK cell homeostasis. *J Exp Med*. 2013;210(6):1179–1187.
202. Gasteiger G, Hemmers S, Firth MA, et al. IL-2-dependent tuning of NK cell sensitivity for target cells is controlled by regulatory T cells. *J Exp Med*. 2013;210(6):1167–1178.
203. Barbi J, Pardoll D, Pan F. Metabolic control of the Treg/Th17 axis. *Immunol Rev*. 2013;252(1):52–77.

CHAPTER SIX

Mechanisms of Surveillance of Dendritic Cells by Regulatory T Lymphocytes

Maxime Dhainaut, Muriel Moser[1]
Laboratory of Immunobiology, Department of Molecular Biology, Université Libre de Bruxelles, Brussel, Belgium
[1]Corresponding author: e-mail address: mmoser@ulb.ac.be

Contents

Abstract

Regulatory T cells (Tregs) are critical to maintain the homeostasis of the organism, i.e., to prevent (excessive) inflammatory reactions against self and nonself. Data in the literature suggest that a variety of suppressive mechanisms exist that may act selectively depending on the tissue environment. An interesting question is whether dendritic cells (DCs), which are involved in immunity and tolerance, are targets of Tregs. We review herein some experimental evidence indicating that some suppressive mechanisms involve DCs, altering their function and inducing an immune control at the very first step of the immune response.

Progress in Molecular Biology and Translational Science, Volume 136
ISSN 1877-1173
http://dx.doi.org/10.1016/bs.pmbts.2015.08.003

Dendritic cells (DCs), discovered by Steinman in 1973,[1] play opposite roles in the immune system: they have the unique capacity to present antigens in an immunogenic or tolerogenic form and therefore control the switch between immunity and tolerance.

The stimulatory capacity of DCs has been amply demonstrated in the years that followed their discovery. They were described as the most potent stimulators of the primary mixed leukocyte reaction in mice.[2] They were considered as sentinels of the immune system, able to detect "danger signals" in the periphery and to activate T lymphocytes in lymphoid organs. Their unique capacity to activate optimally naïve T cells correlated with a specialization of function over time, i.e., they shift from an antigen-presenting mode to a T cell sensitizing mode while migrating to the T cell zones of lymphoid organs. During this process called "maturation," DCs upregulated major histocompatibility complex (MHC) and costimulatory molecule expression and became able to provide both signals to T cells.[3]

This simple view of DCs as inducers of immunity was challenged in 2001 when Hawiger and colleagues used a monoclonal antibody to a DC-restricted endocytic receptor to target an antigen to DCs *in vivo*. Unexpectedly, antigen-specific T cells were initially activated but could not be reactivated when the same antigen was delivered in an immunogenic form. In contrast, coadministration of DC-targeted antigen and anti-CD40 agonistic antibody resulted in prolonged T cell activation. The authors suggested that the primary role of DCs may be to maintain tolerance in the steady state.[4] Subsequent studies confirmed the existence of DCs displaying tolerogenic properties, which appeared critical to maintain immune homeostasis, as constitutively DC-depleted mice or mice in which DCs presented a defect in uptake of apoptotic cellular antigen-developed autoimmunity.[5,6] The molecular mechanisms involved remained elusive and could imply the stage of maturation, the nature of the DC subset, the secretion of inhibitory cytokines, etc.

In addition to DCs, regulatory T cells (Tregs) play a key role in maintaining peripheral tolerance and immune homeostasis. Their importance is best evidenced by the lethal autoimmune syndrome in mice or human lacking a functional Treg population.[7–10] Although Tregs were initially believed to control immune responses to self, there is now clear evidence that they not only prevent autoimmunity but also keep in check responses to foreign antigens, such as microbial components, thereby preventing excessive inflammatory responses.[11,12]

The critical role of DCs in the control of immunity versus tolerance and the close interaction between Tregs and DCs suggested a reciprocal regulation. In this chapter, we will review some mechanisms by which Tregs have been shown to control immune responses by targeting DCs.

1. OF MICE AND MEN: AN HISTORICAL PERSPECTIVE ON Tregs

1.1 Characterization of Suppressor T Cells

Initial observations of a suppressive population of thymus-derived cells were made by two groups in the early seventies. Nishizuka and Sakakura showed that, when performed at day 3, neonatal thymectomy led to organ-specific autoimmunity, which could be rescued by a thymus transplant.[13] Gershon and Kondo investigated the mechanisms of tolerance to large doses of sheep red blood cells[14,15] and noticed that transfer of spleen cells from tolerogenized animals suppressed the antibody response of the secondary recipients to sheep red blood cells only if thymus-derived lymphocytes were present during the course of tolerance induction. This early work prompted Sakaguchi and colleagues, 10 years later, to identify the suppressor T cell population with oophoritis-preventing capacity. They showed that Lyt-1$^+$ cells served as suppressor T cells, as injection of splenocytes depleted of Lyt-1$^+$ cells (i) failed to protect neonatally thymectomized mice from autoimmunity[16] and (ii) caused autoimmunity in nude mice, which could be prevented by a cotransfer of Lyt-1$^+$ cells.[17] A few years later, Powrie and colleagues confirmed the existence of two subsets in the CD4 population which differed by their expression of CD45RB. Injection of CD45RB^{hi}, but not CD45RB^{lo}, T cells in *SCID* mice led to development of colitis, whereas cotransfer of both populations prevented the onset of the disease, implying important regulatory interactions between the two subsets.[18]

A major advance in the field of self-tolerance was made by Sakaguchi and colleagues in 1995. They showed that suppressor T cells which maintained tolerance expressed the IL-2 receptor α-chain, CD25. *Nude* mice injected with CD4$^+$CD25$^-$ T cells displayed manifestations of autoimmunity, which were prevented by CD4$^+$CD25$^+$ cotransfer. Moreover, CD4$^+$CD25$^+$ T cells were able to suppress cellular (directed against allogeneic skin graft) and humoral (production of IgG after HA-BSA immunization) immune responses,[19] suggesting a role in the control of immune responses against self and nonself. Further work by the same group revealed that thymic production of CD25$^+$ T cells began 3 days after birth in mice, in accordance with the autoimmune manifestations associated with early thymectomy.[20]

1.2 Tregs in Mice: From *Scurfy* to Foxp3

In 1949, a new mouse line arose due to a spontaneous mutation in the mice colony of the Oak Ridge National Laboratory (USA). The mice, named *scurfy* after their ruffled and scaly skin, developed an X-linked disease and died at 3 weeks of age.[21] Brunkow *et al.* demonstrated that the *scurfy* phenotype was caused by a frameshift mutation in the gene encoding a new forkhead/winged-helix domain family member, Foxp3, resulting *in fine* in the production of a truncated protein.[8] Foxp3 was specifically expressed in the $CD4^{+}CD25^{+}$ population, which was completely absent in $Foxp3^{-/-}$ mice. As a consequence, these mice succumbed from an aggressive lymphoproliferative autoimmune syndrome comparable to the *scurfy* disease.[9] The link between the suppressor T cells described in the 1990s and Foxp3 was then clarified, but their function in the maintenance of tolerance in adults remained to be addressed. The generation of DEREG (DEpletion of REGulatory T cells)[22] and Foxp3-DTR mice,[23] which expressed the diphtheria toxin (DT) receptor only in the $Foxp3^{+}$ population, allowed to deplete Tregs by DT injections. Treg depletion in adult mice induced the same immunopathology as in $Foxp3^{-/-}$ mice, demonstrating that Tregs prevented autoimmunity throughout the lifespan of the mice, and not only set up mechanisms of tolerance early after birth.[23]

1.3 Tregs in Human: The IPEX Syndrome

First described in 1982,[24] the IPEX syndrome (*I*mmunodysregulation, *P*olyendocrinopathy *E*nteropathy, *X*-linked) is a fatal autoimmune disorder similar to the *scurfy* phenotype. Two groups simultaneously linked the IPEX syndrome with mutations in Foxp3,[7,10] confirming that dysfunction of Tregs were causing the disease. The characterization of human Tregs paralleled the observations made in mice, with some difference in Foxp3 regulation: (i) three different isoforms of human Foxp3 exist whose respective function is still unknown[25,26] and (ii) human effector T cells transiently expressed Foxp3 upon activation,[27,28] albeit at lower levels than in Tregs. Of note, the methylation status of the Foxp3 gene permitted to discriminate between bona fide Tregs and activated effector T cells.[29]

The analysis of the different mutations in IPEX patients contributed to a better understanding of the regulation of Foxp3 function.[28,30,31] An extensive study by Lopes *et al.*, looking at large cohorts of patients, categorized the mutations according to the Foxp3 domain involved. They showed that mutations in the FKH domain impaired nuclear localization of the protein, while the leucine zipper was necessary for its dimerization and the N-terminal domain required for transcription repression.[31]

1.4 The Multiple Faces of Tregs

The past decade has been marked by the description of a number of different Foxp3$^+$ T cells subpopulations (reviewed in Ref. 32).

In addition to thymus-derived, naturally occurring, Tregs (tTregs), Foxp3$^+$ T cells can differentiate from nonregulatory T cells *in vivo* under immunosuppressive conditions, and in particular in the presence of TGF-β.[33,34] These peripheral Tregs (pTregs) differentiated in the mesenteric lymph nodes and gut-associated lymphoid tissues following induction of oral tolerance,[35,36] developed in chronically inflamed tissues,[37] and were found in high numbers in various tumors.[38] Interestingly, there is some evidence that tTregs and pTregs might play distinct roles, preventing autoimmunity and controlling responses to environmental antigens in mucosa-associated lymphoid tissues, respectively.[39] This hypothesis was confirmed by the group of Rudensky, who identified a specific conserved noncoding DNA sequence in the Foxp3 promoter, CNS1, essential for pTreg differentiation but dispensable for tTreg generation.[40] The lack of pTregs in CNS1$^{-/-}$ mice caused a Th2-type pathology at mucosal sites (lung and gastrointestinal tract) and an impaired maternal–fetal tolerance,[41] but no systemic autoimmunity.[42]

In the past years, a functional specialization due to differences in the expression of transcription factors has been described for Tregs and paralleled T helper cell subsets. The transcription factor T-bet has been shown to control Treg homeostasis and function during type 1 inflammation and IFN-γ production.[43,44] Mice in which Tregs were deficient for IRF4 or Stat3, two transcription factors critical for the development of a Th2 or Th17 responses, respectively, failed to control Th2 or Th17 responses, leading to autoimmunity.[45,46] A population of Bcl6$^+$ CXCR5$^+$ Tregs, named follicular Tregs, was found in the germinal centers where it controlled the germinal center response.[47,48] Whether the transcription factors control the migratory capacity of Treg subsets or induced specific mechanisms of suppression remains to be determined.

2. IMMUNOSUPPRESSIVE MECHANISMS OF Tregs

Over the past decade, multiple mechanisms underlying the suppressive function of Tregs have been identified. In accordance with the lymphoproliferative autoimmune disease in Treg-depleted individuals, most suppressive mechanisms targeted directly effector T cells: production of the inhibitory cytokines IL-35[49] and TGF-β,[50] IL-10 secretion in mucosal tissues,[51–53] apoptosis induced by cytokine deprivation,[54] direct killing of effector cells via the perforin-granzyme pathway,[55] intercellular transfer of cAMP via gap junctions,[56] etc.

However, old studies showed that the suppressive activity of Tregs was abrogated by addition in the culture of anti-CD28 mAb[57] or by addition of increased numbers of antigen-presenting cells,[58] suggesting that cognate interactions between $CD25^+$ $CD4^+$, $CD25^-$, $CD4^+$ T cells and antigen-presenting cells may be required for suppression.

A potential target of Tregs was the population of DCs, which constitutes a link between innate and adaptive immune responses and has been shown to display tolerogenic properties. More than 10 years ago, several reports had indeed suggested that tTregs could selectively inhibit immune responses against self-antigens, i.e., presented by DCs at an immature stage. The general belief was that naturally occurring Tregs would mainly induce or maintain the tolerogenic status of immature DCs. However, subsequent studies clearly demonstrated that tTregs also downregulated responses against microbes[11] and innocuous foreign antigens expressed by fully mature DCs,[12] suggesting that the function of fully competent DCs was also under the control of these suppressor cells.

3. Tregs IMPAIR DC FUNCTION

3.1 Tregs Inhibit the Development of DC Populations

Ablation of Tregs in Foxp3-DTR mice by DT injection resulted in a nearly 10-fold increase in the absolute number of $MHCII^+$ $CD11c^{hi}$ DCs in the lymph nodes and potentiated their maturation process, as assessed by elevated CD80 and CD40 expression.[23] Nussenzweig's group analyzed the various precursors of DCs in bone marrow and secondary lymphoid organs and reported that, following Treg ablation, the absolute number of spleen and lymph node (LN) pre-cDCs and cDCs increased by 2- and 12-fold, respectively, which correlated with a higher proportion of proliferating cells (from 5% to 10%). By contrast, the multipotential progenitor (macrophage and DC precursor, MDP) and the bone marrow pre-cDCs remained unchanged, indicating that Tregs controlled DC numbers mainly in spleen and lymph nodes.[59] Flt3L was shown to be required for the expansion of

DCs as only cDCs from Flt3 competent mice were expanded upon DT treatment of mixed bone marrow chimeras.

It is still unclear whether Treg deficiency by itself is sufficient for DC expansion/maturation. Several observations suggest that activated T cells generated after Treg cell elimination may play a role: (i) they express factors that promote DC expansion/activation, such as Flt3L,[60] GM-CSF, RANKL, and CD40L[23]; (ii) no increase in DC numbers was found in lymphosufficient CD28$^{-/-}$ mice or in TCR transgenic Rag2$^{-/-}$ mice, which have a deficiency in Tregs in absence of proliferative disease[23,61]; (iii) reconstitution of Rag$^{-/-}$ mice with effector T cells and IL-2/JES6-1 complexes (leading to sustained IL-2 availability), but not with Tregs and IL-2/JES6-1 complexes, increased DC number in the spleen. Collectively, these data support a model in which Tregs indirectly affect DC maturation by inhibiting T lymphocytes that themselves induce the maturation of DCs.

Some evidence suggests that Treg populations may control selected DCs subsets. In particular, the absence of Tregs in HY-TCRtg-Rag2$^{-/-}$ mice was linked with a twofold expansion of NK cells in the lymph nodes and a specific decrease in the number of immature CD8α^+ DCs, both effects being reverted by injection of Tregs. The authors proposed that Tregs controlled NK cell precursors and maintained (indirectly) DCs in an immature stage, thereby favoring lymph nodes as privileged sites for induction of tolerance in the absence of infection.[62] Ulges *et al.* recently generated mice lacking protein kinase CK2 expression in the Treg compartment (*Csnk2b*$^{fl/fl}$*Foxp3*-Cre).[63] These mice suffered from Th2 lymphoproliferative responses in the lungs and developed asthmatic syndromes. The authors further showed that CK2-deficient (ILT3$^+$) Tregs were unable to control the pro-Th2 PDL-2$^+$ IRF4$^+$ DC subset, thereby identifying a novel Treg subset able to restrain Th2 response-associated diseases by modulating DCs.

3.2 Tregs Inhibit the Migration of DCs to Draining Lymph Nodes

Migration of DC is a critical early step in the priming of an immune response, as DCs sample antigens in the periphery, while naïve T cells are found in lymphoid organs.[3] Some evidence suggests that Tregs modulate DC migration, mainly through production of inhibitory cytokines, during the course of an immune response against self[64] and nonself.[65] Zhang *et al.* analyzed the migration and function of Tregs in a model of transplantation of pancreatic islets from CX3CR1-GFP BALB/c mice underneath the renal capsule of C57BL/6 recipients. Injection of PKH26-labeled *in vitro* TGF-β-induced Tregs prolonged graft survival, but only if these cells migrated sequentially from the graft to the draining lymph nodes. Of note, Tregs and DCs colocalized in the graft and this juxtaposition correlated with a fivefold decrease in DC number in the draining lymph node (dLNs) the day following Treg injection.[65] *In vitro*, addition of Tregs prevented CCL21-dependent DCs egress from islets via IL-10 and TGF-β secretion, and transfer of IL-10-deficient Tregs in transplanted mice had no impact on the DC number in the graft and dLNs, suggesting that Tregs modulated DC migration through IL-10 production. The inhibition of donor DC egress is likely to have a major effect on graft survival, as Swap70$^{-/-}$ grafts (displaying impaired actin rearrangement and cell migration) were prolonged from about 10 to 23 days. More recently, Alissafi *et al.* investigated the function of Tregs in a model of tolerance induction by subimmunogenic delivery of MOG peptide. The amelioration of disease correlated with the induction of MOG-specific Tregs which inhibited the recruitment of inflammatory-type DCs, a subset of DCs which differentiates from monocytes in inflammatory conditions.[64] Analysis of the gene expression pattern of inflammatory DCs from control and tolerized mice revealed a downregulation of genes encoding molecules involved in chemotaxis, and in particular CCR7. This inhibition of DC migration involved IL-10 and the PI3K/Akt signaling pathway. In both models, IL-10 production by Tregs seemed to control the CCL19/CCL21/CCR7 axis promoting DC migration, but neither study provided evidence for a direct mechanism of inhibition by Tregs.

3.3 Tregs Sequester DCs

In the lymph node, naïve T cells scan the DCs coming from the periphery for the presence of their cognate antigen. These interactions are important to allow subsequent activation of the naïve T cells by the DCs. A few studies suggested that Tregs may sequester DCs *in vitro* and *in vivo*, thereby impeding contact-dependent activation of conventional T lymphocytes.

Interactions between Tregs and DCs were first investigated *in vivo* in the context of autoimmune responses.[66,67] The role of Tregs in autoimmune diabetes was examined using as a model transfer of $CD4^+CD25^-$ T cells from BDC2.5 TCR transgenic mice into either NOD, Treg-deficient NOD.CD28$^{-/-}$ mice or Treg-reconstituted NOD.CD28$^{-/-}$.[67] Tregs suppressed the proliferation and differentiation of autoreactive T cells in the lymph nodes and altered the movement dynamics of helper T cells. Imaging by two-photon intravital laser scanning microscopy of explanted lymph nodes revealed arrest and static clustering of autoimmune T cells in prediabetic NOD mice mainly in the absence of Tregs, suggesting that Tregs prevented the formation of stable clusters of T cells with antigen-presenting cells (APCs), a prerequisite for T cell activation. Dustin and colleagues imaged the dynamic interactions between DCs, Tregs, and Teffs in a model of EAE, using the myelin basic protein (MBP) Ac1-11-specific TCR transgenic mice[66,68] and found that (i) autoantigen-specific T cells were more motile in the presence (6–7 μm/min) than in the absence (3 μm/min) of Tregs and that (ii) the contacts between DCs and autoimmune T lymphocytes were of longer duration in the absence of Tregs: from 340 to 995 s.[66] Similar observations were made in a model of graft versus host disease, in which CD11c-DTR-GFP BALB/c mice were irradiated and given labeled T cells from C57BL/6 mice together with T cell-depleted bone marrow.[69] Cotransfer of unlabeled Tregs increased the velocity of the effector T cells and reduced their contact time with DCs in draining lymph nodes in an IL-10-dependent manner. These observations correlated with the outcome of the GVHD reaction: indeed, IL-10-deficiency in Tregs impaired their suppression of GVHD.[70] Collectively, these reports suggested that Tregs attenuated the formation of stable contacts during inflammatory responses.

To test whether a similar mechanism may be involved in maintenance of the homeostasis under basal conditions, Matheu *et al.* examined the dynamics of unperturbed, endogenous Tregs in the steady state using the heterozygous progeny of Foxp3-GFP mice mated with CD11c-YFP reporter mice.[71] Upon transfer of labeled conventional T cells, quantification of the different interactions in the lymph node confirmed that Treg/DC contacts lasted longer than their Tconv/DC counterparts (3.3 vs. 2.5 min) and that Tregs and Tconvs also interacted in the absence of inflammation. Moreover, a population of Tregs was present near the LN capsule and characterized by a decreased motility. When labeled LPS-activated bone marrow-derived dendritic cells (BMDCs) were injected in Foxp3-GFP mice, they were intercepted by the subcapsular Tregs upon entry in the LNs, and established long-lived (up to 40 min) interactions with Tregs.

Insight on the molecular mechanism by which Tregs prevented stable DC/T cell interactions came from *in vitro* studies.[72,73] Both Tregs and Tconvs were forming stable contacts with BMDCs, in an MHCII-dependent manner.[73] Similarly, labeled Tregs or Tconvs from DO11.10 mice were aggregated with splenic DCs, in an antigen-dependent manner.[72] Of note, twice as many Tregs were forming long interactions with BMDCs as compared to Tconvs[73] and simultaneous addition of Treg and Tconvs from DO11.10 mice to OVA-pulsed splenic DCs resulted in an exclusion of Tconvs from the DC/Treg aggregates.[72] Ectopic expression of Foxp3 in Tconvs stabilized their interaction (more than 400 s) with BMDCs, revealing a role for a Foxp3-induced factor. Blocking of neuropilin-1 (Nrp-1) reduced the frequency of Treg/DC interaction, while ectopic Nrp-1 expression in Tconvs had the opposite effect,[73] indicating that Nrp-1 promoted prolonged interactions.[73] LFA-1 seemed to have a similar function, as LFA-1$^{-/-}$ Tregs were unable to outcompete wild-type T cells, whereas CTLA-4$^{-/-}$ Tregs similarly formed preferential clusters with DCs, suggesting that LFA-1 contributed to the preferential Treg aggregation on DCs. Interestingly, the authors further showed that a second mechanism of suppression existed which later downmodulated CD80 and CD86 expression in a CTLA-4- and LFA-1-dependent manner (see section 3.4.2.). Tomiyama *et al.* investigated the role of Mst1, the mammalian homolog of Drosophila Hpo, which interacted with the Rap1-binding protein RAPL and transmitted signals that caused LFA-1-dependent synapse formation of lymphocytes with APCs. They showed that Mst1-deficient Tregs failed to prevent colitis in a T cell transfer model, exhibited defective interactions with DCs and generated defective synapses, characterized by an absence of pMHC clusters and the ICAM-1 ring structure.[74] Consequently, they only partially inhibited the CD86 upregulation, compared to wild-type Tregs.

Finally, Tregs may also directly attract DCs through soluble mediators: they express the CD39 and CD73 ectonucleotidases and are therefore able to generate adenosine from ATP.[75] CD39-deficient Tregs failed to cluster with DCs *in vitro* and DCs migrated toward Tregs, but not CD4^{+} T cells, in Boyden chambers. Adenosine signaling in DCs activated the Epac/Rap1 pathway, leading to a relocalization of Rap1 to the DC plasma membrane, where it associated with the subcorticular actin cytoskeleton, favoring its remodeling and DC migration.[75]

Collectively, these observations suggest that DC-Treg clusters are formed and that this sequestration from responder T cells may (i) form a geographic barrier impeding DC/Tconv interaction and (ii) alter the function of DCs. Although most studies have been conducted in mice, human Tregs have been shown to form clusters with both CD1c^{+} myeloid DCs from peripheral blood and DCs derived from CD14^{+} monocytes.[76] As in mice, DC/Treg conjugates were more stable as compared with DC/Tconvs, suggesting that the mechanisms described above might also be relevant in human (Fig. 1).

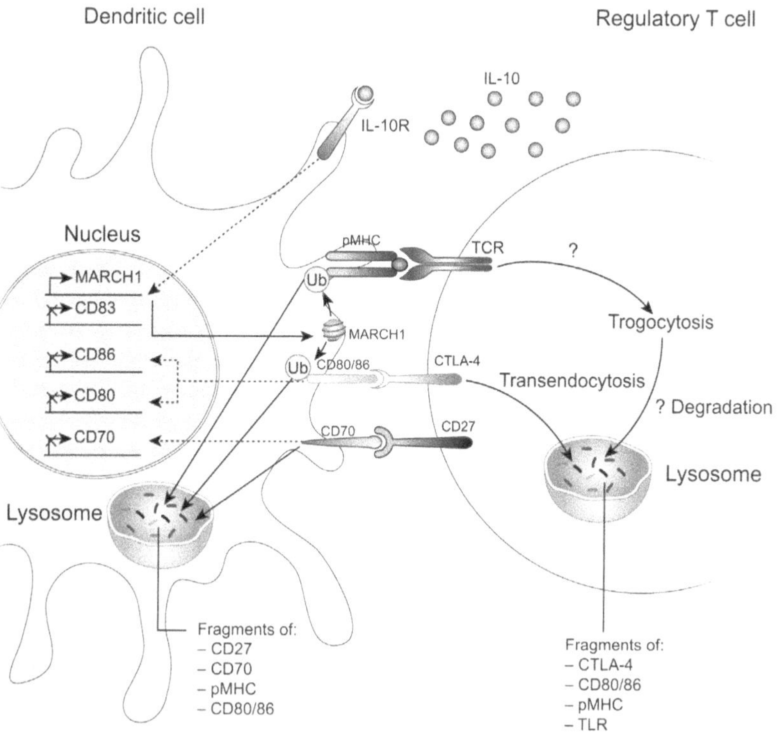

Figure 1 Control of DC stimulatory capacity by Tregs. Tregs closely interact with DCs and have been shown to interfere with their stimulatory capacity via a variety of mechanisms: soluble factors, signaling through transmembrane proteins, membrane exchanges at the immunological synapse (trogocytosis), ubiquitin-mediated degradation of MHC molecules, and control of CD80/CD86 and CD70 expression both at the transcriptional and posttranslational (internalization, degradation, etc.) levels, in a CTLA-4- and CD27-dependent manner, respectively.

3.4 Tregs Downregulate the Stimulatory Capacity of DCs

Data in the literature have shown in various models that Treg-conditioned DCs were less immunogenic, both *in vitro* and *in vivo*, suggesting that Tregs directly affected their ability to prime efficiently naïve T cells. Lewkowich *et al.*[105] reported that Treg depletion enhanced allergen-induced airway hyperresponsiveness in mice only when anti-CD25 was administered before the effector phase of the immune response. The proportion of pulmonary DCs expressing MHCII and costimulatory molecules (CD80 and CD86) was higher in Treg-depleted animals and these cells displayed increased T cell stimulatory capacity. Of note, Tregs appeared to control DC function only in mice resistant (C3H) but not in mice susceptible (A/J) to the development of airway hyperresponsiveness, suggesting that resistance to asthma was mediated by Treg suppression of DC activation. Schildknecht and colleagues investigated the function of DCs upon Treg depletion in DIETER mice (in which tamoxifen-induced presentation of LCMV-derived epitopes, GP_{33-44} and β-$gal_{497-505}$ by DCs).[77] Antigen presentation by steady-state DCs induced peripheral tolerance in DIETER mice but induced protective LCMV-specific immunity in DT-treated DEREG/DIETER mice. In absence of Tregs, DCs were present at increased numbers in lymph nodes and upregulated costimulatory molecules and T cell stimulatory capacity.[78] The impact of *in vitro* culture of DCs with Tregs was assessed by analyzing DC potency *in vivo* in a model of contact hypersensitivity.[79] DCs were purified from the culture, haptenized with TNBS, and injected in mice which were challenged by epicutaneous application of TNCB on the ear. Ear swelling of mice immunized with $DCs_{(Tregs)}$ was reduced by twofold, as compared to the control group. Similarly, $DCs_{(Tregs)}$ pulsed with OVA were less potent in inducing an antigen-specific response, as measured by the percentage of SIINFKL tetramer$^+$ CD8$^+$ T cells.[79]

Some mechanisms involved in this regulation have now been identified and are discussed below (Fig. 1).

3.4.1 Downregulation of pMHC

A few studies suggested that Tregs downmodulated the level of MHC expression on DCs in some conditions. Chattopadhyay and Shevach reported that induced Tregs inhibited MHC class II molecules on DCs in culture (at a high ratio of Tregs to DCs) in an IL-10-dependent manner.[80] IL-10 was produced by iTregs in the coculture and upregulated DC expression of MARCH1, an E3 ubiquitin ligase which induced ubiquitin-dependent internalization and degradation of MHCII.[80] Of note, DCs cocultured with induced Tregs expressed lower levels of CD83, a molecule which has been shown to block MHC class II association with MARCH1.[81] Thus, a major suppressive mechanism of induced Tregs relied on IL-10-dependent reciprocal modulation of MARCH1 and CD83 expression. In humans, Tregs have been also found to downregulate CD83 on monocyte-derived DCs.[82]

Tregs not only controlled MHCII expression but also antigen presentation in the context of MHCII. Indeed, using the C4H3 antibody to detect hen egg lysozyme peptide-MHC complexes, Mahnke *et al.* showed that injection of Tregs reduced by twofold the number of DCs displaying peptide–MHC complexes on their plasma membrane.[79]

3.4.2 Downregulation of the Costimulatory Ligands CD80/CD86

Data in the literature suggest that a major impact of Tregs on DC function could rely on inhibition of costimulatory signals. Fifteen years ago, Cederbom *et al.* reported that $CD4^+$ $CD25^+$ Tregs downregulated CD80 and CD86 surface expression on BMDCs in culture, in a cell-contact-dependent manner.[83] *In vivo* studies in NOD mice showed that $CD4^+$ $CD25^+$ Tregs inhibited $CD8^+$ T cells induced by peptide-pulsed DCs but not peptide/MHC monomers, suggesting that impaired T cell activation was due to DC hypofunction rather than T cell suppression.[106] The authors showed indeed that Tregs could prevent DC maturation *in vitro*. However, mature DCs were refractory to suppression *in vitro* and $CD4^+$ $CD25^+$ Tregs could not inhibit IFN-γ secretion by $CD8^+$ T lymphocytes in response to CpG or LPS-activated DCs *in vivo*. Interestingly, anti-CTLA-4 mAbs abrogated the suppressive activity of Tregs *in vitro* and *in vivo*, whereas blocking IL-4, IL-10, or TGF-β had no effect. The same observations were made using human monocyte-derived DCs.[82]

Subsequent studies highlighted the major role of cytotoxic T lymphocyte-associated antigen 4 (CTLA-4) which is constitutively expressed on Tregs and induced on conventional T cells upon activation. Indeed, blocking CTLA-4 impaired Treg suppression of naïve T cell proliferation *in vitro*[84] and CTLA-4-deficient mice developed a lethal lymphoproliferative condition.[85] The critical role of CTLA-4 in Treg function and control of homeostasis was demonstrated by Sakaguchi and colleagues who generated Balb/c mice with a specific deficiency of CTLA-4 in Tregs. These mice developed spontaneously fatal T cell dependent autoimmune disorders and promoted efficient tumor immunity.[86]

Several CTLA-4-dependent suppressive mechanisms have been identified in the last decade:

(1) In 2003, Fallarino *et al.* described a phenomenon of "reverse signaling" by which Tregs conditioned DCs to express indoleamin 2,3-dioxygenase (IDO) through a CTLA-4- and IFN-γ-dependent mechanism, thereby priming DCs for tolerance induction.[87] In particular, IDO-expressing bone marrow-derived DCs have been shown to skew the differentiation of naïve T cells into Tregs rather than Th17.[88] The suppressive effect of IDO relied on the combined effects of tryptophan starvation and generation of metabolites acting via the aryl hydrocarbon receptor in Tregs, thereby creating a local immune suppression and promoting systemic tolerance by reinforcing Treg function (for review, see Ref. 89).

(2) Cederbom *et al.* observed a decrease in the amount of CD80, but not CD86, encoding mRNA in BMDCs in culture with Tregs.[83] Incubation of BMDCs with CTLA-4Fc fusion proteins decreased CD80 and CD86 levels at the plasma membrane and to a lesser extent at the mRNA levels.[90] The authors proposed a model in which CTLA-4 interaction with CD80 led to increased pSTAT3 levels and decreased NF-κB activity, which accounted for the downregulation of CD80 and CD86 transcription. Of note, phosphorylation of STAT3 was also observed in BMDCs upon culture with *in vitro*-induced Tregs.

(3) CTLA-4 may also indirectly promote MARCH1 expression in DCs, which has been shown to increase ubiquitination-mediated degradation of MHCII, CD80, and CD86.[80] CTLA-4-deficient Tregs produced slightly less IL-10 than their WT counterparts when cultured with splenic DCs, potentially linking the two mechanisms.

(4) The group of David Sansom described an unexpected molecular mechanism by which CTLA-4 stripped CD80 and CD86 from DC membranes.[91] This cell extrinsic function of CTLA-4, referred to as *trans*-endocytosis, involved intracellular membrane exchanges and lysosomal degradation of CTLA-4 ligands. CTLA-4 expression was necessary and sufficient to confer to cells the ability to strip CD86 from DCs, which was acquired by CTLA-4 transfected resting T cells and lost upon addition of a blocking antibody to CTLA-4 in the medium. *Trans*-endocytosis was observed *in vivo* in Rag2$^{-/-}$ mice reconstituted with CD86-GFP-transduced bone marrow and immunized with OVA prior to a transfer of DO11.10 TCR transgenic or DO11.10 CTLA-4$^{-/-}$ T cells. CD86-GFP was detected only in CTLA-4-competent CD4^{+}CD25^{+} T cells upon OVA challenge in the presence of an inhibitor of lysosomal degradation. As expected, CTLA-4-dependent inhibition was strictly dependent on ligand driven T cell activation and was highly effective at low numbers of APCs.[92]

These observations would be consistent with a two-step model of CD80/CD86 downregulation on DCs: a first decrease in protein levels at the plasma membrane by CTLA-4-mediated *trans*-endocytosis and/or MARCH1-dependent degradation, followed by an inhibition of CD80/CD86 transcription.

Finally, a report has suggested a role for LAG-3, a transmembrane CD4-related protein expressed constitutively on Tregs (and transiently on activated T cells) which binds MHCII with higher affinity than CD4. Liang *et al.* have reported that Tregs inhibited DC maturation (CD86 expression) following interaction of LAG-3 with MHCII and ectopic LAG-3 expression conferred regulatory activity in a cell-contact-dependent manner. By analogy with the effect of MHCII cross-linking, the authors proposed that an inhibitory reverse signaling occurred after LAG-3/MHCII interaction, inducing an ITAM-mediated inhibitory pathway, based on FcγRγ and ERK-mediated recruitment of SHP-1.[93]

3.4.3 Downregulation of the Costimulatory Ligand CD70

CD70 is a member of the TNF family whose expression is upregulated on DCs upon maturation. Its only receptor, CD27, is expressed on T cells and NK cells.[94,95] The CD70/CD27 pathway promotes effector $CD8^+$ T cells responses (by sustaining survival of CTLs)[96] and influences polarization of $CD4^+$ T cells, as it represents an alternative IL-12-independent pathway of Th1 priming[97] and suppresses effector Th17 function.[98]

The importance of a control of the CD70/CD27 pathway activity was best evidenced by the phenotype of transgenic mice in which DCs constitutively express CD70 (CD11c-*CD70tg*).[99] These mice exhibited spontaneous activation of T cells leading to fatal immunopathology if not kept on a $CD27^{-/-}$ background. Moreover, the sole expression of CD70 on DCs sufficed to break tolerance and induce immunity in the DIETER mice model. Indeed, while $CD8^+$ T cells did not expand upon tamoxifen-induced presentation of LCMV peptide in DIETER mice, antigen-specific $CD8^+$ T cells proliferated in their *CD70tg*; DIETER counterparts and exhibited effector functions following challenge with LCMV. The authors combined the DIETER model of peripheral tolerance and the DEREG model of Treg ablation and found that Tregs were critical to maintain tolerance: an LCMV GP_{33-41}- and $Gal_{497-504}$-specific response was induced in the DEREG: DIETER model upon Treg depletion. The number of $CD11c^+$ cells in peripheral and mesenteric lymph nodes was increased and DCs displayed a more activated phenotype.[78]

In the same model, Muth *et al.* showed that DCs that lacked MHCII expression, i.e., were unable to interact with Tregs, failed to induce tolerance but induced a LCMV-specific CTL response instead. They conclude from these observations that DC activation in the absence of Tregs was a direct consequence of missing DC/Treg interaction.[107] These MHC class II-deficient DCs expressed higher levels of CD80, CD86, and CD70. Finally, the same group recently reported that CD40-mediated signaling in steady-state DCs was required for the breakdown of tolerance in the absence of suppressive T cells and that CD70 expression on DCs was critical to induce immunity, which was reversed by CD70 blockade *in vivo*.[100] Collectively, these data suggested that a population of CD4$^+$ T cells (presumably Tregs) controlled CD70 expression in DCs, thereby promoting tolerance.

Our own recent data confirmed and extended these observations in an *in vivo* model of Th1 priming, where mice were immunized with KLH-pulsed DCs.[101] Treg depletion in this model led to an increase in Th1 priming, which correlated with an upregulation of CD70 at the plasma membrane of DCs. *In vitro*, Tregs inhibited CD70 expression on DCs in a contact-dependent manner. As for CTLA-4-mediated suppression of CD80/86, CD27 expression on Tregs was required for CD70 downregulation. Interestingly, the molecular mechanism of suppression displayed similar features as the *trans*-endocytosis of CD80/CD86 by CTLA-4, as it involved intercellular transfer of molecules, but occurs at opposite direction, i.e., transfer of the receptor to the DCs.[91] Of note, activated Tconvs downregulated CD70 on DCs *in vitro*, suggesting that they may contribute to the termination of the response *in vivo* by inhibiting CD70 suppression. The group of Jonathan Ashwell similarly observed a CD27-dependent regulation of CD70 expression at the transcriptional level, suggesting that, similarly to CD80/CD86 downregulation, CD70 suppression by Tregs at the protein level could be followed by a reduction of CD70 transcription.

3.4.4 Upregulation of Inhibitory Molecules

A few reports in mouse and humans have suggested that Tregs may not only downregulate costimulatory molecules but also induce expression of inhibitory molecules in DCs. In the mouse, Mahnke *et al.* showed that addition of anti-B7-H3 to MLR culture with Treg-exposed DCs resulted in substantial increase in T cell activation.[79] *In vivo*, CD11c$^+$ DCs present in close proximity of Tregs displayed reduced levels of B7-2 and increased expression of B7-H3. In addition, 25–60% of DCs produced IL-10 after contact with Tregs. These data indicated that Tregs may induce directly an inhibitory phenotype in DCs, although the role of molecules of the B7-H family remains elusive.

3.5 Direct Killing of DCs

Direct killing of effector cells by Tregs has been proposed as one mechanism of suppression of antitumor responses.[55] Granzyme B was not detected in naïve Treg but was highly expressed in 5–30% of CD4$^+$ Foxp3$^+$ Tregs in tumors. Tumor-associated Tregs induced NK and CD8$^+$ T cell death via the granzyme B and perforin pathways. Of note, Amigorena and colleagues provided evidence that Tregs may similarly kill DCs in tumor-draining lymph nodes. They showed that adoptive transfer of OVA-specific OT-I $CD8^+$ T cells and Treg depletion induced rejection of an established fibrosarcoma MCA101-OVA.[102] The surface expression of CD80, ICAM-1, MHC class II, and CD40 was not altered after anti-CD25 treatment, but the absolute number of $CD11c^{hi}$ cells was increased by twofold in tumor-draining lymph nodes. Using two-photon live microscopy after adoptive transfer of DCs, the authors showed a strong correlation between the duration of the contact with Tregs and the death of the DCs. DC death was tumor antigen and perforin dependent. Interestingly, DC killing may be a general mechanism of control by Tregs (not restricted to tumors), as Lin *et al.* have reported recently that infusion of Tregs postallogeneic transplantation led to a decrease (by 50%) in the number of host DCs, compared to the infusion of Tconvs in a model of hematopoietic stem cell transplantation.[69] Tregs disrupted T cell–DC interaction in an IL-10-dependent manner and decreased CD86 and CD54 expression on DCs.

In human, Tregs were cultured with monocyte-derived immature DCs *in vitro* and DCs apoptosis was measured by the percentage of cells incorporating 7-AAD.[103] The authors discriminated between anti-CD3/anti-CD28-activated Tregs (which do not express granzyme A or B) and anti-CD3/anti-CD46-activated Tregs (which contained a subpopulation of $CD4^{lo}$ cells expressing granzyme A). Both Treg populations increased the percentage of $7\text{-}AAD^+$ DCs, although the second one was more efficient in doing so (from 5% to 20% and 35%, respectively). Moreover, perforin activity seemed to be required, as EGTA inhibited DC death. This suggested that human Tregs had the capacity to kill DCs, although the observations were made in a highly artificial experimental set-up.

4. CONCLUDING REMARKS

Tregs probably use multiple suppressive mechanisms depending on the tissue microenvironment and the nature of the immune response. The data summarized here clearly demonstrate that Tregs may target DCs, i.e., exert their suppressive mechanism at the very first step of the immune response (Fig. 2).

Interestingly, three strains of mice, genetically deficient for DCs,[5,6] Tregs,[9,22,23] or CTLA-4[85,86] harbor similar severe autoimmune disorders and die at an early age, suggesting that the control of DC function by

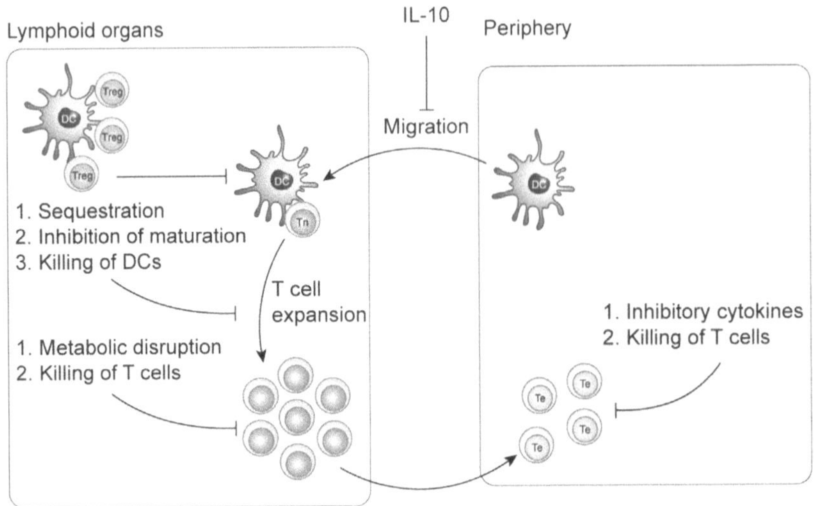

Figure 2 Tregs immunosuppressive mechanisms targeting DCs and T cells. Upon activation by "danger" signals, dendritic cells from the periphery migrate to the draining lymph nodes (dLNs), where they activate antigen-specific T cells, which proliferate, acquire an effector phenotype, and migrate to the periphery. Regulatory T cells may control the immune response by interfering with DC and/or T cell function. Several mechanisms have been described and include: IL-10-dependent inhibition of DC migration to the dLNs, sequestration of DCs in the dLNs and inhibition of their maturation, production of inhibitory cytokines, and killing of T cells and/or DCs.

CTLA-4^+ Tregs is critical to maintain immune homeostasis to self-antigens. Tregs arising from the thymus after birth would control DCs under steady-state conditions, thereby preventing their maturation and ensuring the continuous presentation of self-antigens in a tolerogenic manner. This control by Tregs has been shown to be required throughout the lifespan of mice.[23]

In addition, thymus-derived Tregs control immune responses to foreign antigens, such as infectious agents, by downregulating the adjuvant function of fully mature DCs and by directly inhibiting effector T cells. This control of DCs at a later stage may benefit to the host by preventing excessive inflammatory responses, potentially harmful, and by increasing the avidity, hence the quality of the primary and memory responses. This "quality control" by Tregs was suggested by the observation that Tregs were required for the establishment of high-avidity primary responses and effective memory. The mechanism involved inhibition of stable interactions between DCs and low-avidity T cells through decreased production of chemokines, thereby preventing the overproliferation of low-avidity T cells.[104]

The control of immature versus mature DC function by Tregs would ensure tolerance to self and regulate/restrain immunity to infections, respectively, acting in a complementary manner to maintain homeostasis in steady state and inflammatory conditions.

REFERENCES

1. Steinman RM, Cohn ZA. Identification of a novel cell type in peripheral lymphoid organs of mice. I. Morphology, quantitation, tissue distribution. *J Exp Med.* 1973;137:1142–1162.
2. Steinman RM, Witmer MD. Lymphoid dendritic cells are potent stimulators of the primary mixed leukocyte reaction in mice Immunology. *Proc Natl Acad Sci USA.* 1978;75:5132–5136.
3. De Smedt T, Pajak B, Muraille E, et al. Regulation of dendritic cell numbers and maturation by lipopolysaccharide in vivo. *J Exp Med.* 1996;184:1413–1424.
4. Hawiger D, Inaba K, Dorsett Y, et al. Dendritic cells induce peripheral T cell unresponsiveness under steady state conditions in vivo. *J Exp Med.* 2001;194:769–779.
5. Luckashenak N, Schroeder S, Endt K, et al. Constitutive crosspresentation of tissue antigens by dendritic cells controls CD8+ T cell tolerance in vivo. *Immunity.* 2008;28:521–532.
6. Ohnmacht C, Pullner A, King SBS, et al. Constitutive ablation of dendritic cells breaks self-tolerance of CD4 T cells and results in spontaneous fatal autoimmunity. *J Exp Med.* 2009;206:549–559.
7. Bennett CL, Christie J, Ramsdell F, et al. The immune dysregulation, polyendocrinopathy, enteropathy, X-linked syndrome (IPEX) is caused by mutations of FOXP3. *Nat Genet.* 2001;27:20–21.
8. Brunkow ME, Jeffery EW, Hjerrild KA, et al. Disruption of a new forkhead/winged-helix protein, scurfin, results in the fatal lymphoproliferative disorder of the scurfy mouse. *Nat Genet.* 2001;27:68–73.
9. Fontenot JD, Gavin MA, Rudensky AY. Foxp3 programs the development and function of CD4+CD25+ regulatory T cells. *Nat Immunol.* 2003;4:330–336.
10. Wildin RS, Ramsdell F, Peake J, et al. X-linked neonatal diabetes mellitus, enteropathy and endocrinopathy syndrome is the human equivalent of mouse scurfy. *Nat Genet.* 2001;27:18–20.
11. Belkaid Y, Piccirillo CA, Mendez S, Shevach EM, Sacks DL. CD4+CD25+ regulatory T cells control Leishmania major persistence and immunity. *Nature.* 2002;420:502–507.
12. Oldenhove G, de Heusch M, Urbain-Vansanten G, et al. CD4+ CD25+ regulatory T cells control T helper cell type 1 responses to foreign antigens induced by mature dendritic cells in vivo. *J Exp Med.* 2003;198:259–266.
13. Nishizuka Y, Sakakura T. Thymus and reproduction: sex-linked dysgenesia of the gonad after neonatal thymectomy in mice. *Science.* 1969;166:753–755.
14. Gershon RK, Kondo K. Cell interactions in the induction of tolerance: the role of thymic lymphocytes. *Immunology.* 1970;18:723–737.
15. Gershon RK, Kondo K. Infectious immunological tolerance. *Immunology.* 1971;21:903–914.
16. Sakaguchi S, Takahashi T, Nishizuka Y. Study on cellular events in post-thymectomy autoimmune oophoritis in mice. II. Requirement of Lyt-1 cells in normal female mice for the prevention of oophoritis. *J Exp Med.* 1982;156:1577–1586.
17. Sakaguchi S, Fukuma K, Kuribayashi K, Masuda T. Organ-specific autoimmune diseases induced in mice by elimination of T cell subset. I. Evidence for the active

participation of T cells in natural self-tolerance; deficit of a T cell subset as a possible cause of autoimmune disease. *J Exp Med.* 1985;161:72–87.
18. Powrie F, Leach MW, Mauze S, Caddle LB, Coffman RL. Phenotypically distinct subsets of CD4+ T cells induce or protect from chronic intestinal inflammation in C. B-17 scid mice. *Int Immunol.* 1993;5:1461–1471.
19. Sakaguchi S, Sakaguchi N, Asano M, Itoh M, Toda M. Immunologic self-tolerance maintained by activated T cells expressing IL-2 receptor alpha-chains (CD25). Breakdown of a single mechanism of self-tolerance causes various autoimmune diseases. *J Immunol.* 1995;155:1151–1164.
20. Asano M, Toda M, Sakaguchi N, Sakaguchi S. Autoimmune disease as a consequence of developmental abnormality of a T cell subpopulation. *J Exp Med.* 1996;184:387–396.
21. Russell WL, Russell LB, Gower JS. Exceptional inheritance of a sex-linked gene in the mouse explained on the basis that the X/O sex-chromosome constitution is female. *Proc Natl Acad Sci USA.* 1959;45:554–560.
22. Lahl K, Loddenkemper C, Drouin C, et al. Selective depletion of Foxp3$^+$ regulatory T cells induces a scurfy-like disease. *J Exp Med.* 2007;204:57–63.
23. Kim JM, Rasmussen JP, Rudensky AY. Regulatory T cells prevent catastrophic autoimmunity throughout the lifespan of mice. *Nat Immunol.* 2007;8:191–197.
24. Powell BR, Buist NR, Stenzel P. An X-linked syndrome of diarrhea, polyendocrinopathy, and fatal infection in infancy. *J Pediatr.* 1982;100:731–737.
25. Allan SE, Passerini L, Bacchetta R, et al. The role of 2 FOXP3 isoforms in the generation of human CD4+ Tregs. *J Clin Invest.* 2005;115:3276–3284.
26. Smith EL, Finney HM, Nesbitt AM, Ramsdell F, Robinson MK. Splice variants of human FOXP3 are functional inhibitors of human CD4+ T-cell activation. *Immunology.* 2006;119:203–211.
27. Allan SE, Crome SQ, Crellin NK, et al. Activation-induced FOXP3 in human T effector cells does not suppress proliferation or cytokine production. *Int Immunol.* 2007;19:345–354.
28. Gavin MA, Torgerson TR, Houston E, et al. Single-cell analysis of normal and FOXP3-mutant human T cells: FOXP3 expression without regulatory T cell development. *Proc Natl Acad Sci USA.* 2006;103:6659–6664.
29. Baron U, Floess S, Wieczorek G, et al. DNA demethylation in the human FOXP3 locus discriminates regulatory T cells from activated FOXP3$^+$ conventional T cells. *Eur J Immunol.* 2007;37:2378–2389.
30. d'Hennezel E, Ben-Shoshan M, Ochs HD, et al. FOXP3 forkhead domain mutation and regulatory T cells in the IPEX syndrome. *N Engl J Med.* 2009;361:1710–1713.
31. Lopes JE, Torgerson TR, Schubert LA, et al. Analysis of FOXP3 reveals multiple domains required for its function as a transcriptional repressor. *J Immunol.* 2006;177:3133–3142.
32. Liston A, Gray DHD. Homeostatic control of regulatory T cell diversity. *Nat Rev Immunol.* 2014;14:154–165.
33. Chen W, Jin W, Hardegen N, et al. Conversion of peripheral CD4+CD25- naive T cells to CD4+CD25+ regulatory T cells by TGF-beta induction of transcription factor Foxp3. *J Exp Med.* 2003;198:1875–1886.
34. Yamagiwa S, Gray JD, Hashimoto S, Horwitz DA. A role for TGF-beta in the generation and expansion of CD4+CD25+ regulatory T cells from human peripheral blood. *J Immunol.* 2001;166:7282–7289.
35. Mucida D, Kutchukhidze N, Erazo A, Russo M, Lafaille JJ, Curotto De Lafaille MA. Oral tolerance in the absence of naturally occurring Tregs. *J Clin Invest.* 2005;115:1923–1933.

36. Sun C-M, Hall JA, Blank RB, et al. Small intestine lamina propria dendritic cells promote de novo generation of Foxp3 T reg cells via retinoic acid. *J Exp Med.* 2007;204:1775–1785.
37. Curotto de Lafaille MA, Kutchukhidze N, Shen S, Ding Y, Yee H, Lafaille JJ. Adaptive Foxp3$^+$ regulatory T cell-dependent and -independent control of allergic inflammation. *Immunity.* 2008;29:114–126.
38. Liu VC, Wong LY, Jang T, et al. Tumor evasion of the immune system by converting CD4+CD25- T cells into CD4+CD25+ T regulatory cells: role of tumor-derived TGF-beta. *J Immunol.* 2007;178:2883–2892.
39. Curotto de Lafaille MA, Lafaille JJ. Natural and adaptive Foxp3$^+$ regulatory T cells: more of the same or a division of labor? *Immunity.* 2009;30:626–635.
40. Zheng Y, Josefowicz S, Chaudhry A, Peng XP, Forbush K, Rudensky AY. Role of conserved non-coding DNA elements in the Foxp3 gene in regulatory T-cell fate. *Nature.* 2010;463:808–812.
41. Samstein RM, Josefowicz SZ, Arvey A, Treuting PM, Rudensky AY. Extrathymic generation of regulatory T cells in placental mammals mitigates maternal-fetal conflict. *Cell.* 2012;150:29–38.
42. Josefowicz SZ, Niec RE, Kim HY, et al. Extrathymically generated regulatory T cells control mucosal TH2 inflammation. *Nature.* 2012;482:395–399.
43. Koch MA, Tucker-Heard G, Perdue NR, Killebrew JR, Urdahl KB, Campbell DJ. The transcription factor T-bet controls regulatory T cell homeostasis and function during type 1 inflammation. *Nat Immunol.* 2009;10:595–602.
44. Koch MA, Thomas KR, Perdue NR, Smigiel KS, Srivastava S, Campbell DJ. T-bet(+) Treg cells undergo abortive Th1 cell differentiation due to impaired expression of IL-12 receptor β2. *Immunity.* 2012;37:501–510.
45. Chaudhry A, Rudra D, Treuting P, et al. CD4+ regulatory T cells control TH17 responses in a Stat3-dependent manner. *Science.* 2009;326:986–991.
46. Zheng Y, Chaudhry A, Kas A, et al. Regulatory T-cell suppressor program co-opts transcription factor IRF4 to control T(H)2 responses. *Nature.* 2009;458:351–356.
47. Chung Y, Tanaka S, Chu F, et al. Follicular regulatory T cells expressing Foxp3 and Bcl-6 suppress germinal center reactions. *Nat Med.* 2011;17:983–988.
48. Linterman MA, Pierson W, Lee SK, et al. Foxp3(+) follicular regulatory T cells control the germinal center response. *Nat Med.* 2011;17:975–982.
49. Collison LW, Workman CJ, Kuo TT, et al. The inhibitory cytokine IL-35 contributes to regulatory T-cell function. *Nature.* 2007;450:566–569
50. Li MO, Sanjabi S, Flavell RA. Transforming growth factor-beta controls development, homeostasis, and tolerance of T cells by regulatory T cell-dependent and -independent mechanisms. *Immunity.* 2006;25:455–471.
51. Asseman C, Mauze S, Leach MW, Coffman RL, Powrie F. An essential role for interleukin 10 in the function of regulatory T cells that inhibit intestinal inflammation. *J Exp Med.* 1999;190:995–1004.
52. Atarashi K, Tanoue T, Shima T, et al. Induction of colonic regulatory T cells by indigenous Clostridium species. *Science.* 2011;331:337–341.
53. Chaudhry A, Samstein RM, Treuting P, et al. Interleukin-10 signaling in regulatory T cells is required for suppression of Th17 cell-mediated inflammation. *Immunity.* 2011;34:566–578.
54. Pandiyan P, Zheng L, Ishihara S, Reed J, Lenardo MJ. CD4+CD25+Foxp3+ regulatory T cells induce cytokine deprivation-mediated apoptosis of effector CD4+ T cells. *Nat Immunol.* 2007;8:1353–1362.

55. Cao X, Cai SF, Fehniger TA, et al. Granzyme B and perforin are important for regulatory T cell-mediated suppression of tumor clearance. *Immunity*. 2007;27:635–646.
56. Bopp T, Becker C, Klein M, et al. Cyclic adenosine monophosphate is a key component of regulatory T cell-mediated suppression. *J Exp Med*. 2007;204:1303–1310.
57. Thornton AM, Shevach EM. CD4+CD25+ immunoregulatory T cells suppress polyclonal T cell activation in vitro by inhibiting interleukin 2 production. *J Exp Med*. 1998;188:287–296.
58. Takahashi T, Kuniyasu Y, Toda M, et al. Immunologic self-tolerance maintained by CD25+CD4+ naturally anergic and suppressive T cells: induction of autoimmune disease by breaking their anergic/suppressive state. *Int Immunol*. 1998;10:1969–1980.
59. Liu K, Victora GD, Schwickert TA, et al. In vivo analysis of dendritic cell development and homeostasis. *Science*. 2009;324:392–397.
60. Saito Y, Boddupalli CS, Borsotti C, Manz MG. Dendritic cell homeostasis is maintained by nonhematopoietic and T-cell-produced Flt3-ligand in steady state and during immune responses. *Eur J Immunol*. 2013;43:1651–1658.
61. Bolton HA, Roediger B, Groth BFDS. The effects of IL-2 and Treg cells on dendritic cell homeostasis are mediated indirectly via activation of conventional T cells. *Eur J Immunol*. 2015;45:1141–1147.
62. Giroux M, Yurchenko E, St.-Pierre J, Piccirillo CA, Perreault C. T regulatory cells control numbers of NK cells and CD8alpha+ immature dendritic cells in the lymph node paracortex. *J Immunol*. 2007;179:4492–4502.
63. Ulges A, Klein M, Reuter S, et al. Protein kinase CK2 enables regulatory T cells to suppress excessive TH2 responses in vivo. *Nat Immunol*. 2015;16:267–275.
64. Alissafi T, Hatzioannou A, Ioannou M, et al. De novo-induced self-antigen-specific Foxp3$^+$ regulatory T cells impair the accumulation of inflammatory dendritic cells in draining lymph nodes. *J Immunol*. 2015;194:5812–5824.
65. Zhang N, Schröppel B, Lal G, et al. Regulatory T cells sequentially migrate from inflamed tissues to draining lymph nodes to suppress the alloimmune response. *Immunity*. 2009;30:458–469.
66. Tadokoro CE, Shakhar G, Shen S, et al. Regulatory T cells inhibit stable contacts between CD4+ T cells and dendritic cells in vivo. *J Exp Med*. 2006;203:505–511.
67. Tang Q, Adams JY, Tooley AJ, et al. Visualizing regulatory T cell control of autoimmune responses in nonobese diabetic mice. *Nat Immunol*. 2006;7:83–92.
68. Lafaille JJ, Nagashima K, Katsuki M, Tonegawa S. High incidence of spontaneous autoimmune encephalomyelitis in immunodeficient anti-myelin basic protein T cell receptor transgenic mice. *Cell*. 1994;78:399–408.
69. Lin KL, Fulton LM, Berginski M, et al. Intravital imaging of donor allogeneic effector and regulatory T cells with host dendritic cells during GVHD. *Blood*. 2014;123:1604–1614.
70. Tawara I, Sun Y, Liu C, et al. Donor- but not host-derived interleukin-10 contributes to the regulation of experimental graft-versus-host disease. *J Leukoc Biol*. 2012;91:667–675.
71. Matheu MP, Othy S, Greenberg ML, et al. Imaging regulatory T cell dynamics and CTLA4-mediated suppression of T cell priming. *Nat Commun*. 2015;6:1–11.
72. Onishi Y, Fehervari Z, Yamaguchi T, Sakaguchi S. Foxp3+ natural regulatory T cells preferentially form aggregates on dendritic cells in vitro and actively inhibit their maturation. *Proc Natl Acad Sci USA*. 2008;105:10113–10118.
73. Sarris M, Andersen KG, Randow F, Mayr L, Betz AG. Neuropilin-1 expression on regulatory T cells enhances their interactions with dendritic cells during antigen recognition. *Immunity*. 2008;28:402–413.

74. Tomiyama T, Ueda Y, Katakai T, Kondo N, Okazaki K, Kinashi T. Antigen-specific suppression and immunological synapse formation by regulatory T cells require the Mst1 kinase. *PLoS One*. 2013;8:1–10.
75. Ring S, Pushkarevskaya A, Schild H, et al. Regulatory T cell-derived adenosine induces dendritic cell migration through the Epac-Rap1 pathway. *J Immunol*. 2015;194:3735–3744.
76. Herman S, Krenbek D, Klimas M, et al. Regulatory T cells form stable and long-lasting cell cluster with myeloid dendritic cells (DC). *Int Immunol*. 2012;24:417–426.
77. Probst HC, Lagnel J, Kollias G, Van Den Broek M. Inducible transgenic mice reveal resting dendritic cells as potent inducers of CD8+ T cell tolerance. *Immunity*. 2003;18:713–720.
78. Schildknecht A, Brauer S, Brenner C, et al. FoxP3+ regulatory T cells essentially contribute to peripheral CD8+ T-cell tolerance induced by steady-state dendritic cells. *Proc Natl Acad Sci USA*. 2010;107:199–203.
79. Mahnke K, Ring S, Johnson TS, et al. Induction of immunosuppressive functions of dendritic cells in vivo by CD4+CD25+ regulatory T cells: role of B7-H3 expression and antigen presentation. *Eur J Immunol*. 2007;37:2117–2126.
80. Chattopadhyay G, Shevach E. Antigen-specific induced T regulatory cells impair dendritic cell function via an IL-10/MARCH1-dependent mechanism. *J Immunol*. 2013;191:5875–5884.
81. Tze LE, Horikawa K, Domaschenz H, et al. CD83 increases MHC II and CD86 on dendritic cells by opposing IL-10-driven MARCH1-mediated ubiquitination and degradation. *J Exp Med*. 2011;208:149–165.
82. Misra N, Bayry J, Lacroix-Desmazes S, Kazatchkine MD, Kaveri SV. Cutting edge: human CD4+CD25+ T cells restrain the maturation and antigen-presenting function of dendritic cells. *J Immunol*. 2004;172:4676–4680.
83. Cederbom L, Hall H, Ivars F. CD4+CD25+ regulatory T cells down-regulate co-stimulatory molecules on antigen-presenting cells. *Eur J Immunol*. 2000;30:1538–1543.
84. Takahashi T, Tagami T, Yamazaki S, et al. Immunologic self-tolerance maintained by CD25(+)CD4(+) regulatory T cells constitutively expressing cytotoxic T lymphocyte-associated antigen 4. *J Exp Med*. 2000;192:303–310.
85. Waterhouse P, Penninger JM, Timms E, et al. Lymphoproliferative disorders with early lethality in mice deficient in Ctla-4. *Science*. 1995;270:985–988.
86. Wing K, Onishi Y, Prieto-Martin P, et al. CTLA-4 control over Foxp3+ regulatory T cell function. *Science*. 2008;322:271–275.
87. Fallarino F, Grohmann U, Hwang KW, et al. Modulation of tryptophan catabolism by regulatory T cells. *Nat Immunol*. 2003;4:1206–1212.
88. Nguyen NT, Kimura A, Nakahama T, et al. Aryl hydrocarbon receptor negatively regulates dendritic cell immunogenicity via a kynurenine-dependent mechanism. *Proc Natl Acad Sci USA*. 2010;107:19961–19966.
89. Fallarino F, Grohmann U, Puccetti P. Indoleamine 2,3-dioxygenase: from catalyst to signaling function. *Eur J Immunol*. 2012;42:1932–1937.
90. Kowalczyk A, D'Souza CA, Zhang L. Cell-extrinsic CTLA4-mediated regulation of dendritic cell maturation depends on STAT3. *Eur J Immunol*. 2014;44:1143–1155.
91. Qureshi OS, Zheng Y, Nakamura K, et al. Trans-endocytosis of CD80 and CD86: a molecular basis for the cell-extrinsic function of CTLA-4. *Science*. 2011;332:600–603.
92. Hou TZ, Qureshi OS, Wang CJ, et al. A transendocytosis model of CTLA-4 function predicts its suppressive behavior on regulatory T cells. *J Immunol*. 2015;194:2148–2159.
93. Liang B, Workman C, Lee J, et al. Regulatory T cells inhibit dendritic cells by lymphocyte activation gene-3 engagement of MHC class II. *J Immunol*. 2008;180:5916–5926.

94. Oshima H, Nakano H, Nohara C, et al. Characterization of murine CD70 by molecular cloning and mAb. *Int Immunol.* 1998;10:517–526.
95. Tesselaar K, Xiao Y, Arens R, et al. Expression of the murine CD27 ligand CD70 in vitro and in vivo. *J Immunol.* 2003;170:33–40.
96. Peperzak V, Xiao Y, Veraar EAM, Borst J. CD27 sustains survival of CTLs in virus-infected nonlymphoid tissue in mice by inducing autocrine IL-2 production. *J Clin Invest.* 2010;120:168–178.
97. Soares H, Waechter H, Glaichenhaus N, et al. A subset of dendritic cells induces CD4+ T cells to produce IFN-gamma by an IL-12-independent but CD70-dependent mechanism in vivo. *J Exp Med.* 2007;204:1095–1106.
98. Coquet JM, Middendorp S, van der Horst G, et al. The CD27 and CD70 costimulatory pathway inhibits effector function of T helper 17 cells and attenuates associated autoimmunity. *Immunity.* 2013;38:53–65.
99. Keller AM, Schildknecht A, Xiao Y, van den Broek M, Borst J. Expression of costimulatory ligand CD70 on steady-state dendritic cells breaks CD8+ T cell tolerance and permits effective immunity. *Immunity.* 2008;29:934–946.
100. Muth S, Schütze K, Hain T, Yagita H, Schild H, Probst HC. A CD40/CD40L feedback loop drives the breakdown of CD8+ T-cell tolerance following depletion of suppressive CD4+ T cells. *Eur J Immunol.* 2014;44:1099–1107.
101. Dhainaut M, Coquerelle C, Uzureau S, et al. Thymus-derived regulatory T cells restrain pro-inflammatory Th1 responses by downregulating CD70 on dendritic cells. *EMBO J.* 2015;34:1336–1348. e201490312.
102. Boissonnas A, Scholer-Dahirel A, Simon-Blancal V, et al. Foxp3$^+$ T cells induce perforin-dependent dendritic cell death in tumor-draining lymph nodes. *Immunity.* 2010;32:266–278.
103. Grossman WJ, Verbsky JW, Barchet W, Colonna M, Atkinson JP, Ley TJ. Human T regulatory cells can use the perforin pathway to cause autologous target cell death. *Immunity.* 2004;21:589–601.
104. Pace L, Tempez A, Arnold-Schrauf C, et al. Regulatory T cells increase the avidity of primary CD8+ T cell responses and promote memory. *Science.* 2012;338:532–536.
105. Lewkowich IP, Herman NS, Schleifer KW, et al. CD4+CD25+ T cells protect against experimentally induced asthma and alter pulmonary dendritic cell phenotype and function. *J Exp Med.* 2005;202(11):1549–1561. PMID:16314437.
106. Serra P, Amrani A, Yamanouchi J, et al. CD40 ligation releases immature dendritic cells from the control of regulatory CD4+CD25+ T cells. *Immunity.* 2003;19(6): 877–889. PMID:14670304.
107. Muth S, Schutze K, Schild H, Probst HC. Release of dendritic cells from cognate CD4+ T-cell recognition results in impaired peripheral tolerance and fatal cytotoxic T-cell mediated autoimmunity. *Proc Natl Acad Sci U S A.* 2012;109(23), 9059–9064. PMID:22615402.

CHAPTER SEVEN

Development and Function of Effector Regulatory T Cells

Peggy P. Teh*,†, Ajithkumar Vasanthakumar*,†, Axel Kallies*,†,1
*Walter and Eliza Hall Institute of Medical Research, Parkville, Victoria, Australia
†Department of Medical Biology, University of Melbourne, Parkville, Victoria, Australia
[1]Corresponding author: e-mail address: kallies@wehi.edu.au

Contents

Abstract

Distinguishing self from nonself is a unique feature of the immune system. Although most self-reactive T cells are eliminated in the thymus, a few rogue cells escape the negative selection process and have the potential to mediate autoimmune disease. Over the last decade, there has been a vast improvement in our understanding of the cellular mechanisms that evolved to dampen the deleterious effects of these self-reactive T cells. In particular, T cells expressing the transcription factor FoxP3, known as regulatory T (Treg) cells, play a central role in maintaining immune homeostasis and suppressing autoimmune responses. In addition, Treg cells are endowed with the ability to suppress diverse inflammatory responses both in lymphoid and in nonlymphoid tissues. This requires Treg cells to undergo a peripheral differentiation and specialization program that results in the emergence of effector Treg (eTreg) cells that are characterized by their

Progress in Molecular Biology and Translational Science, Volume 136
ISSN 1877-1173
http://dx.doi.org/10.1016/bs.pmbts.2015.08.005

ability to produce high amounts of immunosuppressive molecules, including IL-10. This chapter discusses the recent advances in our understanding of the mechanisms governing the differentiation, migration, and maintenance of eTreg cells, in particular in nonlymphoid tissues, in health and disease.

1. INTRODUCTION

Immune homeostasis critically depends on the balance between anti- and proinflammatory processes. Excessive and prolonged inflammation, whether in response to infection, tissue injury, or perturbed metabolic homeostasis, is deleterious to the organism. Treg cells, a specialized subset of $CD4^+$ T cells, maintain immunological homeostasis in a nonredundant manner by suppressing the activation of self-reactive and inflammatory T cells. The development, function, and maintenance of Treg cells are dependent on the expression of the transcription factor FoxP3.[1] The essential nature of Treg cells is exemplified in mice or humans with loss-of-function mutations in FoxP3 that result in a fatal, early-onset, T-cell-dependent multiorgan inflammatory disease.[2–4]

Approximately 10–15% of the murine and 5–10% of the human $CD4^+$ T-cell population are comprised of Treg cells. This population can be further subdivided based on their developmental origin into cells that acquire FoxP3 expression in the thymus and those that are induced to differentiate into Tregs in the periphery. Treg cell development in the thymus is a two-stage process with the first stage governed by T-cell receptor (TCR) signaling and the second dependent on the cytokine interleukin (IL)-2.[5–7] In the first stage, transcription factors and epigenetic regulators downstream of the TCR facilitate the development of partially committed Treg precursors, in which the FoxP3 locus is poised for activation. Transcriptional induction of FoxP3 takes place in the second stage, which is driven by the IL-2-dependent transcription factor STAT5 (signal transducer and activator of transcription 5).[8] FoxP3 subsequently establishes the Treg cell transcriptional program, which reinforces stable maintenance of FoxP3 itself and confers functional fitness to the cells. IL-2 and TCR signaling are not only required for the development of Treg cells but are also essential for their continuous survival and homeostasis.[6,9,29]

Treg cells execute their suppressor function by diverse mechanisms (reviewed in Refs. 10–12). These include limiting IL-2 availability to conventional T cells, CTLA-4 (cytotoxic T-lymphocyte-associated protein

4)-mediated inhibition of costimulation, secretion of IL-10 and IL-35, or expression of suppressor molecules such as CD39 and TIGIT (T-cell immunoreceptor with Ig and ITIM domains).[13–18] Although it is clear that Treg cells can utilize multiple pathways that lead to the suppression of immune responses, the relative contribution of each of these mechanisms is unknown.

2. FEATURES OF EFFECTOR T REGULATORY CELLS

Most Treg cells localize to secondary lymphoid organs, and the majority of them have a naïve phenotype. These Treg cells have been termed central Treg (cTreg) cells. They express high amounts of CD25, the alpha chain of the high-affinity IL-2 receptor, and, therefore, can deprive surrounding conventional T cells of IL-2 and limit their proliferation. However, the expression of functionally important molecules such as IL-10, CTLA-4, ICOS (inducible costimulator), and TIGIT is restricted to or greatly increased in a subpopulation of Tregs, which have an activated phenotype. These cells have been termed effector Treg (eTreg) cells.[20] Depending on their localization and surrounding cytokine milieu, eTreg cells are phenotypically diverse. eTreg cell development is regulated by TCR signals that act together with cytokine and tissue-specific factors. T helper (T_H) lineage-specific and other transcription factors cooperate with FoxP3 during this differentiation process to generate functional and spatial diversity, which is critical to contain immune-mediated damage. Mature eTreg cells express the transcription factor B-lymphocyte-induced maturation protein 1 (Blimp-1) and are potent producers of the anti-inflammatory cytokine IL-10.[19,20] Several proinflammatory cytokines are known to upregulate the expression of Blimp-1 in Treg cells. Transcriptional profiling of Blimp-1$^+$ eTreg cells demonstrated their unique transcriptional signature and has provided further insights into key molecular events that underpin the eTreg cell differentiation program.[21] Importantly, the FoxP3-dependent transcriptional program is conserved in eTreg cells. Although Blimp-1 expression was used to initially define eTregs, it is not required for their differentiation. It is, however, essential for IL-10 production and plays a significant role in the migration of eTreg cells to nonlymphoid tissues.[20] While eTreg cells represent a minor population within lymphoid organs, they predominate in nonlymphoid tissues, reflecting their essential role in the resolution of local inflammation and tissue homeostasis.[19]

Analogous to the murine Treg cell differentiation scheme, human cTreg cells are $CD45RA^+FoxP3^{low}$, while eTreg cells are $CD45RO^+CD45RA^-$ $FoxP3^{high}$.[22,23] eTreg cells in humans are also known to express TIGIT, CD39, ICOS, and IL-10.[24–26] Thus, mounting evidence suggests that similar to murine Treg cells, human Treg cells can undergo a peripheral differentiation program critical for the maintenance of immune tolerance.

3. ROLE OF TCR SIGNALING IN EFFECTOR T REGULATORY CELL DIFFERENTIATION

TCR signaling is critical for the first stage of Treg cell development in the thymus. Several TCR-induced transcription factors such as c-Rel and nuclear factor of activated T cells (NFAT) are directly involved in priming the FoxP3 locus and in inducing its transcription. Deletion of these transcription factors results in severe Treg cell deficiency in the thymus and periphery.[27,28] Recently, two studies have shed light on the role of TCR signaling in Treg cell homeostasis demonstrating the continuous requirement of TCR signaling.[29,30] Strikingly, loss of TCRβ specifically affected the eTreg cell population, indicating that TCR signals are required for the development of functionally mature Treg cells. This is consistent with our observations that expression of Nur77, a surrogate marker for TCR signaling, was elevated in eTreg cells in nonlymphoid tissues compared to their cTreg counterparts.[21,31]

We have previously described that the TCR-induced transcription factor interferon regulatory factor (IRF)4 is required for eTreg cell differentiation.[20] IRF4 deficiency severely reduced the number and function of eTreg cells in both lymphoid and nonlymphoid tissues.[20] Although the exact role of IRF4 in eTreg cell differentiation is unknown, evidence from other lymphocyte populations suggests that it affects cell differentiation and function at multiple levels. In $CD4^+$ T cells, IRF4 is required for the differentiation of multiple T_H cell subsets, including T_H2, T_H17, and T follicular helper (T_{FH}) cells,[32–34] while in $CD8^+$ T cells, IRF4 regulates metabolic functions and clonal population expansion in a TCR-affinity-dependent manner.[35] IRF4 is also essential for germinal center B-cell development and plasma cell differentiation, where it is critical in inducing Blimp-1.[36,37] IRF4 also activates expression of Blimp-1 in Treg cells and cooperates with Blimp-1 to activate IL-10[20].

eTreg cells, in particular those residing in nonlymphoid tissues, have distinct TCR repertoires, suggesting that their development can be influenced directly by local cues such as tissue antigens.[38] Differences in TCR-ligand affinity are known to have direct effects on the characteristics and intensity of downstream signaling pathways, which might also influence the differentiation of eTreg cells.[39] Taken together, these observations demonstrate an essential role for TCR signaling in eTreg cell differentiation.

4. CYTOKINE REQUIREMENTS OF EFFECTOR T REGULATORY CELLS

In addition to TCR signaling, cytokines are essential mediators required to shape immune responses. Many cytokines execute their function by inducing the phosphorylation of downstream STAT transcription factors, which independently or in cooperation with other TCR-induced transcription factors reprogram naïve T cells to effector T cells. eTreg cells are dependent upon the local cytokine milieu, suggesting that their differentiation mirrors this paradigm. In line with this idea, STAT3 deficiency in Treg cells disarms their ability to suppress T_H17 responses.[40] Treg cells depend on IL-2 for their development and ongoing survival, which is mediated by the prosurvival molecule Mcl-1.[42] Interestingly, cTregs are particularly responsive to IL-2 and preferentially localize in the T-cell zones of lymphoid organs in a chemokine receptor (CCR)7-dependent manner.[43] Consistent with this, we demonstrated that the Blimp-1$^+$ eTreg cells that do not express CCR7 were excluded from the T-cell zones.[21] Together, these observations suggest that eTreg cells may use alternate cytokines for their survival. Transcriptional profiling revealed that several cytokine receptors including the IL-33 receptor (ST2) were specifically upregulated in eTregs.[21] In line with this finding, we and others have recently demonstrated the specific requirement for IL-33 in the homeostasis of visceral adipose tissue (VAT) and colonic Treg cells.[21,44,45] Similarly, Treg cells present in the muscle during inflammation or injury also express high levels of ST2, although its Treg-specific role in this tissue is unknown.[91] Collectively, these studies support the notion that Treg cells require site-specific cytokines to differentiate to eTreg cells.

5. DISTINCT MIGRATORY PROPERTIES OF EFFECTOR T REGULATORY CELLS

The localization of different Treg populations depends on the expression of distinct chemokine receptors. CCR7 and CD62L (L-selectin) are highly expressed by cTreg cells and allow their efficient trafficking into lymph nodes and the white pulp of the spleen (reviewed in Ref. 46).

Downregulation of CD62L and CCR7 is one of the early events of eTreg cell differentiation that facilitates their egress from lymphoid organs and subsequent migration to sites of inflammation[46–48] (see Table 1). This event is important, as tissue-specific inflammatory disease occurs when eTreg cell migration is impaired.[49] While downregulation of CCR7 is essential in this process, concerted upregulation of other chemokine receptors facilitates the migration and retention of eTreg cells in various peripheral tissue microenvironments. CCR4 expression in Treg cells is required for migration to the skin and lung to suppress aberrant inflammation[49] and promotes Treg cell migration to the gut to limit colitis.[50] In addition to CCR4, Tregs in the small intestinal lamina propria express high amounts of CCR9.[51] Similarly, CCR6 directs trafficking of Treg cells to the intestine and the central nervous system (CNS) during inflammation and allows for the accumulation of Treg cells in mucosal and inflamed sites.[52–54] Furthermore, CXCR3 is specifically expressed on a subset of Treg cells that develop under T_H1 inflammatory conditions,[55] while CXCR5 is expressed on Treg cells in the germinal center.[56]

Integrins also play critical roles by promoting retention of Tregs within inflamed tissues. For example, expression of αE integrin (CD103) on Treg cells poises them for recruitment and retention within nonlymphoid organs during infection or inflammation.[57] Consequently, $CD103^+$ Treg cells can be found in nonlymphoid organs such as the skin, lungs, and the epithelial lining of the gut, and also at sites of inflammation.[49,58] Overall, these observations demonstrate that the migrational properties of Treg cells are intimately linked to their functional specialization and critical for efficient immune suppression.

Table 1 Characteristics of Central (c)Treg Cells Compared to Effector (e)Treg Cells

Central Treg Cells	**Effector Treg Cells**
Dominant in secondary lymphoid organs, including lymph nodes and the white pulp of the spleen	Dominant in nonlymphoid tissues and also present in secondary lymphoid organs, such as the red pulp of the spleen
$FoxP3^+$	$FoxP3^+$
$Blimp\text{-}1^-$	$Blimp\text{-}1^+$
$CD62L^+$	$CD62L^-$
$CCR7^+$	$CCR7^-$
$IL10^-$	$IL10^+$

6. FUNCTIONAL SPECIALIZATION OF EFFECTOR T REGULATORY CELLS

Analogous to other T cells, Treg cells exhibit functional plasticity and have the capacity to adapt to diverse inflammatory conditions.[20,59–63] In line with this notion, Treg cells are able to functionally diversify in response to the same cytokine signals and co-opt the similar transcriptional modules that polarize conventional $CD4^+$ T cells to T_H1, T_H2, T_H17, and T_{FH} cells. This allows Tregs cells to specifically antagonize their conventional counterparts that differentiated in the same inflammatory microenvironment (reviewed in Refs. 19,62,64).

eTreg cells that develop under T_H1 inflammation have been particularly well studied. In addition to FoxP3, these Treg cells rely on the transcription factor T-bet for their differentiation and depend on cytokines such as interferon gamma (IFNγ) and IL-27.[55,65] T-bet expression in Treg cells induces upregulation of CXCR3, thereby promoting colocalization of Tregs and conventional $CD4^+$ T cells during Th1 inflammation.[55] On the other hand, expression of the canonical T_H2 transcription factor GATA3, although not essential for Treg cell homeostasis, is required for T_H2-type Treg cell development during inflammation, particularly at mucosal sites.[66] In the absence of GATA3, Treg cells are unable to control T_H2 responses and instead produce IL-17.[67] Interestingly, a recent study revisiting the role of the transcription factors T-bet and GATA3 has demonstrated that the absence of these transcription factors individually in Treg cells does not result in pathology under homeostatic conditions. However, the combined deletion of both transcription factors in Treg cells resulted in severe autoimmunity, highlighting the importance of these regulators in coordinating the transcriptional program that maintains Treg cell function.[68]

Another well-studied example of codifferentiation of conventional CD4 T cells and Treg cells is the germinal center response, which is required for affinity maturation and immune globulin class switching of B cells. Germinal center development and organization are facilitated by a specialized T_H cell subset called T_{FH} cells.[41] Treg cells can differentiate into a specialized population termed T follicular regulatory (T_{FR}) that modulate the T_{FH} response.[56] Similar to T_{FH} cells, T_{FR} cells depend on the transcription factor Bcl6, and akin to other eTreg populations, T_{FR} cells express Blimp-1.[56] This is in contrast to T_{FH} cells, whose development is inhibited by Blimp-1.[56] Further studies are required to understand commonalities and differences in the differentiation program of T_H cell populations and their corresponding Treg cell subsets.

7. EFFECTOR T REGULATORY CELLS IN NONLYMPHOID ORGANS

Recent murine studies have established the existence of multiple distinct Treg cell subsets residing in nonlymphoid tissues. Depending on their anatomical localization, these Treg cells express distinct molecular signatures and display unique phenotypic features.[61,69,91]

In particular, barrier tissues, such as the gastrointestinal tract and the skin that are constantly exposed to commensal microbiota, have a constitutive eTreg cell presence. Similarly, tissues rich in proinflammatory cytokines, such as the VAT, harbor large numbers of Treg cells. This is in contrast to other nonlymphoid tissues such as the muscle and the CNS, which recruit Treg cells only after injury or inflammation. In the following section, we will discuss some of the well-characterized tissue-resident Treg cells.

7.1 Visceral Adipose Tissue

VAT Treg cells constitute one of the better-known specialized populations of tissue-resident Treg cells. Lean adipose tissue is rich in Treg cells that produce large amounts of IL-10 and are critical for the suppression of local inflammation, which contributes to glucose intolerance and insulin resistance.[70,72] Obesity is a major driver of VAT inflammation and is characterized by infiltration and activation of inflammatory leukocytes resulting in insulin resistance (reviewed in Ref. 73). Genetic or diet-induced obesity leads to a pronounced loss of Treg cells from the VAT,[70] which is also seen in humans, where obesity is directly linked to development of the metabolic syndrome and type 2 diabetes.[74]

VAT Treg cells express high amounts of peroxisome proliferator-activated receptor gamma (PPAR-γ), a transcription factor that regulates adipocyte differentiation and is critical for the development of VAT Treg cells.[75] *In vitro* studies demonstrated that PPAR-γ cooperated with FoxP3 resulting in upregulation of the VAT Treg cell signature.[75] VAT Treg cells have a distinct TCR repertoire and cytokine and chemokine receptor expression pattern compared to cTreg cells.[70] For example, they express high amounts of CCR2 and the killer cell lectin-like receptor G1 (KLRG1), which are not expressed on cTreg and only on some eTregs in other anatomical locations.[70]

We recently discovered that IL-33 is essential for VAT Treg cell differentiation and homeostasis.[21] IL-33, known for its role in enhancing

T_H2-type immunity, is produced by a range of different cell types.[76] It serves as an "alarmin" that signals inflammation and tissue damage to responder cells.[77] Administration of IL-33 to mice resulted in massive VAT Treg population expansion and was able to rescue Treg cell numbers within the VAT of obese mice, leading to attenuated VAT inflammation and improved insulin sensitivity.[21] Induction of both the IL-33 receptor (ST2) and PPAR-γ is controlled by the TCR-induced transcription factor IRF4, and mice lacking either of these factors lacked Treg cells in the VAT.[21,75] This confirms the critical roles of TCR signaling and IRF4 in eTreg cell differentiation.

7.2 Gastrointestinal Tract

The gastrointestinal tract constitutes an important barrier that is constantly exposed to benign commensals, pathological microbes, and food-derived antigens. There is immense interest in understanding the interplay of the various immune cell types and the local microbiota that maintain the balance between pro- and anti-inflammatory stimuli. Gut bacteria modulate the mammalian immune system in a variety of ways, with some bacteria producing metabolites and proteins that enhance the development of induced Treg cells.[78,79] Emerging evidence in the field suggests that disruption to the normal composition of the gut microbiota (dysbiosis) can alter the local immune cells from a tolerogenic state to an inflammatory, pathogenic one, resulting in inflammatory and metabolic diseases (reviewed in Refs. 80,81).

Intestinal Treg cells reside in the lamina propria and gut-associated lymphoid tissues and are essential for regulating local mucosal homeostasis, preventing the development of inflammatory diseases within the gut.[82,113] Intestinal Treg cells secrete high amounts of IL-10, which is critical for the prevention of colitis.[16] In particular, polysaccharide A of the commensal bacterium *Bacteroides fragilis* facilitates the development of IL-10-producing $FoxP3^+$ Treg cells, which can provide protection from experimental colitis.[79] Intestinal Treg cells can originate from either thymic-derived or peripherally induced Treg cells, though there is some uncertainty as to the relative contribution of each group to intestinal Treg pool.[69,83] Interestingly, a subset of Treg cells within the lamina propria coexpress FoxP3 and the transcription factor RORγt, suggesting a unique developmental origin.[84] Similar to other tissue-resident Treg populations, intestinal Treg cells are dependent on the local cytokine milieu for their expansion and maintenance, and analogous to the VAT, IL-33 plays an important role.[85] Disruption of homeostasis in the gastrointestinal system results in the release of

IL-33 from epithelial cells, and IL-33 enhances TGFβ (transforming growth factor beta)-mediated differentiation of Treg cells through serine phosphorylation of GATA3, which is required for maintaining high expression of FoxP3.[85]

7.3 Skin

The skin is a barrier tissue, which offers protection from a variety of environmental insults. It is host to a multitude of commensals but is also frequently exposed to pathogens (reviewed in Ref. 86). A range of immune cells, including Treg cells, reside in the skin. They are required for maintaining immune tolerance but also contribute to wound healing.[71] Similar to results from murine studies, the homeostasis of Treg cells in the human skin is also determined by skin commensals. Human skin Treg cells display unique TCRs, express distinct cell surface markers, produce different cytokines compared to circulating Treg cells, and are preferentially located within hair follicles.[87] Both human and murine skin Treg cells accumulate upon antigenic stimulation and require IL-7 (derived from keratinocytes) for their maintenance.[87,88] They express higher levels of the activation markers such as CTLA-4 and ICOS.[87,89] The recruitment of skin Treg cells is dependent upon the expression of the chemokine receptor CCR4,[49,90] and deficiency in CCR4 expression on Treg cells after an inflammatory insult resulted in the development of severe cutaneous inflammation.[49]

7.4 Muscle

Healthy muscle tissue is almost completely devoid of hematopoietic cells and does not contain Treg cells. However, two recent studies discovered that a specialized population of Treg cells can be found within the muscle after injury and is required for tissue repair processes.[91,92] Muscle Treg cells regulated the differentiation of satellite cells necessary for the regeneration of skeletal muscle.[91] Upon injury, Treg cells accumulated and proliferated within the muscle and were found in close proximity to regenerating muscle fibers.[92] Muscle Treg cells showed enhanced expression of activation and differentiation markers such as glucocorticoid-induced tumor necrosis factor receptor family-related gene, programmed cell death protein 1 (PD-1), T-cell immunoglobulin and mucin protein 3, and KLRG1 and displayed molecules associated with Treg suppressive function such as CTLA-4 and IL-10.[91,92] Interestingly, Treg cells persisted for weeks in the muscle despite the removal of the inciting insult. Analogous to the other tissue-resident

Treg cells, muscle Treg cells show largely distinct TCR repertoires when compared to conventional Treg cells.[91] Muscle Treg cells are characterized by their high production of amphiregulin, an epidermal growth factor-like molecule that is essential for tissue repair (reviewed in Ref. 93). Furthermore, muscle Treg cells upregulate the chemokine receptor CCR2, which is necessary for the recruitment of leucocytes to injured skeletal muscles.[94]

7.5 Central Nervous System

The role of immune cells within the CNS has been intensely studied. While lymphocytes are excluded from the healthy CNS, in inflammatory conditions, Treg cells can enter and accumulate within the CNS and suppress inflammation.[95,97] In line with this, patients diagnosed with multiple sclerosis (MS), a chronic demyelinating inflammatory disorder of the CNS, have increased numbers of Treg cells detected within their cerebrospinal fluid.[96]

In studies utilizing the experimental autoimmune encephalomyelitis (EAE) mouse model of MS, Treg cells are critical to attenuate inflammation.[97] Interestingly, the presence of IL-6 in the local cytokine milieu can abolish Treg cell-mediated suppression of activated T cells in the CNS.[95] Treg cells require CCR6 to infiltrate the CNS compartment and to attenuate the inflammation associated with EAE.[53] In addition, $CD103^+$ Treg cells are found within the CNS during the recovery phase,[98] suggesting that Treg cells are retained in the CNS in a tissue-specific manner.

Intriguingly, although the mechanism of action is not clear, melatonin, a hormone that controls circadian rhythm, was shown to improve EAE and attenuate T-cell responses. Mice that received melatonin showed decreased T_H1- and T_H17 type responses and interestingly displayed increased Treg cell frequencies and IL-10 production in the CNS.[99] Consistent with the eTreg cell model, CNS Treg cells have an activated phenotype and express high levels of IL-10 to facilitate disease remission.[100]

7.6 Kidney

The implications for Treg cells residing in the kidney are important as they can have a significant impact on the outcomes of acute kidney injury (AKI) occurring as a consequence of acute renal allograft rejection, autoimmune diseases, or ischemic reperfusion injuries. Although only low numbers of Treg cells reside in the healthy kidney,[101] their numbers increase after an insult, both in mice and in humans.[114] Kidney Treg cells have been shown to reduce the severity of inflammation and injury related to ischemia reperfusion, a common cause for AKI and a critical cause for early allograft dysfunction in kidney transplantation.[102] These cells express high levels of TGFβ, IL-10, and CXCR3[106] and act to dampen interstitial inflammation by reducing the production of proinflammatory cytokines such as IFNγ and TNFα (tumor necrosis factor alpha), thereby facilitating tissue repair.[103] In line with these findings, Treg cell-deficient mice subjected to ischemia reperfusion injury (IRI), experimental murine glomerulonephritis or nephrotoxic agents show increased renal inflammation, acute glomerular or tubular necrosis, and deterioration in kidney function.[102,104,105] In addition, depletion of Treg cells in mismatched mouse models of kidney transplants resulted in renal allograft rejection.[106]

Although it has been reported that Treg cells upregulate CCR6 in order to home to the kidneys,[107] little is known about the molecular determinants that may facilitate their differentiation and function. Furthermore, while kidney Treg cells are able to secrete IL-10 to ameliorate AKI after IRI[102] and crescenteric glomerulonephritis,[108] it is unclear whether they perform any kidney-specific functions. Thus, more detailed studies are required to define the precise role and nature of kidney-resident Tregs in health and disease.

8. A UNIFIED MODEL OF EFFECTOR T REGULATORY CELL DIFFERENTIATION

At present, the developmental pathways of eTreg cells are under intense study. Although it is clear that they can differentiate from cTreg cells in the periphery, a recent report suggests that tissue-seeding Treg cells develop early during ontogeny in the thymus.[72] Phenotypically, eTreg cell differentiation is characterized by the downregulation of CCR7 and CD62L and the upregulation of suppressor molecules such as IL-10, TIGIT and CTLA-4. As outlined above, eTreg cells also upregulate the transcription factor Blimp-1, which is required for IL-10 production (Fig. 1).[20] Transcriptome analyses of Treg cells isolated from various organs, including lymphoid tissues, VAT, colon, and muscle, have vastly improved our understanding about the specific requirements for eTreg cell differentiation.[59] Furthermore, our direct comparison of Treg cell subsets based on Blimp-1 expression has revealed dramatically different transcriptional profiles of cTregs and eTregs.[21] Importantly, TCR signaling and transcription factors downstream of TCR have emerged as key players in the eTreg cell differentiation process.[29] In particular, IRF4 plays a key role during eTreg differentiation. Loss of IRF4 leads to eTreg cell deficiency in lymphoid tissues and Treg cell deficiencies in nonlymphoid

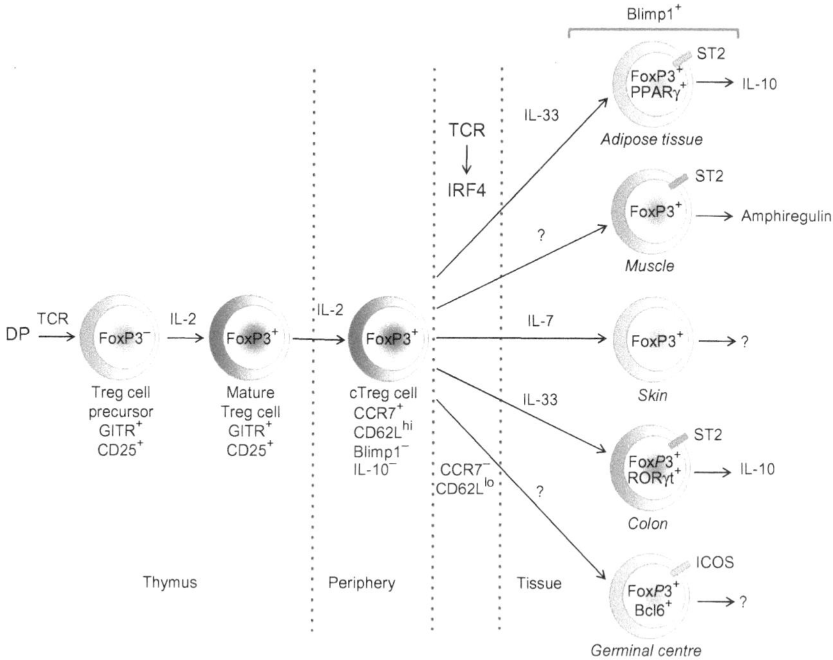

Figure 1 Schematic representation of effector regulatory (eTreg) cell development. Early Treg cell development in the thymus occurs in two stages where T-cell receptor (TCR) and IL-2 signaling cooperate to upregulate the lineage-specific transcription factor FoxP3. Mature Treg cells migrate to the secondary lymphoid tissues where they continue to require IL-2 for their survival. These Treg cells express molecules that promote lymph node homing such as the chemokine receptor CCR7 and are called central (c)Treg cells. cTreg cells do not express IL-10 and display a naïve phenotype (CD62L^{hi}). cTreg cells undergo further differentiation upon receiving TCR stimulation. TCR signaling results in the expression of the transcription factor IRF4, which controls the key events during eTreg cell differentiation such as upregulation of the transcription factor Blimp-1. At this stage, Treg cells downregulate CCR7 and tissue-specific chemokine receptors that guide the migration of eTreg cells to nonlymphoid tissues. IRF4 together with Blimp-1 induce IL-10 production in eTreg cells. IRF4 also regulates the adipose tissue eTreg cell-specific transcription factor PPAR-γ and the receptor for IL-33, ST2.

tissues such as the lung and VAT.[20,21] Specifically, IRF4 is required for the expression of Blimp-1 and IL-10, the correct expression of CCR6, and the upregulation of the costimulatory molecule ICOS.[20,32] Treg cells also depend on IRF4 for the upregulation of the VAT Treg-specific transcription factor PPAR-γ and the IL-33 receptor ST2, a process which requires cooperation with the transcriptional regulator BATF (basic leucine zipper transcription factor, ATF-like).[21,75] This transcriptional module may also be active in other tissue-resident Treg cells as BATF was also found to be required for CCR9 expression and efficient localization of Treg cells to the gut.[109] Together, these findings indicate that IRF4, in cooperation with BATF is central for the differentiation of multiple types of eTreg cells. These data also provide strong support for a model for eTreg differentiation in which IRF4 establishes a common transcriptional program characterized by induction of multiple suppressive molecules, including IL-10, and results in changes to the migratory capacities of eTreg cells. This program is further shaped by a second layer of transcriptional regulators that are directed by distinct inflammatory signals or tissue-specific factors and cytokines. However, precisely how IRF4 establishes this transcriptional framework remains to be determined.

9. FUTURE PERSPECTIVES

Numerous studies over the past few years have addressed the role of immune cells located within nonlymphoid tissues. Accumulating evidence shows that Treg cells are critical to limit or prevent inflammation within these nonlymphoid tissues and have an essential role in maintaining homeostasis. However, the developmental pathways of Treg cells located in nonlymphoid tissues are only now being elucidated, and many intriguing questions remain unanswered. One attractive model would suggest that eTreg cells arise from cTreg cells that are recruited to nonlymphoid organs in a TCR-specific manner, and tissue antigens may play an important role in this process. Once these cells have reached the target organ, they may proliferate and differentiate in response to local tissue-specific or inflammatory signals. Such a model is supported by a number of studies, all of which strongly implicate IL-33 into this process.[21,72,85,110,111] However, it remains to be seen whether other regulatory modules exist that may be responsible for the establishment of organ-specific Treg cell populations. Another, not necessarily exclusive model would predict that eTreg cell differentiation is programmed in the thymus. This is supported by a recent study, which demonstrated that a distinct subset of Treg cells develops in the thymus in an autoimmune regulator (AIRE)-dependent manner and contributes significantly to the maintenance of immune tolerance.[112] It is intriguing to speculate that it is these Treg cells, generated perinatally, that are responsible for "seeding" the tissue-resident Treg cell population. Thus, more studies are required to understand the factors that guide the differentiation program of eTreg cells within nonlymphoid organs.

Much of the success in Treg cell research comes from the study of murine models. While informative, the results of these studies cannot be directly extrapolated to human diseases. They do, however, lead to important considerations, such as the question of how much can be deduced from numbers of Treg cells circulating in the peripheral blood, without assessing the phenotype or differentiation status of these cells. It is conceivable that rather than the absolute number of Treg cells, it is the phenotype or subset of Treg cells that is important for protection from specific insults such inflammation or autoimmune disease. Therefore, much work is required to understand the role of eTreg cells in general, and tissue-resident Treg cells in particular in humans, and novel approaches to identify and manipulate specialized eTreg populations in patients are needed. Defining and distinguishing tissue-resident eTreg cell populations in healthy and diseased organs will enable a better understanding of the cellular defects that result in the development of autoimmune diseases, malignancy, or transplant rejection.

ACKNOWLEDGMENTS

The authors would like to thank the members of the Kallies laboratory, in particular Tom Sidwell, for helpful discussion. We also thank A. Banerjee and G. Grigoriadis for their critical comments on the review. This work was supported by grants and fellowships from the National Health and Medical Research Council of Australia (A.K.), a fellowship from the Sylvia and Charles Viertel Foundation (A.K.), the Victorian State Government Operational Infrastructure Support and Australian Government NHMRC Independent Research Institute Infrastructure Support scheme and Kidney Health Australia (P.T.).

REFERENCES

1. Fontenot JD, Gavin MA, Rudensky AY. Foxp3 programs the development and function of CD4+CD25+ regulatory T cells. *Nat Immunol.* 2003;4(4):330–336.
2. Bennett CL, Christie J, Ramsdell F, et al. The immune dysregulation, polyendocrinopathy, enteropathy, X-linked syndrome (IPEX) is caused by mutations of FOXP3. *Nat Genet.* 2001;27(1):20–21.
3. Brunkow ME, Jeffery EW, Hjerrild KA, et al. Disruption of a new forkhead/winged-helix protein, scurfin, results in the fatal lymphoproliferative disorder of the scurfy mouse. *Nat Genet.* 2001;27(1):68–73.
4. Wildin RS, Ramsdell F, Peake J, et al. X-linked neonatal diabetes mellitus, enteropathy and endocrinopathy syndrome is the human equivalent of mouse scurfy. *Nat Genet.* 2001;27(1):18–20.
5. Sakaguchi S, Setoguchi R, Yagi H, Nomura T. Naturally arising Foxp3-expressing CD25+CD4+ regulatory T cells in self-tolerance and autoimmune disease. *Curr Top Microbiol Immunol.* 2006;305:51–66.
6. Fontenot JD, Rasmussen JP, Gavin MA, Rudensky AY. A function for interleukin 2 in Foxp3-expressing regulatory T cells. *Nat Immunol.* 2005;6(11):1142–1151.
7. Lee HM, Bautista JL, Scott-Browne J, Mohan JF, Hsieh CS. A broad range of self-reactivity drives thymic regulatory T cell selection to limit responses to self. *Immunity.* 2012;37(3):475–486.

8. Burchill MA, Yang J, Vogtenhuber C, Blazar BR, Farrar MA. IL-2 receptor beta-dependent STAT5 activation is required for the development of Foxp3+ regulatory T cells. *J Immunol.* 2007;178(1):280–290.
9. Liston A, Gray DH. Homeostatic control of regulatory T cell diversity. *Nat Rev Immunol.* 2014;14(3):154–165.
10. Tang Q, Bluestone JA. The Foxp3+ regulatory T cell: a jack of all trades, master of regulation. *Nat Immunol.* 2008;9(3):239–244.
11. Sakaguchi S, Yamaguchi T, Nomura T, Ono M. Regulatory T cells and immune tolerance. *Cell.* 2008;133(5):775–787.
12. Wing JB, Sakaguchi S. Multiple treg suppressive modules and their adaptability. *Front Immunol.* 2012;3:178.
13. Joller N, Lozano E, Burkett PR, et al. Treg cells expressing the coinhibitory molecule TIGIT selectively inhibit proinflammatory Th1 and Th17 cell responses. *Immunity.* 2014;40(4):569–581.
14. Pandiyan P, Zheng L, Ishihara S, Reed J, Lenardo MJ. CD4+CD25+Foxp3+ regulatory T cells induce cytokine deprivation-mediated apoptosis of effector CD4+ T cells. *Nat Immunol.* 2007;8(12):1353–1362.
15. Wing K, Onishi Y, Prieto-Martin P, et al. CTLA-4 control over Foxp3+ regulatory T cell function. *Science.* 2008;322(5899):271–275.
16. Rubtsov YP, Rasmussen JP, Chi EY, et al. Regulatory T cell-derived interleukin-10 limits inflammation at environmental interfaces. *Immunity.* 2008;28(4):546–558.
17. Deaglio S, Dwyer KM, Gao W, et al. Adenosine generation catalyzed by CD39 and CD73 expressed on regulatory T cells mediates immune suppression. *J Exp Med.* 2007;204(6):1257–1265.
18. Collison LW, Workman CJ, Kuo TT, et al. The inhibitory cytokine IL-35 contributes to regulatory T-cell function. *Nature.* 2007;450(7169):566–569.
19. Cretney E, Kallies A, Nutt SL. Differentiation and function of Foxp3(+) effector regulatory T cells. *Trends Immunol.* 2013;34(2):74–80.
20. Cretney E, Xin A, Shi W, et al. The transcription factors Blimp-1 and IRF4 jointly control the differentiation and function of effector regulatory T cells. *Nat Immunol.* 2011;12(4):304–311.
21. Vasanthakumar A, Moro K, Xin A, et al. The transcriptional regulators IRF4, BATF and IL-33 orchestrate development and maintenance of adipose tissue-resident regulatory T cells. *Nat Immunol.* 2015;16(3):276–285.
22. Seddiki N, Santner-Nanan B, Martinson J, et al. Expression of interleukin (IL)-2 and IL-7 receptors discriminates between human regulatory and activated T cells. *J Exp Med.* 2006;203(7):1693–1700.
23. Sakaguchi S, Miyara M, Costantino CM, Hafler DA. FOXP3+ regulatory T cells in the human immune system. *Nat Rev Immunol.* 2010;10(7):490–500.
24. Ferraro A, D'Alise AM, Raj T, et al. Interindividual variation in human T regulatory cells. *Proc Natl Acad Sci USA.* 2014;111(12):E1111–E1120.
25. Fuhrman CA, Yeh WI, Seay HR, et al. Divergent phenotypes of human regulatory T cells expressing the receptors TIGIT and CD226. *J Immunol.* 2015;195(1):145–155.
26. Ito T, Hanabuchi S, Wang YH, et al. Two functional subsets of FOXP3+ regulatory T cells in human thymus and periphery. *Immunity.* 2008;28(6):870–880.
27. Isomura I, Palmer S, Grumont RJ, et al. c-Rel is required for the development of thymic Foxp3+ CD4 regulatory T cells. *J Exp Med.* 2009;206(13):3001–3014.
28. Ruan Q, Kameswaran V, Tone Y, et al. Development of Foxp3(+) regulatory T cells is driven by the c-Rel enhanceosome. *Immunity.* 2009;31(6):932–940.
29. Levine AG, Arvey A, Jin W, Rudensky AY. Continuous requirement for the TCR in regulatory T cell function. *Nat Immunol.* 2014;15(11):1070–1078.

30. Vahl JC, Drees C, Heger K, et al. Continuous T cell receptor signals maintain a functional regulatory T cell pool. *Immunity*. 2014;41(5):722–736.
31. Moran AE, Holzapfel KL, Xing Y, et al. T cell receptor signal strength in Treg and iNKT cell development demonstrated by a novel fluorescent reporter mouse. *J Exp Med*. 2011;208(6):1279–1289.
32. Zheng Y, Chaudhry A, Kas A, et al. Regulatory T-cell suppressor program co-opts transcription factor IRF4 to control T(H)2 responses. *Nature*. 2009;458(7236): 351–356.
33. Brustle A, Heink S, Huber M, et al. The development of inflammatory T(H)-17 cells requires interferon-regulatory factor 4. *Nat Immunol*. 2007;8(9):958–966.
34. Bollig N, Brustle A, Kellner K, et al. Transcription factor IRF4 determines germinal center formation through follicular T-helper cell differentiation. *Proc Natl Acad Sci USA*. 2012;109(22):8664–8669.
35. Man K, Miasari M, Shi W, et al. The transcription factor IRF4 is essential for TCR affinity-mediated metabolic programming and clonal expansion of T cells. *Nat Immunol* 2013;14(11):1155–1165.
36. Willis SN, Good-Jacobson KL, Curtis J, et al. Transcription factor IRF4 regulates germinal center cell formation through a B cell-intrinsic mechanism. *J Immunol*. 2014;192(7):3200–3206.
37. Kallies A, Hasbold J, Tarlinton DM, et al. Plasma cell ontogeny defined by quantitative changes in blimp-1 expression. *J Exp Med*. 2004;200(8):967–977.
38. Lathrop SK, Santacruz NA, Pham D, Luo J, Hsieh CS. Antigen-specific peripheral shaping of the natural regulatory T cell population. *J Exp Med*. 2008;205(13):3105–3117.
39. Gottschalk RA, Corse E, Allison JP. TCR ligand density and affinity determine peripheral induction of Foxp3 in vivo. *J Exp Med*. 2010;207(8):1701–1711.
40. Chaudhry A, Rudra D, Treuting P, et al. CD4+ regulatory T cells control TH17 responses in a Stat3-dependent manner. *Science*. 2009;326(5955):986–991.
41. Linterman MA, Vinuesa CG. T follicular helper cells during immunity and tolerance. *Prog Mol Biol Transl Sci*. 2010;92:207–248.
42. Pierson W, Cauwe B, Policheni A, et al. Antiapoptotic Mcl-1 is critical for the survival and niche-filling capacity of Foxp3(+) regulatory T cells. *Nat Immunol*. 2013;14(9):959–965.
43. Smigiel KS, Richards E, Srivastava S, et al. CCR7 provides localized access to IL-2 and defines homeostatically distinct regulatory T cell subsets. *J Exp Med*. 2014;211(1):121–136.
44. Duan L, Chen J, Zhang H, et al. Interleukin-33 ameliorates experimental colitis through promoting Th2/Foxp3(+) regulatory T-cell responses in mice. *Mol Med*. 2012;18:753–761.
45. Sedhom MA, Pichery M, Murdoch JR, et al. Neutralisation of the interleukin-33/ST2 pathway ameliorates experimental colitis through enhancement of mucosal healing in mice. *Gut*. 2013;62(12):1714–1723.
46. Forster R, Davalos-Misslitz AC, Rot A. CCR7 and its ligands: balancing immunity and tolerance. *Nat Rev Immunol*. 2008;8(5):362–371.
47. Menning A, Hopken UE, Siegmund K, Lipp M, Hamann A, Huehn J. Distinctive role of CCR7 in migration and functional activity of naive- and effector/memory-like Treg subsets. *Eur J Immunol*. 2007;37(6):1575–1583.
48. Huehn J, Siegmund K, Lehmann JC, et al. Developmental stage, phenotype, and migration distinguish naive- and effector/memory-like CD4+ regulatory T cells. *J Exp Med*. 2004;199(3):303–313.
49. Sather BD, Treuting P, Perdue N, et al. Altering the distribution of Foxp3(+) regulatory T cells results in tissue-specific inflammatory disease. *J Exp Med*. 2007;204(6):1335–1347.

50. Yuan Q, Bromley SK, Means TK, et al. CCR4-dependent regulatory T cell function in inflammatory bowel disease. *J Exp Med*. 2007;204(6):1327–1334.
51. Guo Z, Jang MH, Otani K, et al. CD4+CD25+ regulatory T cells in the small intestinal lamina propria show an effector/memory phenotype. *Int Immunol*. 2008;20(3):307–315.
52. Kitamura K, Farber JM, Kelsall BL. CCR6 marks regulatory T cells as a colon-tropic, IL-10-producing phenotype. *J Immunol*. 2010;185(6):3295–3304.
53. Villares R, Cadenas V, Lozano M, et al. CCR6 regulates EAE pathogenesis by controlling regulatory CD4+ T-cell recruitment to target tissues. *Eur J Immunol*. 2009;39(6):1671–1681.
54. Yamazaki T, Yang XO, Chung Y, et al. CCR6 regulates the migration of inflammatory and regulatory T cells. *J Immunol*. 2008;181(12):8391–8401.
55. Koch MA, Tucker-Heard G, Perdue NR, Killebrew JR, Urdahl KB, Campbell DJ. The transcription factor T-bet controls regulatory T cell homeostasis and function during type 1 inflammation. *Nat Immunol*. 2009;10(6):595–602.
56. Linterman MA, Pierson W, Lee SK, et al. Foxp3+ follicular regulatory T cells control the germinal center response. *Nat Med*. 2011;17(8):975–982.
57. Suffia I, Reckling SK, Salay G, Belkaid Y. A role for CD103 in the retention of CD4+CD25+ Treg and control of Leishmania major infection. *J Immunol*. 2005;174(9):5444–5455.
58. Lehmann J, Huehn J, de la Rosa M, et al. Expression of the integrin alpha Ebeta 7 identifies unique subsets of CD25+ as well as CD25- regulatory T cells. *Proc Natl Acad Sci USA*. 2002;99(20):13031–13036.
59. Feuerer M, Hill JA, Mathis D, Benoist C. Foxp3+ regulatory T cells: differentiation, specification, subphenotypes. *Nat Immunol*. 2009;10(7):689–695.
60. Campbell DJ, Koch MA. Phenotypical and functional specialization of FOXP3(+) regulatory T cells. *Nat Rev Immunol*. 2011;11(2):119–130.
61. Burzyn D, Benoist C, Mathis D. Regulatory T cells in nonlymphoid tissues. *Nat Immunol*. 2013;14(10):1007–1013.
62. Chaudhry A, Rudensky AY. Control of inflammation by integration of environmental cues by regulatory T cells. *J Clin Invest*. 2013;123(3):939–944.
63. Liston A, Gray DH. Homeostatic control of regulatory T cell diversity. *Nat Rev Immunol*. 2014;14:154–165.
64. Lazarevic V, Glimcher LH, Lord GM. T-bet: a bridge between innate and adaptive immunity. *Nat Rev Immunol*. 2013;13(11):777–789.
65. Hall AO, Beiting DP, Tato C, et al. The cytokines interleukin 27 and interferon-gamma promote distinct Treg cell populations required to limit infection-induced pathology. *Immunity*. 2012;37(3):511–523.
66. Wohlfert EA, Grainger JR, Bouladoux N, et al. GATA3 controls Foxp3(+) regulatory T cell fate during inflammation in mice. *J Clin Invest*. 2011;121(11):4503–4515.
67. Rudra D, deRoos P, Chaudhry A, et al. Transcription factor Foxp3 and its protein partners form a complex regulatory network. *Nat Immunol*. 2012;13(10):1010–1019.
68. Yu F, Sharma S, Edwards J, Feigenbaum L, Zhu J. Dynamic expression of transcription factors T-bet and GATA-3 by regulatory T cells maintains immunotolerance. *Nat Immunol*. 2015;16(2):197–206.
69. Lathrop SK, Bloom SM, Rao SM, et al. Peripheral education of the immune system by colonic commensal microbiota. *Nature*. 2011;478(7368):250–254.
70. Feuerer M, Herrero L, Cipolletta D, et al. Lean, but not obese, fat is enriched for a unique population of regulatory T cells that affect metabolic parameters. *Nat Med*. 2009;15(8):930–939.
71. Mueller SN, Zaid A, Carbone FR. Tissue-resident T cells: dynamic players in skin immunity. *Front Immunol*. 2014;5:332.
72. Kolodin D, van Panhuys N, Li C, et al. Antigen- and cytokine-driven accumulation of regulatory T cells in visceral adipose tissue of lean mice. *Cell Metab*. 2015;21(4):543–557.

73. Han JM, Levings MK. Immune regulation in obesity-associated adipose inflammation. *J Immunol.* 2013;191(2):527–532.
74. Alvehus M, Buren J, Sjostrom M, Goedecke J, Olsson T. The human visceral fat depot has a unique inflammatory profile. *Obesity.* 2010;18(5):879–883.
75. Cipolletta D, Feuerer M, Li A, et al. PPAR-gamma is a major driver of the accumulation and phenotype of adipose tissue Treg cells. *Nature.* 2012;486(7404):549–553.
76. Schmitz J, Owyang A, Oldham E, et al. IL-33, an interleukin-1-like cytokine that signals via the IL-1 receptor-related protein ST2 and induces T helper type 2-associated cytokines. *Immunity* 2005;23(5):479–490.
77. Pichery M, Mirey E, Mercier P, et al. Endogenous IL-33 is highly expressed in mouse epithelial barrier tissues, lymphoid organs, brain, embryos, and inflamed tissues: in situ analysis using a novel Il-33-LacZ gene trap reporter strain. *J Immunol.* 2012;188(7):3488–3495.
78. Arpaia N, Rudensky AY. Microbial metabolites control gut inflammatory responses. *Proc Natl Acad Sci USA.* 2014;111(6):2058–2059.
79. Round JL, Mazmanian SK. Inducible Foxp3+ regulatory T-cell development by a commensal bacterium of the intestinal microbiota. *Proc Natl Acad Sci USA.* 2010;107(27):12204–12209.
80. Round JL, Mazmanian SK. The gut microbiota shapes intestinal immune responses during health and disease. *Nat Rev Immunol.* 2009;9(5):313–323.
81. Manichanh C, Borruel N, Casellas F, Guarner F. The gut microbiota in IBD. *Nat Rev Gastroenterol Hepatol.* 2012;9(10):599–608.
82. Josefowicz SZ, Niec RE, Kim HY, et al. Extrathymically generated regulatory T cells control mucosal TH2 inflammation. *Nature.* 2012;482(7385):395–399.
83. Cebula A, Seweryn M, Rempala GA, et al. Thymus-derived regulatory T cells contribute to tolerance to commensal microbiota. *Nature.* 2013;497(7448):258–262.
84. Zhou L, Lopes JE, Chong MM, et al. TGF-beta-induced Foxp3 inhibits T(H)17 cell differentiation by antagonizing RORgammat function. *Nature.* 2008;453(7192): 236–240.
85. Schiering C, Krausgruber T, Chomka A, et al. The alarmin IL-33 promotes regulatory T-cell function in the intestine. *Nature.* 2014;513(7519):564–568.
86. Grice EA, Segre JA. The skin microbiome. *Nat Rev Microbiol.* 2011;9(4):244–253.
87. Sanchez Rodriguez R, Pauli ML, Neuhaus IM, et al. Memory regulatory T cells reside in human skin. *J Clin Invest.* 2014;124(3):1027–1036.
88. Gratz IK, Truong HA, Yang SH, et al. Cutting edge: memory regulatory T cells require IL-7 and not IL-2 for their maintenance in peripheral tissues. *J Immunol.* 2013;190(9):4483–4487.
89. Rosenblum MD, Gratz IK, Paw JS, Lee K, Marshak-Rothstein A, Abbas AK. Response to self antigen imprints regulatory memory in tissues. *Nature.* 2011;480(7378):538–542.
90. Campbell JJ, O'Connell DJ, Wurbel MA. Cutting edge: chemokine receptor CCR4 is necessary for antigen-driven cutaneous accumulation of CD4 T cells under physiological conditions. *J Immunol.* 2007;178(6):3358–3362.
91. Burzyn D, Kuswanto W, Kolodin D, et al. A special population of regulatory T cells potentiates muscle repair. *Cell.* 2013;155(6):1282–1295.
92. Villalta SA, Rosenthal W, Martinez L, et al. Regulatory T cells suppress muscle inflammation and injury in muscular dystrophy. *Sci Transl Med.* 2014;6(258):258ra142.
93. Zaiss DM, Gause WC, Osborne LC, Artis D. Emerging functions of amphiregulin in orchestrating immunity, inflammation, and tissue repair. *Immunity.* 2015;42(2): 216–226.
94. Warren GL, Hulderman T, Mishra D, et al. Chemokine receptor CCR2 involvement in skeletal muscle regeneration. *FASEB J.* 2005;19(3):413–415.
95. Korn T, Reddy J, Gao W, et al. Myelin-specific regulatory T cells accumulate in the CNS but fail to control autoimmune inflammation. *Nat Med.* 2007;13(4):423–431.

96. Venken K, Hellings N, Thewissen M, et al. Compromised CD4+ CD25(high) regulatory T-cell function in patients with relapsing-remitting multiple sclerosis is correlated with a reduced frequency of FOXP3-positive cells and reduced FOXP3 expression at the single-cell level. *Immunology*. 2008;123(1):79–89.
97. O'Connor RA, Malpass KH, Anderton SM. The inflamed central nervous system drives the activation and rapid proliferation of Foxp3+ regulatory T cells. *J Immunol*. 2007;179(2):958–966.
98. McGeachy MJ, Stephens LA, Anderton SM. Natural recovery and protection from autoimmune encephalomyelitis: contribution of CD4+CD25+ regulatory cells within the central nervous system. *J Immunol*. 2005;175(5):3025–3032.
99. Alvarez-Sanchez N, Cruz-Chamorro I, Lopez-Gonzalez A, et al. Melatonin controls experimental autoimmune encephalomyelitis by altering the T effector/regulatory balance. *Brain Behav Immun*. 2015. Available at http://dx.doi.org/10.1016/j.bbi.2015.06.021.
100. Zhang X, Koldzic DN, Izikson L, et al. IL-10 is involved in the suppression of experimental autoimmune encephalomyelitis by CD25+CD4+ regulatory T cells. *Int Immunol*. 2004;16(2):249–256.
101. Ascon DB, Ascon M, Satpute S, et al. Normal mouse kidneys contain activated and CD3+CD4- CD8- double-negative T lymphocytes with a distinct TCR repertoire. *J Leukoc Biol*. 2008;84(6):1400–1409.
102. Kinsey GR, Sharma R, Huang L, et al. Regulatory T cells suppress innate immunity in kidney ischemia-reperfusion injury. *J Am Soc Nephrol*. 2009;20(8):1744–1753.
103. Gandolfo MT, Jang HR, Bagnasco SM, et al. Foxp3+ regulatory T cells participate in repair of ischemic acute kidney injury. *Kidney Int*. 2009;76(7):717–729.
104. Ooi JD, Snelgrove SL, Engel DR, et al. Endogenous foxp3(+) T-regulatory cells suppress anti-glomerular basement membrane nephritis. *Kidney Int*. 2011;79(9):977–986.
105. Lee H, Nho D, Chung HS, et al. CD4+CD25+ regulatory T cells attenuate cisplatin-induced nephrotoxicity in mice. *Kidney Int*. 2010;78(11):1100–1109.
106. Hu M, Wang C, Zhang GY, et al. Infiltrating Foxp3(+) regulatory T cells from spontaneously tolerant kidney allografts demonstrate donor-specific tolerance. *Am J Transplant*. 2013;13(11):2819–2830.
107. Turner JE, Paust HJ, Steinmetz OM, et al. CCR6 recruits regulatory T cells and Th17 cells to the kidney in glomerulonephritis. *J Am Soc Nephrol*. 2010;21(6):974–985.
108. Ostmann A, Paust HJ, Panzer U, et al. Regulatory T cell-derived IL-10 ameliorates crescentic GN. *J Am Soc Nephrol*. 2013;24(6):930–942.
109. Wang C, Thangamani S, Kim M, et al. BATF is required for normal expression of gut-homing receptors by T helper cells in response to retinoic acid. *J Exp Med*. 2013;210(3):475–489.
110. Han JM, Wu D, Denroche HC, Yao Y, Verchere CB, Levings MK. IL-33 reverses an obesity-induced deficit in visceral adipose tissue ST2+ T regulatory cells and ameliorates adipose tissue inflammation and insulin resistance. *J Immunol*. 2015;194(10):4777–4783.
111. Molofsky AB, Van Gool F, Liang HE, et al. Interleukin-33 and interferon-gamma counter-regulate group 2 innate lymphoid cell activation during immune perturbation. *Immunity*. 2015;43(1):161–174.
112. Yang S, Fujikado N, Kolodin D, Benoist C, Mathis D. Immune tolerance. Regulatory T cells generated early in life play a distinct role in maintaining self-tolerance. *Science*. 2015;348(6234):589–594.
113. Makita S, Kanai T, Nemoto Y, et al. Intestinal lamina propria retaining CD4+CD25+ regulatory T cells is a suppressive site of intestinal inflammation. *J Immunol*. 2007;178(8):4937–4946.
114. Kinsey GR, Huang L, Vergis AL, Li L, Okusa MD. Regulatory T cells contribute to the protective effect of ischemic preconditioning in the kidney. *Kidney Int*. 2010;77(9):771–780.

CHAPTER EIGHT

Treg Cell Differentiation: From Thymus to Peripheral Tissue

David M. Richards*[,1], Michael Delacher*[,1], Yael Goldfarb[†,1], Danny Kägebein*, Ann-Cathrin Hofer*, Jakub Abramson[†,2], Markus Feuerer*[,2,3]

*Immune Tolerance, Tumor Immunology Program, German Cancer Research Center (DKFZ), Heidelberg, Germany
[†]Faculty of Biology, Department of Immunology, Weizmann Institute of Science, Rehovot, Israel
[3]Corresponding author: e-mail address: m.feuerer@dkfz.de

Contents

Abstract

Regulatory T cells (Tregs) are crucial mediators of self-tolerance in the periphery. They differentiate in the thymus, where interactions with thymus-resident antigen-presenting cells, an instructive cytokine milieu, and stimulation of the T cell receptor lead to the selection into the Treg lineage and the induction of *Foxp3* gene expression. Once mature, Treg cells leave the thymus and migrate into either the secondary lymphoid tissues, e.g., lymph nodes and spleen, or peripheral nonlymphoid tissues. There is growing evidence that Treg cells go beyond the classical modulation of immune responses and also play important functional roles in nonlymphoid peripheral tissues. In this review, we summarize recent findings about the thymic Treg lineage differentiation as well as the further specialization of Treg cells in the secondary lymphoid and in the peripheral nonlymphoid organs.

[1] Equally contributed first authors.
[2] Equally contributed last authors.

Progress in Molecular Biology and Translational Science, Volume 136
ISSN 1877-1173
http://dx.doi.org/10.1016/bs.pmbts.2015.07.014

1. INTRODUCTION

$Foxp3^+$ regulatory T (Treg) cells are a subpopulation of $CD4^+$ T cells, critical for the maintenance and regulation of immunological homeostasis and self-tolerance. Although Treg cells comprise only 3–5% of the peripheral blood T cell pool, they are important for regulating the activation of the immune system and preventing pathological reactivity to self (autoimmunity) and/or harmless antigens (allergy). The critical role of Treg cells within the immune system is best demonstrated by the lethal phenotypes resulting from their functional deficiency.[1,2] This is well illustrated by the scurfy mice carrying a mutated *Foxp3* gene, which consequently exhibit lymphoproliferation, develop fatal multiorgan autoimmunity, and die by the age of 24 days.[3,4] Similarly, humans with an X-linked *FOXP3* mutation suffer from the fatal immune dysregulation polyendocrinopathy enteropathy syndrome.[5] Moreover, there is growing evidence that Treg cells go beyond the classical modulation of immune responses and also play important functional roles in various peripheral nonlymphoid tissues.[6] In fact, it is becoming clear that tissue-specific subsets of Treg cells exist and seem to develop in response to organ-specific environmental signals. In this review, we summarize recent findings about the origin, phenotype and function of thymic, secondary lymphoid, and peripheral nonlymphoid tissue-resident Treg cells.

2. THYMIC TREG CELLS

It is now well established that there are two main pathways for the generation of Treg cells *in vivo*. The majority of functionally mature Treg cells are produced in the thymus, where recognition of self-antigen by certain clones leads to their deviation into the thymus-derived $Foxp3^+$ Treg (tTreg) cell lineage. The second pathway of Treg cell generation is in the periphery, where naïve $CD4^+Foxp3^-$ T cells encounter their cognate antigens and, under certain conditions, differentiate into periphery-derived $CD4^+Foxp3^+$ Treg (pTreg) cells.[7]

2.1 Which Signals Are Important for tTreg Cell Generation in the Thymus?

The past two decades have seen a true explosion of research into the development of tTreg cells. The early studies highlighted a central role for high-affinity T cell receptor (TCR) interactions with self-peptides in tTreg cell development.[8,9] These and several other subsequent reports set the stage for the still prominent hypothesis that developing thymocytes recognize self-antigen with high affinity, leading to clonal deviation from the conventional T cell (Tconv) fate, and differentiate into tTreg cells.[10] Although interactions between the TCR and self-peptides presented by the major histocompatibility complex (MHC) are required for tTreg cell development, as demonstrated by the lack of tTreg cells in mice lacking conventional MHC molecules, the exact mechanisms that determine the choice between tTreg cell generation and negative selection of self-reactive T cells are still not defined.[11] Moreover, the requirement of a high-affinity interaction has recently been put into question. Specifically, tTreg cells can be generated from thymocytes with varying TCR affinities for their cognate antigen, suggesting that the spectrum of affinities required for tTreg cell generation is much broader than previously thought.[12] Nevertheless, the idea that productive tTreg cell differentiation ensues from interactions that lie between the signaling strength required for positive selection on the one side and clonal deletion on the other side still remains the prevalent view underlying the generation of Treg cells in the thymus.

Given that signaling strength plays a pivotal role in tTreg cell induction, it seems reasonable that costimulatory signals would also be required in this process. Indeed, in parallel to TCR activation, signaling via the CD28 coreceptor has also been shown to play an important part in the initiation of the tTreg cell differentiation program. Specifically, deficiency in CD28 or its corresponding ligands, CD80/CD86 (B7-1/B7-2), was shown to result in an about 80% decrease in the frequency of Treg cells.[13–15] Interestingly, these mice did not develop autoimmunity, presumably because of the corresponding impact of diminished costimulation on Tconv cell activation, including self-reactive T cells.[16] Moreover, CD28 was shown to have a cell-intrinsic role in the induction of *Foxp3*, as well as *Gitr* and *Ctla-4* in double-positive (DP) thymocytes, indicating that it regulates tTreg cell development and function at multiple levels.[14] Other costimulatory molecules have also been implicated in tTreg cell differentiation. Specifically, deletion of either CD154 (CD40L) or CD40 was shown to result in a threefold decrease in tTreg cell frequency.[17] However, this decrease is not attributed to the direct role of CD154/CD40 in tTreg cell generation, but rather to the maintenance of tTreg cell homeostasis and survival. Another recent report suggested that tumor-necrosis factor receptor (TNFR) family members, such as GITR and OX40, could couple TCR signal strength to tTreg cell differentiation and thus support tTreg cell generation as well.[18]

Cytokine signaling is another important element that regulates development of hematopoietic cells in general and T cells in particular. Although many of the initial studies focused on pTreg cell differentiation, it has now become evident that cytokines also play an important role in tTreg cell differentiation. Since both tTreg and pTreg cells are characterized by high expression of IL-2Rα (CD25), it has been hypothesized that IL-2 plays a pivotal role in all Treg cell development. Initial studies, however, demonstrated that IL-2 is not absolutely required for tTreg development, as the frequency of tTreg cells was reduced by only 50% in IL-2- or CD25-deficient mice.[19,20] Nevertheless, more recent studies have begun to shed light on the role that IL-2 signaling plays in tTreg cell development. First, in mixed bone marrow (BM) chimera experiments, the frequency of tTreg cells is decreased by about fivefold in CD25-deficient, compared to CD25-sufficient donor cells, suggesting that IL-2 is particularly important in a competitive environment with normal thymocytes.[20] Second, it is likely that the absence of IL-2 can be compensated for by other common γ-chain cytokines in the thymus, such as IL-7 or IL-15. Indeed, analysis of common γ-chain receptor-, IL-2-, IL-7-, IL-15-, STAT5-, or IL-2Rβ (CD122)-deficient mice revealed further declines in the frequency of tTreg cells.[19–22] Because neither IL-7 nor IL-15 deficiency by itself affects tTreg cell production, it is the prevailing view that IL-2 is the principal common γ-chain cytokine required for tTreg cell development. However, it seems that IL-7 and IL-15 can, at least in part, compensate for its loss. This notion is supported by the *in vitro* finding that both IL-2 and IL-15 can support Treg differentiation from tTreg precursors.[23]

Another cytokine that seems to play an essential role in Treg cell development is transforming growth factor-β (TGF-β). Although TGF-β is known to be important for the conversion of $CD4^+Foxp3^-$ cells into pTreg cells in the periphery, it was for a long time believed to be dispensable for tTreg cell generation. This notion was mostly based on the observation that adult mice lacking TGF-βRII demonstrate fairly normal tTreg cell frequencies, while the pTreg cell pool was severely diminished.[24–26] Thus, these data set the prevailing view that although TGF-β is crucial for pTreg cell homeostasis and maintenance of *Foxp3* expression, it is not absolutely necessary for tTreg cell differentiation. However, subsequent studies demonstrated that T cell-specific ablation of the TGF-β-receptor results in a significant diminution of the first wave of neonatal tTreg cell production, around day 4 after birth, suggesting that Treg cell differentiation in the neonatal and adult thymus might differ in their requirement for TGF-β.[27] Moreover, mice deficient for both TGF-β and IL-2 are completely deprived of tTreg cells, suggesting that TGF-β might compensate for IL-2 deficiency and induce *Foxp3* expression.[27] How and when IL-2 and TGF-β signaling pathways intersect in the thymus to generate Foxp3^{+} cells, however, remains to be further elucidated.

The downstream mechanisms critical for tTreg cell development involve an orchestrated action of various transcription factors, which activate expression of the master regulator transcription factor Foxp3. Specifically, TCR-CD28 costimulation induces various signaling pathways that culminate in the activation of transcription factors including NFκB, AP-1, and NFAT. Indeed, inactivation of genes involved in NFκB activation such as protein kinase C-θ (PKC-θ), CARD-containing MAGUK protein 1 (CARMA1), TAK1, and IkappaB kinase (IKKb) leads to defective tTreg cell generation.[28–31] Of the five NFκB family members (i.e., NFκB1, NFκB2, RelA, RelB, and c-Rel), c-Rel has been pinpointed as having the most central role in tTreg cell development.[32] First, c-Rel is highly expressed in tTreg cells. Second, c-Rel-deficient mice show severe deficiencies in tTreg cell frequencies compared with normal frequencies in NFκB1-null mice and an intermediate reduction in RelA-deficient mice.[33,34] Paradoxically, activation of the PI3K–Akt pathway, which is also downstream of the TCR-CD28 signaling, was shown to repress tTreg cell differentiation.[35–37] Specifically, the inhibition of the Akt pathway was recently found to be critical for activation of the Foxo1 and Foxo3a transcription factors, which in turn translocate to the nucleus and collaborate with NFκB and other factors to induce *Foxp3* transcription.[38,39] Therefore, tTreg cell development seems to strongly depend on the balance between the NFkB and the PI3K–Akt signals, which are likely determined by the quality (i.e., strength and duration) of antigen stimulation.[40] Parallel to the transcription factors directly activated by TCR-CD28 costimulation, additional transcription factors were found to operate as secondary modulators of *Foxp3* expression. As already mentioned above, STAT5, a key transcription factor activated by γ-chain cytokine signaling, is essential for tTreg cell differentiation.[22] Moreover, members of the nuclear receptor 4a (Nr4a) family of orphan nuclear receptors (*Nr4a1, Nr4a2*, and *Nr4a3*) were also found to be critical for *Foxp3* expression, as triple-deficient mice were found to display a diminished tTreg cell repertoire and lethal autoimmune phenotypes.[41] The role of TGF-β-induced transcription factors, like the Smad family, in tTreg cell development still remains controversial. The current prevalent view suggests that TGF-β/Smad signaling promotes tTreg cell generation by restraining negative selection rather than direct transcriptional control of *Foxp3*.[40]

In conclusion, several models have been proposed to explain the development of tTreg cells. Based on the current knowledge, the following model seems to reflect the known aspects of Treg cell development in the thymus. In the first step, TCR stimulation of CD4 single-positive (SP) thymocytes results in the generation of $CD25^+Foxp3^-$ CD4SP Treg cell precursors. As such, the TCR affinity can be high (but below the threshold that would induce negative selection), medium, or low (but not too low as to insufficiently activate the TCR). The critical element determining the tTreg cell fate decision, however, seems to be the balance between NFkB and PI3K/Akt signaling, which are both triggered by the TCR-CD28 costimulation. In the second step, the $CD25^+Foxp3^-$ CD4SP Treg cell precursors need to respond to common γ-chain cytokines, predominantly IL-2, as well as to other stimuli (e.g., TGF-β), in order to initiate the transcriptional machinery controlling *Foxp3* gene expression and the subsequent tTreg cell generation (Fig. 1). Interestingly, while IL-2, TGF-β, and costimulation via CD28 are all required for the development of tTreg cells, they are dispensable for Tconv cell development in the thymus, thus highlighting the unique role of these factors in controlling the tTreg cell developmental program.

2.2 Which Cells Are Important for tTreg Cell Generation?

Developing tTreg cells require interactions with other thymic populations to complete their differentiation process. As tTreg cells are mostly located within the thymic medulla, it has been proposed that the unique microenvironment provided by medulla-resident stromal cells is required for tTreg cell generation.[11,42] Specifically, both medullary thymic epithelial cells (mTECs) and thymic dendritic cells (tDCs) have been identified as the key players in this process by providing both antigens for TCR stimulation and the necessary costimulatory signals required for tTreg cell development.[18,43]

The importance of the thymic medulla, and specifically mTECs, for controlling the tTreg cell developmental program is well exemplified by studies in which an enlarged mTEC compartment correlated with higher tTreg cell frequencies.[44,45] In contrast, a severely diminished mTEC compartment, due to impaired noncanonical NFκB signaling, correlated with significantly reduced tTreg cell frequencies.[46,47] Furthermore, it has been demonstrated that the $CD25^+Foxp3^-$ Treg cell precursors, at the $CD69^+CCR7^+CCR9^-$ stage, require physical interaction with $RelB^+$ mTECs in order to develop into mature $Foxp3^+$ tTreg cells.[48] Moreover, reduced MHC-II expression

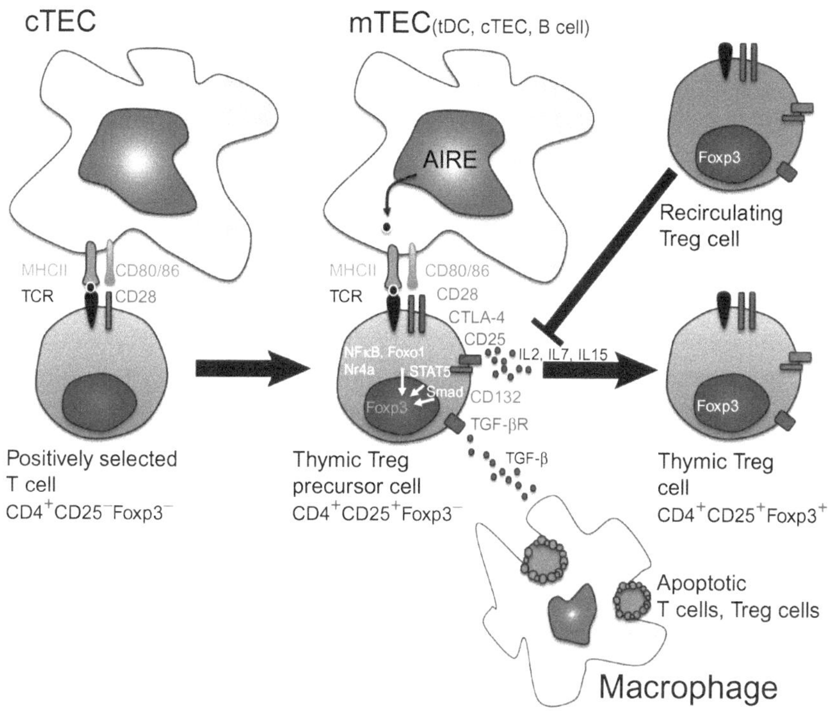

Figure 1 Development of tTreg cells is a multistep process. In the first step, TCR stimulation (with low, mid, or high affinities) on CD4$^+$CD8$^-$ single-positive thymocytes results in the generation of CD25$^+$Foxp3$^-$ CD4SP Treg cell precursors. In the second step, the CD25$^+$Foxp3$^-$ CD4SP Treg cell precursors respond to common γ-chain cytokines, predominantly IL-2, as well as to other stimuli (e.g., TGF-β), which are provided in part by macrophages phagocytosing apoptotic T cell clones. This, together with TCR-CD28 signals, initiates transcriptional machinery involving multiple transcription factors (e.g., NFkB, Nr4a, Foxo1, STAT5) controlling *Foxp3* gene expression and the subsequent tTreg cell generation. A critical element determining the tTreg cell fate decision is the balance between the NFkB and the PI3K/Akt pathways triggered by the TCR-CD28 costimulation. The development of tTreg cells does not require a dedicated APC, but rather entails a high degree of flexibility in the stromal cell types involved. The cooperative action of mTECs and tDCs, however, plays a pivotal role in this process. Aire-mediated expression of tissue-restricted antigens in mTECs is critical for the induction of a unique subpopulation of tTreg cells during the perinatal period.

on mTECs resulted in a smaller proportion of CD4 SP thymocytes being deleted and a higher differentiation rate into the tTreg cell fate.[49] Nevertheless, the optimal environment for tTreg cell development seems to be provided by the cooperative action of both mTECs and tDCs. Specifically, it has been demonstrated that mTECs not only supply antigen to tDCs for subsequent cross-presentation to developing tTreg cells, but themselves act as professional antigen-presenting cells (APCs), responsible for tTreg cell generation.[50] More recently, it was shown that mTECs and BM-derived thymic APCs contribute nonredundantly to tTreg cell generation, implying that each population presents a different array of self-antigens to developing thymocytes. In particular, Batf3-dependent $CD8\alpha^{+}CD11c^{+}$ DCs were found to be responsible for the cross-presentation of about 50% of the Aire-dependent antigens, demonstrating an intimate and nonredundant crosstalk between thymic populations.[51]

Other thymic populations have also been implicated in controlling tTreg cell development, including cortical thymic epithelial cells (cTECs),[52,53] thymic B cells,[54] and apoptotic thymocytes.[24,55] Specifically, DP $Foxp3^{+}$ cells, which express high levels of CCR7, localize in the cortex, whereas CD4 SP $Foxp3^{+}$ Treg cells localize in the medulla. Specifically, experiments utilizing K14-$A\beta^{b}$ $Ab1^{-/-}$ mice, with expression of MHC-II restricted to cTECs, were able to demonstrate that the thymic cortex is sufficient for supporting the generation of $Foxp3^{+}$ tTreg cells and that these "cortical" tTreg cells rapidly migrate to the medulla via a CCR7-dependent mechanism.[53] The role of thymic B cells in tTreg cell development was shown using BAFF-transgenic mice, which have a twofold increase in the number and frequency of thymic tTreg cells and an increase in thymic output as measured by $Helios^{+}Foxp3^{+}$ Treg cells in the periphery.[54] Moreover, tTreg cell frequencies are decreased in B cell-deficient mice and BM chimeras from B cell-deficient donors into wild-type or BAFF-transgenic hosts also demonstrate a reduction in the tTreg cell population.[54] Finally, a unique role for apoptotic thymocytes in tTreg cell development was delineated in an effort to explain the perplexing observation of delayed tTreg cell emergence on day 3 after birth.[55] Although some previous studies suggested that the late tTreg cell onset is due to the incomplete structural organization of the thymic medulla immediately after birth, some noticed that tTreg cell emergence coincides with the previously reported massive rise in thymocyte apoptosis beginning 2 days postpartum.[56] Hence, it was hypothesized that thymic apoptosis may play a critical role in tTreg cell development. Indeed, it has been found that TGF-β, which has been shown to be secreted by phagocytes following the engulfment of apoptotic cells,[57,58] has an indispensable role at the early tTreg cell generation (discussed above). This is well illustrated by experiments where augmented apoptosis resulted in higher tTreg cell numbers and frequencies, while inhibition of apoptosis had the opposite effect, both in adult and in neonatal thymi.

In summary, tTreg cell differentiation does not seem to require a dedicated APC, but rather entails a high degree of flexibility of the stromal cell types involved. However, the cooperative action of mTECs and tDCs in the thymic medulla seems to play a pivotal role in this process.

2.3 Is Aire Important for tTreg Cell Generation?

The central role of the Autoimmune regulator (*Aire*) gene in the negative selection of self-reactive thymocytes is well established; however, its contribution to the selection and generation of tTreg cells has remained controversial for over a decade.[59] Several early studies demonstrated that Aire does not impinge on Foxp3^{+} Treg cell generation as *Aire*-deficient mice exhibit unremarkable changes in Treg cell number and frequency or their proliferative and suppressive capabilities.[60–62] Moreover, crossing *Foxp3*-mutant scurfy mice with *Aire*-deficient mice did not exacerbate lymphoproliferation nor did it change the profile of target organs afflicted by autoimmune attack in the scurfy mice.[63] Therefore, the contention that Aire's primary function is to mediate negative selection of autoreactive T cells rather than positive selection or clonal diversion of tTreg cells became the prevalent view in the field.

Nevertheless, these original studies were somewhat contrasted by the observed defects in the Treg cell compartment of human patients with autoimmune polyendocrine syndrome type 1 (APS1) who carry a mutated *AIRE* gene. Although these patients were found to have normal frequencies of circulating CD25^{hi}CD4^{+} Treg cells, they had reduced expression of FOXP3, resulting in compromised Treg cell suppressive capacity.[64,65] Moreover, several subsequent reports challenged the original studies in mice[60–62] by showing mild, though significant, reduction in tTreg cell frequencies and numbers in *Aire*-deficient mice.[66–69] The first evidence supporting the view that Aire-dependent expression of tissue-restricted antigens (TRAs) in mice regulates tTreg cell development was demonstrated using an Aire-hemagglutinin (HA) transgenic mouse, which expresses an HA peptide under the control of the Aire promoter in mTECs.[50] Specifically, in experiments using double transgenic mice, where Aire-driven HA peptides presented on MHC-II molecules of mTECs are recognized by transgenic HA-specific T cell clones, it was demonstrated that Aire-driven expression of the HA antigen in mTECs is sufficient to generate antigen-specific tTreg cells. Moreover, HA-expressing mTECs were shown to be more effective inducers of Treg cells *ex vivo* than HA-expressing tDCs, implicating that

Aire-expressing cells and Aire-dependent expression of target genes in mTEC are critical for the generation of tTreg cells in mouse models.[50] In addition, the direct involvement of Aire-dependent expression of a specific TRA in the selection of tTreg cells was recently described by the identification of a specific recurring tumor-infiltrating tTreg cell clone termed MJ23.[68] These MJ23 Treg cells were found to develop in the thymus and to be specific for an Aire-dependent prostate antigen, rather than a tumor-specific antigen, implying that a subpopulation of Aire-dependent Treg cells may be co-opted by tumors developing within the corresponding organ. Moreover, most recent studies based on deep sequencing of the Treg cell TCRα chain repertoire in a fixed TCRβ chain model, in both *Aire*-sufficient and *Aire*-deficient mice, demonstrated that Aire is involved in both the selection process and deletion of Treg cells.[51] Aire shaped the Treg cell TCR repertoire at the polyclonal stage and specifically affected lower frequency TCR clones, an effect that could be easily overlooked when the entire polyclonal Treg cell population, albeit restricted, is examined.[51] These data challenged an earlier report that suggested the Aire-dependent TRA expression in mTECs is dispensable for TCR diversity and selection of dominant TCRs on Treg cells.[70]

Most recently, the role of Aire in shaping the tTreg cell repertoire was further elucidated in a very elegant study by the Benoist–Mathis group who identified a unique subpopulation of Aire-dependent perinatal tTreg cells that are critical for the induction of immune self-tolerance.[69] Although the expression of Aire has previously been shown to be critical early in life for the establishment of central tolerance, the mechanism underlying this phenomenon remained unclear.[71] Recently, it was demonstrated that Treg cell ablation using the NOD.Foxp3-DTR system during the first 10 days of life resulted in multiorgan autoimmunity typical of *Aire*-deficient mice on the NOD background, accompanied by severe weight loss and ultimately mortality by 24 days of age. Interestingly, such a dramatic outcome was not evident in mice depleted of Treg cells at a later time-point.[69] To demonstrate Aire dependence of this autoimmune phenotype, Treg cells from *Aire*-sufficient and -deficient mice were transferred into Treg cell-depleted perinates. Indeed, only Treg cells from *Aire*-sufficient mice were capable of protecting against the development of Aire-like autoimmunity, indicating Aire-dependent generation of Treg cells during a short perinatal window. Lineage tracing of the Treg cells using the NOD.Foxp3eGFP-Cre-ERT2xR26Y reporter mouse, in which tamoxifen injection labels existent GFP^+ Treg cells with YFP, allowed further investigation of the perinatal

Treg cell population. These double-labeled cells were shown to persist for over 2 months, and when peripheral Treg cells from perinates were transferred into *Aire*-deficient recipients, they were capable of significantly reversing the autoimmune phenotype. In contrast, this therapeutic affect was not achieved following transfer of GFP^+YFP^+ Treg cells from older mice or GFP^+YFP^- Treg cells of either age, demonstrating that the perinatal Treg cell population has long-term functional significance.[69]

In conclusion, Aire seems to be critical for the induction of a unique subpopulation of tTreg cells during the perinatal period.[69] This, together with its well-defined role in clonal deletion of self-reactive thymocytes, extends the notion of a "layered" immune system and further blurs the virtual border between central and peripheral modes of immunological self-tolerance.

3. FURTHER DIFFERENTIATION OF TREG CELLS IN THE PERIPHERY

Following development in the thymus, Treg cells enter the periphery where they circulate through the secondary lymphoid tissues as well as the peripheral nonlymphoid tissues. Interestingly, circulating peripheral Treg cells can also reenter the thymus and control thymic tTreg output (Fig. 1).[72] Treg cells in the periphery have heterogeneous expression patterns of activation/memory markers, adhesion molecules, and homing receptors. Originally, Treg cells were subdivided into subsets based on these expression patterns; however, the current Treg cell subset concept arose from evidence describing the diversity of Treg cell suppressor mechanisms. These mechanisms are dynamic and depend on the target cells, the location, and the inflammatory context. Collectively, it is becoming clear that the environmental signals present during Treg cell activation contribute to a further differentiation into unique Treg cell subsets that express genetic signatures mimicking that of their target cells or tissues (Table 1).[6,73,74]

3.1 Do Treg Cells Undergo Further Differentiation in the Secondary Lymphoid Tissues?

The first Treg cell subsets were identified in the secondary lymphoid tissues, i.e., spleen and lymph nodes, where they make up about 10–15% of the $CD4^+$ T cell pool. Treg cells were initially divided into naïve and memory subsets based on their function and differential expression of the surface molecules CD62L and CD103.[75,76] It has long been known that Tconv cells develop into distinct subsets depending on the cytokine signals that are

Table 1 Distribution, Frequency, and Function of Tissue-Resident Treg Cells

Location	Frequency (%)	Function(s)	Subset(s)
Spleen and LN	10–15	Systemic autoimmune disease	Follicular Treg cells T_H-specific subsets: T-bet$^+$ Treg cells GATA3$^+$IFR4$^+$ Treg cells STAT3$^+$ Treg cells Bcl6$^+$ Treg cells
Adipose tissue	10–50	Metabolic control	PPAR-γ^+ Treg cells
Skin	30–90	Psoriasis Contact dermatitis	
Lung	<10	Asthma Pulmonary fibrosis Transplantation	
Liver	5–40	Autoimmune hepatitis HCV infection	
Pancreas	5–15	Type 1 diabetes	
Muscle	10–50	Postinjury muscle regeneration	Areg$^+$ Treg cells
Gut	10–50	Intestinal barrier homeostasis Tolerance to commensal bacteria	FFAR2$^+$ Treg cells

Frequency: relative fraction of Foxp3$^+$CD25$^+$ Treg cells of the CD4$^+$ T cell pool.
Functions: physiological or pathological situations in which these cells have been implicated.

present during their activation. For example, T helper 1 (TH1) cells develop in the presence of IFN-γ and IL-12, TH2 cells in the presence of IL-4, and TH17 in the presence of TGF-β and IL-6. These Tconv cell subsets are stabilized (and can be identified) by the differential expression of master transcription factors, such T-bet and GATA3. The potential for matched symmetry between Tconv cell subsets and Treg cell subsets has led many investigators to examine the importance of these transcription factors in Treg cell function. In fact, Treg cell-specific deletion of certain transcription factors resulted in the impaired suppression of only certain Tconv cell subsets. For example, Treg cell-specific deletion of IRF4 or GATA3 resulted in inability to suppress TH2 responses.[77–79] In addition, Treg cell-specific deletion of STAT3 or T-bet resulted in the inability to suppress TH17 and TH1 responses, respectively.[80,81] Finally, in germinal centers located in the secondary lymphoid tissues, the follicular Treg cell (TFR) and follicular Tconv cell (TFH) subsets develop and survive due to their expression of the transcription factors Blimp-1 and Bcl-6.[82,83] In fact, Treg cell-specific deletion of Bcl-6 resulted in inability to suppress TFH responses.[82] In summary, it is clear that specific subsets of Treg cells in the secondary lymphoid tissues are able to regulate immune responses and maintain homeostasis by specifically controlling T cell and B cell responses.

3.2 Do Treg Cells Undergo Further Differentiation in Peripheral Nonlymphoid Tissues?

Recent evidence suggests that environmental signals found in peripheral nonlymphoid tissues are also responsible for the development of tissue-specific Treg cell subsets. The proportion of tissue-specific Treg cell subsets within tissues is difficult to determine due to differences between inflammatory and steady-state conditions as well as the presence of both long-term (resident) and short-term (migratory) Treg cells.

The identification of the first tissue-specific Treg cell subset in visceral adipose tissue (VAT) opened the possibility that each tissue might harbor a distinct Treg cell subset that is responsible for controlling inflammation and maintaining tissue homeostasis.[84] These VAT-specific Treg cells exhibited a distinct gene signature compared to Treg cells isolated from secondary lymphoid tissues. More recent reports have shown that peripheral nonlymphoid tissue-specific Treg cells can be distinguished from secondary lymphoid tissue Treg cells based on their phenotype and function. In the following sections, we will address the identification, differentiation, and function of these peripheral tissue-specific Treg cell subsets in a number of nonlymphoid tissues.

3.2.1 Treg Cells in the Adipose Tissue

Most studies about "fat" Treg cells focus on VAT-resident Treg cells. They are a dynamic population of Treg cells that increase in frequency between 5 and 25 weeks of age in mice. After reaching a plateau of about 50% of all $CD4^+$ T cells, their frequency declines to about 10% in 40-week-old animals.[84,85] In contrast, this population dynamic has not been observed in the splenic Treg cell compartment. The origin of these VAT-specific Treg cells was investigated through the analysis of the TCR repertoire and little overlap was identified between secondary lymphoid tissue- and VAT-specific Treg cells, which excluded a random influx of short-term (migratory) Treg cells.[84,86] The VAT Treg cell TCR signatures were also unique compared to VAT-resident Tconv cells, ruling out peripheral conversion as a significant source of VAT Treg cells.[84,86] Currently, it is hypothesized that an antigen-specific signal based on TCR specificity leads to retention of specific Treg cells in the VAT, followed by the expansion of these clones *in situ*. It has recently been reported that VAT Treg colonization happens within the first weeks of age and is later independent of thymic input. Additionally, the accumulation of VAT Treg cells depends on antigen presentation by MHC-II and soluble mediators, specifically IL-33.[86] It is still not entirely clear how and where the specialized phenotype is generated, but it is tempting to speculate that once Treg cells reside in the VAT, they initiate a specific gene expression prolife leading to a discrete VAT-specific Treg cell signature, with more than 2000 genes being differently expressed between secondary lymphoid tissue- and VAT-specific Treg cells, whereby still 60% of the "typical" Treg cell signature remains intact (most notably CD25, GITR, CTLA-4, OX40, Foxp3, and KLRG1).[84,87] This translates into the expression of specific chemokine receptors, namely CCR1, CCR2, CCR3, CCR5, CCR9, and CXCR6, while CCR6, CCR7, and CXCR3 are downregulated. Furthermore, VAT Treg cells acquire new functional abilities, such as taking up fat droplets via the expression of CD36, and producing more IL-10 than their splenic counterparts.[88] Upon close examination of the VAT Treg cell gene expression signature, peroxisome proliferator-activated receptor-γ (PPAR-γ) was identified as the key driver of VAT Treg cell accumulation, phenotype, and function.[88] This definitive study demonstrated that selective deletion of PPAR-γ in Treg cells caused only a VAT-specific decrease in Treg cell frequency and severely impinged on the specialized phenotype of VAT-specific Treg cells. Similarly, treatment of wild-type animals with a PPAR-γ inhibitor caused a decline only in VAT-resident GATA3$^+$ Treg cells, not in splenic Treg cell frequency.[88]

While all these changes occur under physiological (i.e., homeostatic) conditions, several studies focused on the importance of VAT-resident Treg cells in dysregulated metabolic conditions. Feeding animals a high-fat diet (HFD) leads to diet-induced obesity (DIO) with body weight increase and insulin resistance. During disease progression, an increase in the frequency and activation status of proinflammatory macrophages as well as CD4$^+$ and CD8$^+$ T cells can be observed. In contrast, the frequency of VAT Treg cells declined.[84,88,89] In two sensitive loss-of-function experiments, Treg cell-depleted Foxp3-DTR mice[84] and Treg-specific PPAR-γ-deficient mice[88] showed an increase in proinflammatory cytokine expression, as well as glucose and insulin levels, which indicated an emerging insulin resistance. Similar conclusions can be drawn from studies with both DIO- and leptin-deficient mice.[90] Mitogenic anti-CD3 monoclonal antibody (mAb) was injected into HFD-fed mice and a selective increase in $CD4^+Foxp3^+$ Treg cell number was observed in VAT, almost to a level comparable to lean animal controls. After this treatment, improved glucose tolerance and insulin sensitivity, transient weight loss, and an increase in both antiinflammatory T cell and macrophage populations in VAT were observed. In leptin-deficient mice (ob/ob), the complete absence of functional leptin leads to increased body weight followed by severe insulin resistance and elevated liver enzymes. Additionally, oral and nasal administration of anti-CD3 mAb plus β-glucosylceramide showed fat-specific induction of $CD4^+Foxp3^+$ Treg cells, without systemic Treg cell expansion.[91] Decreased amounts of proinflammatory cytokines (e.g., TNFα) were observed in fat tissue, along with decreased cellular infiltrates. On a broader scale, long-term decreases in blood glucose levels, liver enzyme levels, and blood cholesterol were also observed. Recently, another approach was used to study the impact of VAT Treg cell depletion in obesity models.[92] It took advantage of the db/db mouse which lacks leptin receptor signaling, leading to obesity, hyperphagia, hyperglycemia, and diabetic nephropathy. Treg cells were systemically depleted with an anti-CD25 mAb, and increased fasting blood glucose levels, insulin resistance, increased proinflammatory cytokine expression, and renal cell infiltration were observed. When adoptively transferring Treg cells, insulin sensitivity was improved, $CD8^+CD69^+$ T cell infiltrates decreased, and damage to renal organs was reduced.[92]

In summary, these studies have shown a profound ability of VAT-resident Treg cells to modulate obesity-associated insulin resistance, proinflammatory cytokine secretion, and cell infiltration. The definitive study identifying PPAR-γ also provided insight into the mechanisms of action. When treating obese animals with pioglitazone, a PPAR-γ agonist and insulin-sensitizing agent, Treg cells in VAT, but not in spleen or liver, were selectively enriched. Their gene expression profile shifted more toward the VAT-specific signature and they showed enhanced expression of CD36, enabling them to take up lipids.[88] From these data, it is possible to infer that one of the pioglitazone-specific targets is VAT-resident Treg cells, probably enhancing their protective phenotype by increasing their frequency and potency and thereby contributing to reduced insulin resistance. This study showed, as a proof of concept, that organ-specific targeting of Treg cells is possible. Taken together, these findings strengthen the hypothesis that VAT-resident Treg cells are pivotal agents controlling physiology and pathology of VAT. A recent study analyzed Treg cells in the brown adipose tissue (BAT).[93] In contrast to the energy-storing white adipose tissue, the BAT and the inducible brown-in-white (brite) adipocytes are specialized in the dissipation of energy in the form of heat in a process called uncoupling thermogenesis. This study defined a BAT-specific Treg subset with implications for the regulation of energy homeostasis in response to environmental stress.[93]

3.2.2 Treg Cells in the Skin

Skin-resident Treg cells, also called cutaneous Treg cells, represent a high percentage of $CD4^+$ T cells in the skin. In humans, they can account for 40–90% of all $CD4^+$ T cells[94] and they are similarly high in mice.[95] This population is maintained through migration of Treg cells into the dermis via E- and P-selectin binding in a CCR4-dependent manner[96] as well as local induction of pTreg cells. For example, UV irradiation causes RANK-L induction in keratinocytes, which in turn drives the activation of Langerhans cells (LC). These activated LC then induce pTreg cell differentiation in an IL-10- and OX-40-dependent manner.[97,98] Interestingly, there are few overlapping TCR sequences between Treg cells and Tconv cells in the skin.[99,100] This is consistent with data from the VAT but opens questions as to the origin and antigen specificity of the converted pTreg cells. Skin Treg cells have a unique surface expression profile. They express CD44 and CD103, along with high levels of the chemokine receptors CCR4, CCR5, CCR6, and CCR7.[101] In addition, they express high levels of activation markers (e.g., CTLA-4, CD25, and ICOS).[99,100]

The influence of cutaneous Treg cells on the development of skin disease has been studied in both humans and mice. Several studies have investigated the importance of Treg cells during the development of contact hypersensitivity (CHS) in mice, a model for contact dermatitis (reviewed in Ref. 101). CHS begins with the "sensitization phase," where hapten-linked epidermal peptides are presented to naïve T cells in skin-draining lymph nodes. This is followed by the "elicitation phase," where reexposure of the skin to haptens causes antigen-specific inflammation. Treg cell depletion results in increased skin inflammation and it has been demonstrated that Treg cells can influence both phases of CHS development. For example, Treg cells can control the magnitude of T cell activation during sensitization. Furthermore, several studies showed that intravenous administration of Treg cells 1 day before elicitation can suppress inflammatory cell infiltration in an

IL-10-dependent manner. Finally, the selective depletion of Treg cells during elicitation resulted in enhanced and prolonged pathology, again indicating the importance of Treg cells in CHS.

Another very prevalent immune-mediated skin disease in humans is psoriasis and it is characterized by thickening of the epidermis and massive infiltration of both myeloid and lymphoid cells.[102] There are conflicting reports about the role of Treg cells in this disease, but a general correlation has been established from patient studies showing increasing numbers of infiltrating Treg cells and disease severity.[103] It is hypothesized that, even though enriched in psoriatic lesions, these Treg cells have impaired regulatory function[104] and/or effector cells can resist their suppressive effects.[105] Finally, it has also been observed that patient-derived psoriatic Treg cells differentiate easily into Th17 cells *in vitro*, in contrast control Treg cells, which raises questions about the origin and stability of this population.[106] In the study of another human autoimmune skin disease, Bullous pemphigoid (BP), a significant reduction of Treg cells in patients has been observed in the peripheral blood.[107] This is similar to findings in patients suffering from systemic sclerosis and morphea, a connective tissue disease with autoimmune involvement. These patients have fewer Treg cells in both skin lesions and peripheral blood, fewer IL-10- and TGF-β-producing cells in skin, and lower serum IL-10 and TGF-β levels.[94] The same observation has been made in dermatomyositis, a rare disease affecting muscles and skin. Here, Treg cell frequency and overall TGF-β-producing cell numbers are reduced in peripheral blood and skin lesions.[108]

Taken together, these findings in both mouse models (CHS) and human patients (psoriasis, BP, system sclerosis, and dermatomyositis) show an important role of Treg cells for skin-related physiology and pathology. Whereas, in most cases, the sheer reduction in Treg cell frequency in skin might promote a more severe autoimmune phenotype, psoriasis disputes this dogma by displaying an even increased frequency of Treg cells in skin lesions. It will be important to determine a specific phenotype of cutaneous Treg cells to understand their molecular makeup and identify key transcription factors responsible for their adaption to the cutaneous environment. This might enable a deeper understanding of their pathophysiological role in healthy and autoimmune-inflamed skin disease.

3.2.3 Treg Cells in the Lung

Respiration causes a constant influx of many immunogenic antigens such as plant and animal proteins. Therefore, the lung depends on complex systems to ensure tolerance to harmless antigens while guaranteeing protection against pathogens. Anatomically, the lung can be separated into conducting airways and lung parenchyma. Ciliated and secretory epithelial cells, DCs, macrophages, and $CD8^+$ T cells are abundant in these tissues. In addition, the presence of bronchial-associated lymphoid tissue (BALT) as local means to support B cell and T cell proliferation has been described.[109] In this secondary lymphoid tissue, homing of Treg cells in a model of pulmonary fibrosis (PF) in CCR7-deficient animals has been observed.[110] But the steady-state presence, frequency, and function of Treg cells in BALT remain elusive. The alveolar space is mostly populated by macrophages. The underlying parenchyma also contains macrophages, along with DCs, B cells, T cells, and mast cells.[109] When measuring Treg cell frequency in the lung, many studies do not discriminate between airway and parenchyma. Hence, it has been observed that Treg cells account for only a small percentage of CD4 T cells in the murine (<10% of CD4 T cells[109]) and human lung (<3%[111]) under physiologic conditions. They have been reported to highly express CCR4, with less than 15% being $CD103^+$ as well.[109]

Treg cells in lung tissue have mostly been studied in the context of PF, asthma, and transplantation. In PF, repeated inflammatory and regenerative phases cause excessive collagen deposition and lung scarring, which decreases oxygen-uptake capacity. CCR7-deficient mice challenged with bleomycin to induce PF showed increased numbers of Treg cells, indicating that loss of CCR7 might retain Treg cells in the inflamed lung and thereby decrease inflammation and pathology.[110] Lung-resident Treg cells have also been implicated in asthma. Here, genetic and environmental factors produce chronic inflammation of the lung. If there is a failure of Treg cell recruitment, excessive tissue repair and wound healing indicative of chronic asthma prevail.[112] In asthmatic children, a decrease in Treg cell frequency in the bronchoalveolar lavage (BAL) fluid has been described, along with an impaired ability of Treg cells to suppress TH2 responder cells.[113] Controversially, other studies show increased Treg cell numbers in patients with severe allergic pneumonitis, but they show impaired suppressive capacity.[114] Interestingly, treatment with steroids has been shown to promote Treg cell activity and might partially explain the therapeutic benefit of this treatment.[113] Next to chronic pulmonary disease, Treg cells have also been studied in the context of lung transplantation. It has been reported that increased numbers of Treg cells in the transplanted tissue correlate with protection from acute rejection. Once a rejection episode is resolved, Treg cell numbers decrease in frequency. Furthermore, it has been noted that Treg cells infiltrate lung tissue before or at the same time as effector T cells, enabling Treg cell frequency in BAL fluid to be used as a prognostic marker of rejection and bronchitis obliterans syndrome.[114] In summary, lung-resident Treg cells are an attractive target to modulate disease severity in common airway diseases such as PF and asthma. In a recent study, the protein kinase CK2 has been implicated to be specifically required for lung-associated Treg cells to prevent excessive TH2 responses in the lung.[115] This example highlights the need for a more detailed profiling of the lung-associated Treg cell populations to evaluate their detailed molecular phenotype and to develop novel treatment strategies.

3.2.4 Treg Cells in the Liver

Immune tolerance is an essential requirement for the homeostasis of the liver for two main reasons. First, the liver is permanently exposed to food and microbial products from the intestine.[116] Second, as a highly metabolic organ, the liver produces a multitude of self-antigens. Therefore, a well-performing self-tolerance system mediated by, among others, Treg cells is important.

The observation that the liver is a unique organ in regard to immune tolerance was first shown by the "liver tolerance effect" in 1967 when it was demonstrated that liver allografts were accepted even with MHC mismatches without immunosuppression.[117] Furthermore, nonliver allografts were accepted if cotransplanted with a liver from the same donor.[118] It has also been shown that expression of extra-hepatic antigens in the liver results in conversion of Tconv cells to highly suppressive pTreg cells that inhibit immune reactions to the antigen in affected organs.[119]

The frequencies of Treg cells among $CD4^+$ T cells range from 5% to 10% in mice[95,120] and can reach up to 30–40% in chronic hepatitis C virus (HCV)-infected patients.[121] Liver Treg cells were shown to express higher levels of CTLA-4, GITR, and CD103 than splenic Treg cells, thus resembling an effector/memory Treg cell phenotype. Additionally, they were able to suppress hepatitis development in a TGF-β-dependent manner.[122] It was also observed that frequencies of liver-specific Treg cells increased up to 20% in C57BL/6 mice after concanavalin A (ConA) treatment in a ConA-induced hepatitis model.

Studies have shown an important role of Treg cells in the prevention of autoimmune hepatitis (AIH).[123,124] For example, a reduction of Treg cell frequencies was identified in patients with AIH, in addition to an inverse correlation between Treg cell numbers and clinic markers such as titers of liver-related autoantibodies.[125] The functional properties of Treg cell-mediated suppression have also been analyzed and showed that direct contact of Treg cells to target cells leads to secretion of IL-10 and TGF-β, both of which are important for the prevention of AIH.[126]

Besides AIH, Treg cells are also crucial during the progression of hepatic cancer and chronic infections of HCV. In hepatic cancer, several studies showed that higher numbers of Treg cells in the lesion are associated with cancer progression and poor prognosis for patients.[127,128] In fact, the balance between $CD8^+$ T cells and Treg cells seems to be important for overall and disease-free survival. Comparable tendencies have been observed during chronic HCV infections. Here, high numbers of Treg cells are associated with virus persistence.[129] It was shown that Treg cells directly suppress HCV-specific $CD8^+$ T cells in an IL-10- and TGF-β-independent, but cell contact-dependent, manner.[130] In summary, Treg cells are essential players in the immune tolerance of the liver. Although their roles under pathological conditions have been well studied, it will be important to further investigate their function under physiological conditions.

3.2.5 Treg Cells in the Pancreas

The prominent role of Treg cells in the pancreas is highlighted by the observation that patients and mice lacking Treg cells (e.g., due to mutations in the *Foxp3* gene) develop Type 1 diabetes (T1D).[5,131,132] Self-reactive T cells destroy the insulin-producing β-cells of the pancreas and thereby initiate T1D. Most studies about pancreatic Treg cells focus on this situation.

It was shown that the frequency of Treg cells in the pancreas of NOD mice, a mouse strain that spontaneously develops T1D, was at about 10% of $CD4^+$ T cells.[132–134] Transcriptional analysis of pancreatic Treg cells revealed a higher expression of IL-10, GITR, and ICOS compared to lymph node Treg cells.[135] Furthermore, an upregulation of CD103, IL1R2, S100a6, and the chemokine receptors CCR5, CXCR3, and CCR2 were found.[136] Chemokine receptors, like the aforementioned, play a pivotal role in the homing and circulation of immune cells. For CXCR3, it was observed that it is important for homing of T cells into inflamed islets in NOD mice.[137] Similarly, it was shown that double-stranded adeno-associated virus serotype 8 (dsAAV8) vector-mediated expression of CCL22 (ligand of CCR4) in islets leads to Treg cell recruitment and prevention of diabetes onset.[138]

Besides the transfer of *ex vivo* expanded Treg cells, strategies targeting the IL-2 signaling pathway are promising for the treatment of T1D. In several studies, it was shown that low-dose IL-2 treatment or IL-2 produced by activated effector T cells can boost Treg cell activation/expansion and thus reverse established diabetes in NOD mice.[139,140]

Another disease setting in which Treg cells are involved is the development of pancreatic ductal carcinoma. A significantly higher number of Treg cells were found in ductal carcinoma lesions compared to normal pancreas tissue, and poor prognosis was associated with high Treg frequency in the lesion.[141] Moreover, these tumor-infiltrating Treg cells secrete TGF-β and IL-10.[142]

In summary, pancreatic Treg cells play a decisive role in the immune homeostasis of the pancreas and imbalances in their frequencies or function can lead to severe pathologies such as T1D. Currently, most observations are derived from disease models, where migratory Treg cells are also being recruited. The specialization and function of long-term tissue-resident Treg cells in the pancreas are still unclear.

3.2.6 Treg Cells in the Muscle Tissue

Treg cells are only a small cell population in healthy muscle tissue, about 10% of all muscle-resident $CD4^+$ T cells. Since they are very limited, no molecular characterization is yet available for muscle-resident Treg cells under physiological conditions. Therefore, this specific Treg cell population is mostly studied under pathological conditions, such as muscular dystrophies or induced muscle injuries, where accumulation of proinflammatory $CD4^+$ and $CD8^+$ T cells seems to promote disease progression.[143,144] In a recent study, acute muscle injury was induced via the administration of cardiotoxin, which causes myofiber necrosis. Interestingly, Treg cells rapidly accumulated in skeletal muscle and increased in frequency to about 50% of all CD4 T cells within 7 weeks and showed a distinct gene expression profile.[145] Similarly, Treg cells are also elevated in dystrophic muscles of mice harboring a muscular dystrophy (mdx) mutation, leaving them deficient of dystrophin, and human Duchenne muscular dystrophy patients. Under these conditions, muscle Treg cells display an activated phenotype in both mice and humans, and their selective depletion in mdx mutant mice leads to enhancement of immune cell infiltration and type 1 inflammatory response, resulting in even more impaired myofiber repair.[145] In contrast to this, treatment with IL-2/anti-IL-2 mAb complexes increased Treg cell frequency and IL-10 concentrations in muscle tissue, decreasing myofiber injury.[146]

Moreover, osteopontin seems to regulate Treg cell number and/or stability since muscles from osteopontin-deficient and mdx mutant mice display increased Treg cell numbers along with decreased fibrosis.[147] Manipulation of mdx-dystrophic mice to increase or decrease muscle Treg cell numbers resulted in improved or impaired muscle repair, respectively. The muscle-repair factor amphiregulin was identified to be expressed specifically by muscle-specific Treg cells, enhancing the differentiation of muscle cell precursors and lowering the expression of proteins associated with fibrosis.[145] In conclusion, muscle-specific Treg cells play an important role once muscle damage has occurred. They can limit damage by suppressing the development of type 1 inflammatory responses and express factors to actively enhance muscle repair and thereby promote faster recovery.[145]

3.2.7 Treg Cells in the Gastrointestinal Tract

The gut contains a large reservoir of both secondary lymphoid tissue-resident Treg cells as well as nonlymphoid tissue-resident Treg cells. In addition, it is the main location of pTreg cell generation. Both tTreg and pTreg cells are necessary for maintaining intestinal barrier homeostasis, especially in the colon.[148] Lamina propria-resident Treg cells can be distinguished from secondary lymphoid tissue Treg cells by the expression of free fatty acid receptor 2 (FFAR2, also known as GPR43).[149] The development of these gut tissue-specific Treg cells is dependent on unique environmental signals, which require products produced by commensal bacteria.[150–152] For example, short-chain fatty acids (SCFA), products of bacterial metabolism, are required for the maintenance of intestinal Treg cell homeostasis, including homing and proliferation potential. Importantly, it was shown that these effects were dependent on Treg cell expression of FFAR2, the receptor for SCFA.[149] In summary, these reports show that specialized Treg cells in the gut and commensal bacteria have developed unique communication ways to limit immune activation and establish gut homeostasis.

4. FUTURE PERSPECTIVES

The differentiation of Treg cells in the thymus depends on the antigen repertoire that is available at a given time-point. This concept is highlighted by a recent study showing that a unique subpopulation of Aire-dependent perinatal tTreg cells, critical for the induction of immune self-tolerance, can only be generated during the perinatal phase.[69] It is not entirely clear if this first wave of Treg cells is indeed critically to seed all peripheral tissues. It is tempting to speculate that such a perinatal wave would once colonize the peripheral tissues in an antigen-specific fashion and could be self-sustained. This would be in analogy to the tissue-resident macrophages, which are believed to be of yolk sac and fetal liver origin and fairly independent of additional input of monocyte-derived macrophages under steady-state homeostasis. The difference, of course, is the antigen specificity of the TCR. On the other hand, tissue-resident memory $CD8^+$ T cells can be established after, e.g., an infection in the skin during adulthood. Therefore, tissue-resident Treg cells might have a dual origin, a perinatal one and some tissue Treg cells may be established as a result of a specific tissue need in the adulthood.

The proportion of tissue-specific Treg cells within tissues is difficult to determine and characterize, especially in pathological conditions due to the presence of both long-term (resident) and short-term (migratory) Treg cells. In the future, the field needs to establish more precisely the molecular mechanisms that are required to differentiate into different subsets of tissue-resident Treg cells. PPAR-γ and VAT Treg cells are a good example for such a tissue-specific modifier. This information will yield specific markers and novel targets to modify Treg cells in a tissue-specific manner.

ACKNOWLEDGMENTS

This review was supported by grants from the Helmholtz Association of German Research Centers (HGF; HZ-NG-505) to M.F.; the German-Israeli Helmholtz Research School in Cancer Biology to M.D.; the DKFZ Postdoctoral Fellowship Program to D.K.; the Cooperation Program in Cancer Research of the DKFZ and Israel's Ministry of Science, Technology and Space (MOST) to M.F., A.-C.H., and J.A.; Israel Science Foundation and Sy Syms foundations to J.A.

REFERENCES

1. Kim JM, Rasmussen JP, Rudensky AY. Regulatory T cells prevent catastrophic autoimmunity throughout the lifespan of mice. *Nat Immunol.* 2007;8(2):191–197.
2. Lahl K, Loddenkemper C, Drouin C, et al. Selective depletion of Foxp3+ regulatory T cells induces a scurfy-like disease. *J Exp Med.* 2007;204(1):57–63.
3. Brunkow ME, Jeffery EW, Hjerrild KA, et al. Disruption of a new forkhead/winged-helix protein, scurfin, results in the fatal lymphoproliferative disorder of the scurfy mouse. *Nat Genet.* 2001;27(1):68–73.
4. Godfrey VL, Wilkinson JE, Russell LB. X-linked lymphoreticular disease in the scurfy (sf) mutant mouse. *Am J Pathol.* 1991;138(6):1379–1387.
5. Bennett CL, Christie J, Ramsdell F, et al. The immune dysregulation, polyendocrinopathy, enteropathy, X-linked syndrome (IPEX) is caused by mutations of FOXP3. *Nat Genet.* 2001;27(1):20–21.
6. Burzyn D, Benoist C, Mathis D. Regulatory T cells in nonlymphoid tissues. *Nat Immunol.* 2013;14(10):1007–1013.

7. Abbas AK, Benoist C, Bluestone JA, et al. Regulatory T cells: recommendations to simplify the nomenclature. *Nat Immunol.* 2013;14(4):307–308.
8. Apostolou I, Sarukhan A, Klein L, von Boehmer H. Origin of regulatory T cells with known specificity for antigen. *Nat Immunol.* 2002;3(8):756–763.
9. Jordan MS, Boesteanu A, Reed AJ, et al. Thymic selection of CD4+CD25+ regulatory T cells induced by an agonist self-peptide. *Nat Immunol.* 2001;2(4):301–306.
10. Benoist C, Mathis D. Treg cells, life history, and diversity. *Cold Spring Harb Perspect Biol.* 2012;4(9):a007021.
11. Stephens GL, Andersson J, Shevach EM. Distinct subsets of FoxP3+ regulatory T cells participate in the control of immune responses. *J Immunol.* 2007;178(11):6901–6911.
12. Lee HM, Bautista JL, Scott-Browne J, Mohan JF, Hsieh CS. A broad range of self-reactivity drives thymic regulatory T cell selection to limit responses to self. *Immunity.* 2012;37(3):475–486.
13. Salomon B, Lenschow DJ, Rhee L, et al. B7/CD28 costimulation is essential for the homeostasis of the CD4+CD25+ immunoregulatory T cells that control autoimmune diabetes. *Immunity.* 2000;12(4):431–440.
14. Tai X, Cowan M, Feigenbaum L, Singer A. CD28 costimulation of developing thymocytes induces Foxp3 expression and regulatory T cell differentiation independently of interleukin 2. *Nat Immunol.* 2005;6(2):152–162.
15. Vang KB, Yang J, Pagan AJ, et al. Cutting edge: CD28 and c-Rel-dependent pathways initiate regulatory T cell development. *J Immunol.* 2010;184(8):4074–4077.
16. Lohr J, Knoechel B, Kahn EC, Abbas AK. Role of B7 in T cell tolerance. *J Immunol.* 2004;173(8):5028–5035.
17. Spence PJ, Green EA. Foxp3+ regulatory T cells promiscuously accept thymic signals critical for their development. *Proc Natl Acad Sci USA.* 2008;105(3):973–978.
18. Mahmud SA, Manlove LS, Schmitz HM, et al. Costimulation via the tumor-necrosis factor receptor superfamily couples TCR signal strength to the thymic differentiation of regulatory T cells. *Nat Immunol.* 2014;15(5):473–481.
19. Bayer AL, Lee JY, de la Barrera A, Surh CD, Malek TR. A function for IL-7R for CD4+CD25+Foxp3+ T regulatory cells. *J Immunol.* 2008;181(1):225–234.
20. Fontenot JD, Rasmussen JP, Gavin MA, Rudensky AY. A function for interleukin 2 in Foxp3-expressing regulatory T cells. *Nat Immunol.* 2005;6(11):1142–1151.
21. Vang KB, Yang J, Mahmud SA, Burchill MA, Vegoe AL, Farrar MA. IL-2, -7, and -15, but not thymic stromal lymphopoeitin, redundantly govern CD4+Foxp3+ regulatory T cell development. *J Immunol.* 2008;181(5):3285–3290.
22. Yao Z, Kanno Y, Kerenyi M, et al. Nonredundant roles for Stat5a/b in directly regulating Foxp3. *Blood.* 2007;109(10):4368–4375.
23. Lio CW, Hsieh CS. A two-step process for thymic regulatory T cell development. *Immunity.* 2008;28(1):100–111.
24. Chen W, Konkel JE. Development of thymic Foxp3(+) regulatory T cells: TGF-beta matters. *Eur J Immunol.* 2015;45(4):958–965.
25. Li MO, Sanjabi S, Flavell RA. Transforming growth factor-beta controls development, homeostasis, and tolerance of T cells by regulatory T cell-dependent and -independent mechanisms. *Immunity.* 2006;25(3):455–471.
26. Marie JC, Liggitt D, Rudensky AY. Cellular mechanisms of fatal early-onset autoimmunity in mice with the T cell-specific targeting of transforming growth factor-beta receptor. *Immunity.* 2006;25(3):441–454.
27. Liu Y, Zhang P, Li J, Kulkarni AB, Perruche S, Chen W. A critical function for TGF-beta signaling in the development of natural CD4+CD25+Foxp3+ regulatory T cells. *Nat Immunol.* 2008;9(6):632–640.
28. Barnes MJ, Krebs P, Harris N, et al. Commitment to the regulatory T cell lineage requires CARMA1 in the thymus but not in the periphery. *PLoS Biol.* 2009;7(3):e51.

29. Rudd CE, Taylor A, Schneider H. CD28 and CTLA-4 coreceptor expression and signal transduction. *Immunol Rev*. 2009;229(1):12–26.
30. Schmidt-Supprian M, Tian J, Grant EP, et al. Differential dependence of CD4+CD25 + regulatory and natural killer-like T cells on signals leading to NF-kappaB activation. *Proc Natl Acad Sci USA*. 2004;101(13):4566–4571.
31. Wan YY, Chi H, Xie M, Schneider MD, Flavell RA. The kinase TAK1 integrates antigen and cytokine receptor signaling for T cell development, survival and function. *Nat Immunol*. 2006;7(8):851–858.
32. Engel M, Sidwell T, Vasanthakumar A, Grigoriadis G, Banerjee A. Thymic regulatory T cell development: role of signalling pathways and transcription factors. *Clin Dev Immunol*. 2013;2013:617595.
33. Isomura I, Palmer S, Grumont RJ, et al. c-Rel is required for the development of thymic Foxp3+ CD4 regulatory T cells. *J Exp Med*. 2009;206(13):3001–3014.
34. Ruan Q, Kameswaran V, Tone Y, et al. Development of Foxp3(+) regulatory t cells is driven by the c-Rel enhanceosome. *Immunity*. 2009;31(6):932–940.
35. Haxhinasto S, Mathis D, Benoist C. The AKT-mTOR axis regulates de novo differentiation of CD4+Foxp3+ cells. *J Exp Med*. 2008;205(3):565–574.
36. Liu G, Burns S, Huang G, et al. The receptor S1P1 overrides regulatory T cell-mediated immune suppression through Akt-mTOR. *Nat Immunol*. 2009;10(7):769–777.
37. Patton DT, Garden OA, Pearce WP, et al. Cutting edge: the phosphoinositide 3-kinase p110 delta is critical for the function of CD4+CD25+Foxp3+ regulatory T cells. *J Immunol*. 2006;177(10):6598–6602.
38. Kerdiles YM, Stone EL, Beisner DR, et al. Foxo transcription factors control regulatory T cell development and function. *Immunity*. 2010;33(6):890–904.
39. Ouyang W, Beckett O, Ma Q, Paik JH, DePinho RA, Li MO. Foxo proteins cooperatively control the differentiation of Foxp3+ regulatory T cells. *Nat Immunol*. 2010;11(7):618–627.
40. Luo CT, Li MO. Transcriptional control of regulatory T cell development and function. *Trends Immunol*. 2013;34(11):531–539.
41. Sekiya T, Kashiwagi I, Yoshida R, et al. Nr4a receptors are essential for thymic regulatory T cell development and immune homeostasis. *Nat Immunol*. 2013; 14(3):230–237.
42. Cowan JE, Jenkinson WE, Anderson G. Thymus medulla fosters generation of natural Treg cells, invariant gammadelta T cells, and invariant NKT cells: what we learn from intrathymic migration. *Eur J Immunol*. 2015;45(3):652–660.
43. Hinterberger M, Wirnsberger G, Klein L. B7/CD28 in central tolerance: costimulation promotes maturation of regulatory T cell precursors and prevents their clonal deletion. *Front Immunol*. 2011;2:30.
44. Akiyama N, Shinzawa M, Miyauchi M, et al. Limitation of immune tolerance-inducing thymic epithelial cell development by Spi-B-mediated negative feedback regulation. *J Exp Med*. 2014;211(12):2425–2438.
45. Hauri-Hohl M, Zuklys S, Hollander GA, Ziegler SF. A regulatory role for TGF-beta signaling in the establishment and function of the thymic medulla. *Nat Immunol*. 2014;15(6):554–561.
46. Bonito AJ, Aloman C, Fiel MI, et al. Medullary thymic epithelial cell depletion leads to autoimmune hepatitis. *J Clin Invest*. 2013;123(8):3510–3524.
47. Mouri Y, Nishijima H, Kawano H, et al. NF-kappaB-inducing kinase in thymic stroma establishes central tolerance by orchestrating cross-talk with not only thymocytes but also dendritic cells. *J Immunol*. 2014;193(9):4356–4367.
48. Cowan JE, Parnell SM, Nakamura K, et al. The thymic medulla is required for Foxp3+ regulatory but not conventional CD4+ thymocyte development. *J Exp Med*. 2013;210(4):675–681.

49. Hinterberger M, Aichinger M, Prazeres da Costa O, Voehringer D, Hoffmann R, Klein L. Autonomous role of medullary thymic epithelial cells in central CD4(+) T cell tolerance. *Nat Immunol.* 2010;11(6):512–519.
50. Aschenbrenner K, D'Cruz LM, Vollmann EH, et al. Selection of Foxp3+ regulatory T cells specific for self antigen expressed and presented by Aire+ medullary thymic epithelial cells. *Nat Immunol.* 2007;8(4):351–358.
51. Perry JS, Lio CW, Kau AL, et al. Distinct contributions of Aire and antigen-presenting-cell subsets to the generation of self-tolerance in the thymus. *Immunity.* 2014;41(3):414–426.
52. Bensinger SJ, Bandeira A, Jordan MS, Caton AJ, Laufer TM. Major histocompatibility complex class II-positive cortical epithelium mediates the selection of CD4(+)25(+) immunoregulatory T cells. *J Exp Med.* 2001;194(4):427–438.
53. Liston A, Nutsch KM, Farr AG, et al. Differentiation of regulatory Foxp3+ T cells in the thymic cortex. *Proc Natl Acad Sci USA.* 2008;105(33):11903–11908.
54. Walters SN, Webster KE, Daley S, Grey ST. A role for intrathymic B cells in the generation of natural regulatory T cells. *J Immunol.* 2014;193(1):170–176.
55. Konkel JE, Jin W, Abbatiello B, Grainger JR, Chen W. Thymocyte apoptosis drives the intrathymic generation of regulatory T cells. *Proc Natl Acad Sci USA.* 2014;111(4):E465–E473.
56. Surh CD, Sprent J. T-cell apoptosis detected in situ during positive and negative selection in the thymus. *Nature.* 1994;372(6501):100–103.
57. Fadok VA, Bratton DL, Konowal A, Freed PW, Westcott JY, Henson PM. Macrophages that have ingested apoptotic cells in vitro inhibit proinflammatory cytokine production through autocrine/paracrine mechanisms involving TGF-beta, PGE2, and PAF. *J Clin Invest.* 1998;101(4):890–898.
58. Xiao YQ, Freire-de-Lima CG, Schiemann WP, Bratton DL, Vandivier RW, Henson PM. Transcriptional and translational regulation of TGF-beta production in response to apoptotic cells. *J Immunol.* 2008;181(5):3575–3585.
59. Mathis D, Benoist C. Aire. *Annu Rev Immunol.* 2009;27:287–312.
60. Anderson MS, Venanzi ES, Chen Z, Berzins SP, Benoist C, Mathis D. The cellular mechanism of Aire control of T cell tolerance. *Immunity.* 2005;23(2):227–239.
61. Anderson MS, Venanzi ES, Klein L, et al. Projection of an immunological self shadow within the thymus by the aire protein. *Science.* 2002;298(5597):1395–1401.
62. Kuroda N, Mitani T, Takeda N, et al. Development of autoimmunity against transcriptionally unrepressed target antigen in the thymus of Aire-deficient mice. *J Immunol.* 2005;174(4):1862–1870.
63. Chen Z, Benoist C, Mathis D. How defects in central tolerance impinge on a deficiency in regulatory T cells. *Proc Natl Acad Sci USA.* 2005;102(41):14735–14740.
64. Kekalainen E, Tuovinen H, Joensuu J, et al. A defect of regulatory T cells in patients with autoimmune polyendocrinopathy-candidiasis-ectodermal dystrophy. *J Immunol.* 2007;178(2):1208–1215.
65. Laakso SM, Laurinolli TT, Rossi LH, et al. Regulatory T cell defect in APECED patients is associated with loss of naive FOXP3(+) precursors and impaired activated population. *J Autoimmun.* 2010;35(4):351–357.
66. Aricha R, Feferman T, Scott HS, Souroujon MC, Berrih-Aknin S, Fuchs S. The susceptibility of Aire(−/−) mice to experimental myasthenia gravis involves alterations in regulatory T cells. *J Autoimmun.* 2011;36(1):16–24.
67. Lei Y, Ripen AM, Ishimaru N, et al. Aire-dependent production of XCL1 mediates medullary accumulation of thymic dendritic cells and contributes to regulatory T cell development. *J Exp Med.* 2011;208(2):383–394.
68. Malchow S, Leventhal DS, Nishi S, et al. Aire-dependent thymic development of tumor-associated regulatory T cells. *Science.* 2013;339(6124):1219–1224.

69. Yang S, Fujikado N, Kolodin D, Benoist C, Mathis D. Immune tolerance. Regulatory T cells generated early in life play a distinct role in maintaining self-tolerance. *Science*. 2015;348(6234):589–594.
70. Daniely D, Kern J, Cebula A, Ignatowicz L. Diversity of TCRs on natural Foxp3+ T cells in mice lacking Aire expression. *J Immunol*. 2010;184(12):6865–6873.
71. Guerau-de-Arellano M, Martinic M, Benoist C, Mathis D. Neonatal tolerance revisited: a perinatal window for Aire control of autoimmunity. *J Exp Med*. 2009;206(6):1245–1252.
72. Thiault N, Darrigues J, Adoue V, et al. Peripheral regulatory T lymphocytes recirculating to the thymus suppress the development of their precursors. *Nat Immunol*. 2015;16(6):628–634.
73. Feuerer M, Hill JA, Kretschmer K, von Boehmer H, Mathis D, Benoist C. Genomic definition of multiple ex vivo regulatory T cell subphenotypes. *Proc Natl Acad Sci USA*. 2010;107(13):5919–5924.
74. Feuerer M, Hill JA, Mathis D, Benoist C. Foxp3+ regulatory T cells: differentiation, specification, subphenotypes. *Nat Immunol*. 2009;10(7):689–695.
75. Huehn J, Siegmund K, Lehmann JC, et al. Developmental stage, phenotype, and migration distinguish naive- and effector/memory-like CD4+ regulatory T cells. *J Exp Med*. 2004;199(3):303–313.
76. Siegmund K, Feuerer M, Siewert C, et al. Migration matters: regulatory T-cell compartmentalization determines suppressive activity in vivo. *Blood*. 2005;106(9):3097–3104.
77. Wang Y, Su MA, Wan YY. An essential role of the transcription factor GATA-3 for the function of regulatory T cells. *Immunity*. 2011;35(3):337–348.
78. Wohlfert EA, Grainger JR, Bouladoux N, et al. GATA3 controls Foxp3(+) regulatory T cell fate during inflammation in mice. *J Clin Invest*. 2011;121(11):4503–4515.
79. Zheng Y, Chaudhry A, Kas A, et al. Regulatory T-cell suppressor program co-opts transcription factor IRF4 to control T(H)2 responses. *Nature*. 2009;458(7236):351–356.
80. Chaudhry A, Rudra D, Treuting P, et al. CD4+ regulatory T cells control TH17 responses in a Stat3-dependent manner. *Science*. 2009;326(5955):986–991.
81. Koch MA, Tucker-Heard G, Perdue NR, Killebrew JR, Urdahl KB, Campbell DJ. The transcription factor T-bet controls regulatory T cell homeostasis and function during type 1 inflammation. *Nat Immunol*. 2009;10(6):595–602.
82. Chung Y, Tanaka S, Chu F, et al. Follicular regulatory T cells expressing Foxp3 and Bcl-6 suppress germinal center reactions. *Nat Med*. 2011;17(8):983–988.
83. Linterman MA, Pierson W, Lee SK, et al. Foxp3+ follicular regulatory T cells control the germinal center response. *Nat Med*. 2011;17(8):975–982.
84. Feuerer M, Herrero L, Cipolletta D, et al. Lean, but not obese, fat is enriched for a unique population of regulatory T cells that affect metabolic parameters. *Nat Med*. 2009;15(8):930–939.
85. Cipolletta D, Cohen P, Spiegelman BM, Benoist C, Mathis D. Appearance and disappearance of the mRNA signature characteristic of Treg cells in visceral adipose tissue: age, diet, and PPARgamma effects. *Proc Natl Acad Sci USA*. 2015;112(2):482–487.
86. Kolodin D, van Panhuys N, Li C, et al. Antigen- and cytokine-driven accumulation of regulatory T cells in visceral adipose tissue of lean mice. *Cell Metab*. 2015;21(4):543–557.
87. Cipolletta D, Kolodin D, Benoist C, Mathis D. Tissular T(regs): a unique population of adipose-tissue-resident Foxp3+CD4+ T cells that impacts organismal metabolism. *Semin Immunol*. 2011;23(6):431–437.
88. Cipolletta D, Feuerer M, Li A, et al. PPAR-gamma is a major driver of the accumulation and phenotype of adipose tissue Treg cells. *Nature*. 2012;486(7404):549–553.
89. Deiuliis J, Shah Z, Shah N, et al. Visceral adipose inflammation in obesity is associated with critical alterations in tregulatory cell numbers. *PLoS One*. 2011;6(1). e16376.

90. Winer S, Chan Y, Paltser G, et al. Normalization of obesity-associated insulin resistance through immunotherapy. *Nat Med*. 2009;15(8):921–929.
91. Ilan Y, Maron R, Tukpah AM, et al. Induction of regulatory T cells decreases adipose inflammation and alleviates insulin resistance in ob/ob mice. *Proc Natl Acad Sci USA*. 2010;107(21):9765–9770.
92. Eller K, Kirsch A, Wolf AM, et al. Potential role of regulatory T cells in reversing obesity-linked insulin resistance and diabetic nephropathy. *Diabetes*. 2011;60(11):2954–2962.
93. Medrikova D, Sijmonsma TP, Sowodniok K, et al. Brown adipose tissue harbors a distinct sub-population of regulatory T cells. *PLoS One*. 2015;10(2). e0118534.
94. Antiga E, Quaglino P, Bellandi S, et al. Regulatory T cells in the skin lesions and blood of patients with systemic sclerosis and morphoea. *Br J Dermatol*. 2010;162(5):1056–1063.
95. Sather BD, Treuting P, Perdue N, et al. Altering the distribution of Foxp3(+) regulatory T cells results in tissue-specific inflammatory disease. *J Exp Med*. 2007;204(6):1335–1347.
96. Dudda JC, Perdue N, Bachtanian E, Campbell DJ. Foxp3+ regulatory T cells maintain immune homeostasis in the skin. *J Exp Med*. 2008;205(7):1559–1565.
97. Loser K, Beissert S. Regulation of cutaneous immunity by the environment: an important role for UV irradiation and vitamin D. *Int Immunopharmacol*. 2009;9(5):587–589.
98. Yoshiki R, Kabashima K, Sakabe J, et al. The mandatory role of IL-10-producing and OX40 ligand-expressing mature Langerhans cells in local UVB-induced immunosuppression. *J Immunol*. 2010;184(10):5670–5677.
99. Rosenblum MD, Gratz IK, Paw JS, Lee K, Marshak-Rothstein A, Abbas AK. Response to self antigen imprints regulatory memory in tissues. *Nature*. 2011;480(7378):538–542.
100. Sanchez Rodriguez R, Pauli ML, Neuhaus IM, et al. Memory regulatory T cells reside in human skin. *J Clin Invest*. 2014;124(3):1027–1036.
101. Honda T, Miyachi Y, Kabashima K. Regulatory T cells in cutaneous immune responses. *J Dermatol Sci*. 2011;63(2):75–82.
102. Gaspari AA. Innate and adaptive immunity and the pathophysiology of psoriasis. *J Am Acad Dermatol*. 2006;54(3 suppl 2):S67–S80.
103. Zhang L, Yang XQ, Cheng J, Hui RS, Gao TW. Increased Th17 cells are accompanied by FoxP3(+) Treg cell accumulation and correlated with psoriasis disease severity. *Clin Immunol*. 2010;135(1):108–117.
104. Zhang K, Li X, Yin G, Liu Y, Niu X, Hou R. Functional characterization of CD4+CD25+ regulatory T cells differentiated in vitro from bone marrow-derived haematopoietic cells of psoriasis patients with a family history of the disorder. *Br J Dermatol*. 2008;158(2):298–305.
105. Sugiyama H, Gyulai R, Toichi E, et al. Dysfunctional blood and target tissue CD4+CD25high regulatory T cells in psoriasis: mechanism underlying unrestrained pathogenic effector T cell proliferation. *J Immunol*. 2005;174(1):164–173.
106. Bovenschen HJ, van de Kerkhof PC, van Erp PE, Woestenenk R, Joosten I, Koenen HJ. Foxp3+ regulatory T cells of psoriasis patients easily differentiate into IL-17A-producing cells and are found in lesional skin. *J Invest Dermatol*. 2011;131(9):1853–1860.
107. Quaglino P, Antiga E, Comessatti A, et al. Circulating CD4+ CD25brightFOXP3+ regulatory T-cells are significantly reduced in bullous pemphigoid patients. *Arch Dermatol Res*. 2012;304(8):639–645.
108. Antiga E, Kretz CC, Klembt R, et al. Characterization of regulatory T cells in patients with dermatomyositis. *J Autoimmun*. 2010;35(4):342–350.

109. Joe B, Letwin NE, Garrett MR, et al. Transcriptional profiling with a blood pressure QTL interval-specific oligonucleotide array. *Physiol Genomics*. 2005;23(3): 318–326.
110. Trujillo G, Hartigan AJ, Hogaboam CM. T regulatory cells and attenuated bleomycin-induced fibrosis in lungs of CCR7 −/− mice. *Fibrogenesis Tissue Repair*. 2010;3:18.
111. Purwar R, Campbell J, Murphy G, Richards WG, Clark RA, Kupper TS. Resident memory T cells (T(RM)) are abundant in human lung: diversity, function, and antigen specificity. *PLoS One*. 2011;6(1). e16245.
112. Busse WW, Lemanske Jr RF. Asthma. *N Engl J Med*. 2001;344(5):350–362.
113. Hartl D, Koller B, Mehlhorn AT, et al. Quantitative and functional impairment of pulmonary CD4+CD25hi regulatory T cells in pediatric asthma. *J Allergy Clin Immunol*. 2007;119(5):1258–1266.
114. Pesenacker AM, Broady R, Levings MK. Control of tissue-localized immune responses by human regulatory T cells. *Eur J Immunol*. 2015;45(2):333–343.
115. Ulges A, Klein M, Reuter S, et al. Protein kinase CK2 enables regulatory T cells to suppress excessive TH2 responses in vivo. *Nat Immunol*. 2015;16(3):267–275.
116. Tiegs G, Lohse AW. Immune tolerance: what is unique about the liver. *J Autoimmun*. 2010;34(1):1–6.
117. Calne RY, Sells RA, Pena JR, et al. Induction of immunological tolerance by porcine liver allografts. *Nature*. 1969;223(5205):472–476.
118. Benseler V, McCaughan GW, Schlitt HJ, Bishop GA, Bowen DG, Bertolino P. The liver: a special case in transplantation tolerance. *Semin Liver Dis*. 2007;27(2):194–213.
119. Luth S, Huber S, Schramm C, et al. Ectopic expression of neural autoantigen in mouse liver suppresses experimental autoimmune neuroinflammation by inducing antigen-specific Tregs. *J Clin Invest*. 2008;118(10):3403–3410.
120. Ma X, Hua J, Mohamood AR, Hamad AR, Ravi R, Li Z. A high-fat diet and regulatory T cells influence susceptibility to endotoxin-induced liver injury. *Hepatology*. 2007;46(5):1519–1529.
121. Ward SM, Fox BC, Brown PJ, et al. Quantification and localisation of FOXP3+ T lymphocytes and relation to hepatic inflammation during chronic HCV infection. *J Hepatol*. 2007;47(3):316–324.
122. Wei HX, Chuang YH, Li B, et al. CD4+ CD25+ Foxp3+ regulatory T cells protect against T cell-mediated fulminant hepatitis in a TGF-beta-dependent manner in mice. *J Immunol*. 2008;181(10):7221–7229.
123. Breous E, Somanathan S, Vandenberghe LH, Wilson JM. Hepatic regulatory T cells and Kupffer cells are crucial mediators of systemic T cell tolerance to antigens targeting murine liver. *Hepatology*. 2009;50(2):612–621.
124. Erhardt A, Biburger M, Papadopoulos T, Tiegs G. IL-10, regulatory T cells, and Kupffer cells mediate tolerance in concanavalin A-induced liver injury in mice. *Hepatology*. 2007;45(2):475–485.
125. Longhi MS, Ma Y, Bogdanos DP, Cheeseman P, Mieli-Vergani G, Vergani D. Impairment of CD4(+)CD25(+) regulatory T-cells in autoimmune liver disease. *J Hepatol*. 2004;41(1):31–37.
126. Longhi MS, Hussain MJ, Mitry RR, et al. Functional study of CD4+CD25+ regulatory T cells in health and autoimmune hepatitis. *J Immunol*. 2006;176(7):4484–4491.
127. Gao Q, Qiu SJ, Fan J, et al. Intratumoral balance of regulatory and cytotoxic T cells is associated with prognosis of hepatocellular carcinoma after resection. *J Clin Oncol*. 2007;25(18):2586–2593.
128. Kobayashi N, Hiraoka N, Yamagami W, et al. FOXP3+ regulatory T cells affect the development and progression of hepatocarcinogenesis. *Clin Cancer Res*. 2007;13(3):902–911.

129. Rushbrook SM, Ward SM, Unitt E, et al. Regulatory T cells suppress in vitro proliferation of virus-specific CD8+ T cells during persistent hepatitis C virus infection. *J Virol.* 2005;79(12):7852–7859.
130. Boettler T, Spangenberg HC, Neumann-Haefelin C, et al. T cells with a CD4+CD25 + regulatory phenotype suppress in vitro proliferation of virus-specific CD8+ T cells during chronic hepatitis C virus infection. *J Virol.* 2005;79(12):7860–7867.
131. Wildin RS, Smyk-Pearson S, Filipovich AH. Clinical and molecular features of the immunodysregulation, polyendocrinopathy, enteropathy, X linked (IPEX) syndrome. *J Med Genet.* 2002;39(8):537–545.
132. Feuerer M, Shen YL, Littman DR, Benoist C, Mathis D. How punctual ablation of regulatory T cells unleashes an autoimmune lesion within the pancreatic islets. *Immunity.* 2009;31(4):654–664.
133. Grinberg-Bleyer Y, Baeyens A, You S, et al. IL-2 reverses established type 1 diabetes in NOD mice by a local effect on pancreatic regulatory T cells. *J Exp Med.* 2010;207(9):1871–1878.
134. Kornete M, Sgouroudis E, Piccirillo CA. ICOS-dependent homeostasis and function of Foxp3+ regulatory T cells in islets of nonobese diabetic mice. *J Immunol.* 2012;188(3):1064–1074.
135. Herman AE, Freeman GJ, Mathis D, Benoist C. CD4+CD25+ T regulatory cells dependent on ICOS promote regulation of effector cells in the prediabetic lesion. *J Exp Med.* 2004;199(11):1479–1489.
136. Chen Z, Herman AE, Matos M, Mathis D, Benoist C. Where CD4+CD25+ T reg cells impinge on autoimmune diabetes. *J Exp Med.* 2005;202(10):1387–1397.
137. van Halteren AG, Kardol MJ, Mulder A, Roep BO. Homing of human autoreactive T cells into pancreatic tissue of NOD-scid mice. *Diabetologia.* 2005;48(1):75–82.
138. Montane J, Bischoff L, Soukhatcheva G, et al. Prevention of murine autoimmune diabetes by CCL22-mediated Treg recruitment to the pancreatic islets. *J Clin Invest.* 2011;121(8):3024–3028.
139. Baeyens A, Saadoun D, Billiard F, et al. Effector T cells boost regulatory T cell expansion by IL-2, TNF, OX40, and plasmacytoid dendritic cells depending on the immune context. *J Immunol.* 2015;194(3):999–1010.
140. Tang Q, Adams JY, Penaranda C, et al. Central role of defective interleukin-2 production in the triggering of islet autoimmune destruction. *Immunity.* 2008;28(5):687–697.
141. Hiraoka N, Onozato K, Kosuge T, Hirohashi S. Prevalence of FOXP3+ regulatory T cells increases during the progression of pancreatic ductal adenocarcinoma and its premalignant lesions. *Clin Cancer Res.* 2006;12(18):5423–5434.
142. Liyanage UK, Moore TT, Joo HG, et al. Prevalence of regulatory T cells is increased in peripheral blood and tumor microenvironment of patients with pancreas or breast adenocarcinoma. *J Immunol.* 2002;169(5):2756–2761.
143. Eghtesad S, Jhunjhunwala S, Little SR, Clemens PR. Rapamycin ameliorates dystrophic phenotype in mdx mouse skeletal muscle. *Mol Med.* 2011;17(9-10):917–924.
144. Foulds CE, Tsimelzon A, Long W, et al. Research resource: expression profiling reveals unexpected targets and functions of the human steroid receptor RNA activator (SRA) gene. *Mol Endocrinol.* 2010;24(5):1090–1105.
145. Burzyn D, Kuswanto W, Kolodin D, et al. A special population of regulatory T cells potentiates muscle repair. *Cell.* 2013;155(6):1282–1295.
146. Villalta SA, Rosenthal W, Martinez L, et al. Regulatory T cells suppress muscle inflammation and injury in muscular dystrophy. *Sci Transl Med.* 2014;6(258):258ra142.
147. Vetrone SA, Montecino-Rodriguez E, Kudryashova E, et al. Osteopontin promotes fibrosis in dystrophic mouse muscle by modulating immune cell subsets and intramuscular TGF-beta. *J Clin Invest.* 2009;119(6):1583–1594.

148. Littman DR, Rudensky AY. Th17 and regulatory T cells in mediating and restraining inflammation. *Cell*. 2010;140(6):845–858.
149. Smith PM, Howitt MR, Panikov N, et al. The microbial metabolites, short-chain fatty acids, regulate colonic Treg cell homeostasis. *Science*. 2013;341(6145):569–573.
150. Round JL, Mazmanian SK. Inducible Foxp3+ regulatory T-cell development by a commensal bacterium of the intestinal microbiota. *Proc Natl Acad Sci USA*. 2010;107(27):12204–12209.
151. Atarashi K, Tanoue T, Shima T, et al. Induction of colonic regulatory T cells by indigenous Clostridium species. *Science*. 2011;331(6015):337–341.
152. Geuking MB, Cahenzli J, Lawson MA, et al. Intestinal bacterial colonization induces mutualistic regulatory T cell responses. *Immunity*. 2011;34(5):794–806.

CHAPTER NINE

Regulatory T cells in Arthritis

Noriko Komatsu*, Hiroshi Takayanagi*,†,1
*Department of Immunology, Graduate School of Medicine and Faculty of Medicine, The University of Tokyo, Bunkyo-ku, Tokyo, Japan
†Japan Science and Technology Agency, Exploratory Research for Advanced Technology Program, Takayanagi Osteonetwork Project, Bunkyo-ku, Tokyo, Japan
[1]Corresponding author: e-mail address: takayana@m.u-tokyo.ac.jp

Contents

Abstract

Rheumatoid arthritis (RA) is one of the most common autoimmune diseases and is characterized by inflammation and subsequent bone destruction in multiple joints. In mice, depletion of regulatory T (Treg) cells results in the onset of a variety of autoimmune diseases including arthritis, while replenishment of Treg cells alleviates arthritic symptoms. The importance of Treg cells in RA is supported by the effectiveness of CTLA4-Ig therapy, an increased Treg cell/effector T cell ratio after anti-IL-6R or anti-TNF-α treatment and the identification of CTLA 4 as an RA-associated gene. Thus, Treg cells constitute a useful target in the treatment of RA. $Foxp3^+$ T cells consist of heterogeneous populations in terms of their surface markers, suppressive function, and plasticity. Plastic $Foxp3^+$ T cells are able to convert into pathogenic Th17 cells, which have been shown to exacerbate arthritis in mice. Therefore, it is important to identify a stable suppressive $Foxp3^+$ Treg cell subpopulation along with suppressive molecules and surface markers. In addition, considering the recent studies on the identification of arthritic antigens, the generation of antigen-specific Treg cells from naïve $CD4^+$ T cells or effector T cells is now feasible, along with the induction of Foxp3 and stabilization of the suppressive function by epigenetic modification of Treg cell signature genes. These approaches will lead to the establishment of new therapeutic strategies against arthritis that work by increasing the Treg cell/effector T cell ratio in favor of Treg cells. Here, we summarize our understanding of the role of Treg cells in arthritis based on recent human and murine studies.

Progress in Molecular Biology and Translational Science, Volume 136
ISSN 1877-1173
http://dx.doi.org/10.1016/bs.pmbts.2015.07.021

1. INTRODUCTION

Rheumatoid arthritis (RA) is an autoimmune disease with a high degree of physical morbidity characterized by joint inflammation which ultimately leads to bone destruction. Although the etiology of RA has not been well elucidated, both genetic and environmental factors are believed to be involved.[1] T cells are known to play an important role in RA pathogenesis for a number of reasons, as follows: (1) T cells accumulate in inflamed joints; (2) autoantibodies are detected in RA patients; (3) SNPs in HLA-DR- and T cell-related genes, such as CCR6 and PTPN22, are associated with RA; (4) treatments targeting T cell molecules are effective in RA; and (5) studies in mouse models of arthritis have shown that T cells are indispensable for arthritis.[1–5] In mice, depletion of regulatory T (Treg) cells causes multiple organ-specific autoimmune diseases, including arthritis.[6]

Treg cells express the transcription factor Foxp3 and play a key role in immunological tolerance. Foxp3 is essential for the differentiation and function of Treg cells as Foxp3 mutants, e.g., immunodysregulation, polyendocrinopathy, enteropathy X-linked (IPEX) syndrome patients and scurfy mice, develop multiple fatal autoimmune diseases due to the lack of functional Treg cells.[7–11] In mice, the depletion of Treg cells during adulthood also results in the development of fatal autoimmune diseases, indicating that Treg cells are necessary for immune homeostasis throughout the lifespan.[12] Furthermore, Treg cells have been shown to be protective against arthritis using mouse models of arthritis that recapitulate RA.[13,14] In humans, recent reports from RA patients also suggest the importance of Treg cells in RA. Here, we summarize recent advances in the understanding of Treg cells in RA and discuss the potential use of Treg cell-targeted therapies in the future.

2. TREG CELLS IN THE INFLAMMATION THAT OCCURS IN RA

Treg cells highly express CD25 and CTLA-4, which are thus commonly regarded as Treg cell-marker genes. Early candidate gene association studies in RA suggested CTLA-4 to be an RA susceptibility gene,[15] while defects in CTLA-4 are associated with abnormal Treg cell function in RA.[16] Furthermore, recent genome-wide association studies with a combination of epigenetic analyses have suggested that among the immune cell subsets,

Treg cells exhibit the greatest change in methylation status of RA risk loci.[17] These studies suggest that Treg cells play an important role in the suppression of onset and/or the progression of RA.

In mice, Treg cells were shown to be protective against arthritis, since the transfer of $CD4^+CD25^+$ Treg cells inhibited arthritis, while the deletion of $Foxp3^+$ Treg cells resulted in exacerbation of the disease.[13,14] To date, there have been numerous studies investigating the number and function of Treg cells obtained from the peripheral blood or synovial fluid of RA patients. Although the frequency of Treg cells in the peripheral blood in RA patients varies among the reports, most studies have consistently reported that Treg cells are abundant in synovial fluid.[18] Arthritis, however, does worsen in the presence of Treg cells, possibly due to the higher number of effector T cells present. It is reported that inflammatory cytokines, such as TNF-α, inhibit Treg cell function,[19] and Treg cells from RA patients treated with anti-TNF-α therapy recover their previously lost suppressive function.[20] TNF-α is reported to inhibit Treg cell function by increasing PKC-θ recruitment to the immunological synapse in Treg cells,[21] while a recent study also demonstrated that TNF-α inhibits the phosphorylation of FOXP3,[22] resulting in the inactivation of FOXP3 under arthritic conditions. In addition, under inflammatory conditions, T effector cells acquire resistance to Treg cell-mediated suppression in a PKB/c-Akt-dependent manner.[23] These studies suggest that the suppressive function of Treg cells may not be sufficient at the site of active inflammation. Although many studies reported that Treg cells retain their suppressive function *in vitro*,[24–27] this may be due to a lack of the proper inflammatory cytokines under these non-physiological conditions.

CTLA-4 prevents the CD28-mediated activation of T cells by competitively binding to CD80/CD86 or downregulating CD80/CD86 on dendritic cells.[28–30] Treg cell-specific deletion of CTLA-4 leads to the onset of fatal autoimmune diseases, resembling the symptoms of both Foxp3-deficient and CTLA-4-deficient mice, indicating that CTLA-4 is indispensable for Treg cell suppressive function.[29–32] The efficacy of CTLA4-Ig in RA treatment may be explained by the complementary effect of Treg cell function.

Additionally, several studies have reported that anti-IL-6 and anti-TNF-α treatments are accompanied by an increased Treg cell/effector T cell ratio, suggesting that these therapies work in part because they enable Treg cells to recover their suppressive function.[33,34] Taken together, Treg cells represent an attractive therapeutic target in RA.

3. TREG CELLS IN THE BONE DESTRUCTION IN RA

Inflammation in the joints of RA patients is frequently accompanied by severe bone destruction. Osteoclasts, which are differentiated from the monocyte/macrophage lineages in the presence of receptor activator of NF-κB ligand (RANKL) and macrophage colony-stimulating factor (M-CSF), are responsible for the bone destruction that occurs in RA.[35,36] Compared to inflammation, fewer studies have reported a role for Treg cells in the control of bone destruction. Treg cells suppress RANKL-induced osteoclastogenesis through several different mechanisms that may be dependent on cytokines such as IL-10 or dependent on cell–cell contact via CTLA-4.[37,38] A recent study reported that CTLA-4 induced the apoptosis of osteoclast precursors though CD80/86 and indoleamine 2,3-dioxygenase (IDO)-dependent pathways. In addition, osteoclast precursor cells were decreased in the peripheral blood of RA patients treated with CTLA4-Ig.[39,40] Thus, one of the protective mechanisms of CTLA4-Ig may be the direct inhibition of osteoclastogenesis. *In vivo* investigation demonstrated that the bone loss in arthritic mice is suppressed by the transfer of bone marrow cells from Foxp3-Tg mice and augmented by the transfer of Foxp3-deficient bone marrow cells.[41] Additionally, adoptive transfer of Treg cells into T cell-deficient mice leads to an increase in bone volume.[42] Collectively, these studies suggest that an increased number of Treg cells results in the suppression of osteoclastogenesis and increased bone volume *in vivo*. Recently, it has been shown that the adaptive immune system regulates bone metabolism through the direct effect of immune complexes on osteoclasts.[43] Given that Treg cells suppress the T cell production of autoantibodies, Treg cells may also protect bone by reducing the amount of immune complexes.

4. THE PLASTICITY OF FOXP3$^+$ T CELL IN ARTHRITIS

Treg cells consist of thymus-derived Treg (tTreg) cells and peripherally derived Treg (pTreg) cells. Since tTreg cells are known to contain self-reactive T cells, the transdifferentiation of Treg cells may have pathogenic consequences.[44] Thus, the stability of Foxp3 in Treg cells under inflammatory conditions has attracted considerable attention recently, but is still controversial.[45–47] By cell transfer of Foxp3$^+$ T cells and tracking of Foxp3$^+$ T cell fate *in vivo*, Foxp3$^+$ T cells, particularly those with low CD25 expression, were shown to convert into Th17 cells in mice afflicted with collagen-induced arthritis (CIA). Th17 cells differentiated from Foxp3$^+$ T cells (termed exFoxp3Th17 cells) accumulated in the inflamed joints, where the conversion was promoted by the IL-6 produced by arthritic synovial fibroblasts.[48] Epigenetic analysis demonstrated that exFoxp3 T cells are not derived from tTreg cells or conventional T cells, and it is suggested that they are derived from pTreg cells, but further analysis is required to derive a clear conclusion. Microarray analysis revealed that exFoxp3Th17 cells expressed genes distinct from the Th17 cells differentiated from naïve CD4$^+$ T cells. In particular, exFoxp3Th17 cells expressed higher levels of Sox4, CCR6, CCL20, IL-23R, and RANKL, some of which have been shown to contribute to the pathogenesis of arthritis. Interestingly, in the presence of synovial fibroblasts, exFoxp3Th17 cells were able to induce osteoclast generation more potently than any other T cell subset previously reported. Although exFoxp3Th17 cells express more RANKL than any other T cell subset, the RANKL expressed on fibroblasts was recently shown to play a more important role in arthritic bone destruction than RANKL on T cells.[49] Collagen-exposed CD25^{lo}Foxp3$^+$ T cells exhibited exacerbated inflammation and bone destruction in arthritis, suggesting that CD25^{lo}Foxp3$^+$ T cells contain T cells which recognize self-antigens and convert into self-reactive arthritogenic T cells after losing Foxp3. Thus, exFoxp3Th17 cells constitute a newly identified arthritogenic Th17 population.[48]

In humans, it is not possible to chase the fate of Foxp3$^+$ T cells. Instead, Foxp3$^+$IL-17$^+$ T cells, which are considered to be in a transient state during conversion into Th17 cells, can be evaluated. Foxp3$^+$IL-17$^+$ T cells are commonly observed in the active RA synovium, supporting the notion that plastic FOXP3$^+$ T cells may also convert into arthritogenic T cells in RA.

5. TREG CELL-TARGETED THERAPIES AGAINST RA

In general, increasing the Treg cell number or enhancing Treg cell suppressive function may prove to be beneficial in the suppression of autoimmune diseases, including arthritis. However, increasing the Foxp3$^+$ T cell number may not be efficient, since it is well known that Foxp3 can be expressed by non-Treg cells and thus Foxp3$^+$ T cells are comprised of both functionally suppressive Treg cells and conventional T cells.[50,51] Therefore, it is important to determine the contribution of each Foxp3-expressing population and to identify the stably suppressive Treg subpopulation along with its functional molecules or surface markers.

The generation of antigen-specific Treg cells from naïve $CD4^+$ T cells or effector T cells would be a highly useful approach. A recent study identified the 60S ribosomal protein L23a (RPL23A) as an arthritis-causative antigen using SKG mice, which spontaneously develop arthritis due to defective T cell selection in the thymus. Notably, a T cell response and autoantibodies toward this antigen have been detected in RA patients.[52] It is a matter of great importance that methods for the identification of causative antigen(s) in individual RA patients be established in the future. Another recent interesting study has shown that Th17 cells transdifferentiate into Treg cells under inflammatory conditions.[53] In this report, using fate-mapping mice, authors chased the fate of IL-17-producing cells, demonstrating that Th17 cells lost IL-17 expression and were converted into Tr1 cells, a Treg cell subset that produces a large amount of IL-10, but expresses neither CD25 nor Foxp3. In addition, the Tr1 cells that were converted from Th17 cells were shown to be functional *in vivo*, although the precise mechanism for this conversion remains to be determined. *Bona fide* Treg cells are characterized by the expression of Foxp3 together with demethylation of the Treg cell-marker genes. Thus, it is feasible that antigen-specific Treg cells that stably exert suppressive function can be generated from naïve $CD4^+$ T cells or even effector T cells using arthritogenic antigens along with the induction of Foxp3 and epigenetic modifications.

Finally, the fate of the plastic $Foxp3^+$ T cells may be a critical determinant of both amelioration and progression in arthritis. The therapeutic effects of anti-IL-6 and anti-TNF-α may be partly attributable to the regulation of the transition of plastic $Foxp3^+$ T cells into Treg cells. The molecular basis underlying the differentiation and function of arthritogenic exFoxp3Th17 cells will provide critically important information on advancing therapeutic strategies against arthritis.

6. CONCLUSIONS

Accumulating evidence from human and murine studies suggests that Treg cells play a protective role in arthritis. However, in inflamed joints, where proinflammatory cytokines are abundant, $Foxp3^+$ Treg cells may not be able to adequately suppress ongoing symptoms because of pathogenic conversion, Foxp3 dysfunctionality, or resistance of effector T cells to Treg cell suppression. Why is it that exFoxp3Th17 cells, potentially "harmful" plastic cells, are needed? Continuous immunosuppression can have obviously deleterious effects, such as the progression of cancer and infectious diseases. Thus, it is possible that the immune system may be equipped with mechanistic flexibility in order to adapt to changes in microenvironments, hence, the need for plasticity. Further studies focusing on Treg cell biology will improve our understanding of their role in health and disease as well as help in the establishment of new therapies against various autoimmune diseases, including RA.

ACKNOWLEDGMENT

We would like to thank Stephanie Win for helpful discussion.

REFERENCES

1. Firestein GS. Evolving concepts of rheumatoid arthritis. *Nature*. 2003;423:356–361.
2. Kochi Y, Okada Y, Suzuki A, et al. A regulatory variant in CCR6 is associated with rheumatoid arthritis susceptibility. *Nat Genet*. 2010;42:515–519.
3. Begovich AB, Carlton VE, Honigberg LA, et al. A missense single-nucleotide polymorphism in a gene encoding a protein tyrosine phosphatase (PTPN22) is associated with rheumatoid arthritis. *Am J Hum Genet*. 2004;75:330–337.
4. Lee AT, Li W, Liew A, et al. The PTPN22 R620W polymorphism associates with RF positive rheumatoid arthritis in a dose-dependent manner but not with HLA-SE status. *Genes Immun*. 2005;6:129–133.
5. Moreland LW, Alten R, Van den Bosch F, et al. Costimulatory blockade in patients with rheumatoid arthritis: a pilot, dose-finding, double-blind, placebo-controlled clinical trial evaluating CTLA-4Ig and LEA29Y eighty-five days after the first infusion. *Arthritis Rheum*. 2002;46:1470–1479.
6. Sakaguchi S, Sakaguchi N, Asano M, et al. Immunologic self-tolerance maintained by activated T cells expressing IL-2 receptor alpha-chains (CD25). Breakdown of a single mechanism of self-tolerance causes various autoimmune diseases. *J Immunol*. 1995;155:1151–1164.
7. Sakaguchi S, Yamaguchi T, Nomura T, Ono M. Regulatory T cells and immune tolerance. *Cell*. 2008;133:775–787.
8. Hori S, Nomura T, Sakaguchi S. Control of regulatory T cell development by the transcription factor Foxp3. *Science*. 2003;299:1057–1061.
9. Fontenot JD, Gavin MA, Rudensky AY. Foxp3 programs the development and function of CD4+CD25+ regulatory T cells. *Nat Immunol*. 2003;4:330–336.
10. Khattri R, Cox T, Yasayko SA, Ramsdell F. An essential role for Scurfin in CD4+ CD25+ T regulatory cells. *Nat Immunol*. 2003;4:337–342.
11. Godfrey VL, Wilkinson JE, Russell LB. X-linked lymphoreticular disease in the scurfy (sf) mutant mouse. *Am J Pathol*. 1991;138:1379–1387.
12. Kim JM, Rasmussen JP, Rudensky AY. Regulatory T cells prevent catastrophic autoimmunity throughout the lifespan of mice. *Nat Immunol*. 2007;8:191–197.
13. Morgan ME, Flierman R, van Duivenvoorde LM, et al. Effective treatment of collagen-induced arthritis by adoptive transfer of CD25+ regulatory T cells. *Arthritis Rheum*. 2005;52:2212–2221.
14. Nguyen LT, Jacobs J, Mathis D, Benoist C. Where FoxP3-dependent regulatory T cells impinge on the development of inflammatory arthritis. *Arthritis Rheum*. 2007;56:509–520.

15. Plenge RM, Padyukov L, Remmers EF, et al. Replication of putative candidate-gene associations with rheumatoid arthritis in >4,000 samples from North America and Sweden: association of susceptibility with PTPN22, CTLA4, and PADI4. *Am J Hum Genet.* 2005;77:1044–1060.
16. Flores-Borja F, Jury EC, Mauri C, Ehrenstein MR. Defects in CTLA-4 are associated with abnormal regulatory T cell function in rheumatoid arthritis. *Proc Natl Acad Sci USA.* 2008;105:19396–19401.
17. Okada Y, Wu D, Trynka G, et al. Genetics of rheumatoid arthritis contributes to biology and drug discovery. *Nature.* 2014;506:376–381.
18. Miyara M, Gorochov G, Ehrenstein M, et al. Human FoxP3+ regulatory T cells in systemic autoimmune diseases. *Autoimmun Rev.* 2011;10:744–755.
19. Valencia X, Stephens G, Goldbach-Mansky R, et al. TNF down modulates the function of human CD4+CD25hi T-regulatory cells. *Blood.* 2006;108:253–261.
20. Ehrenstein MR, Evans JG, Singh A, et al. Compromised function of regulatory T cells in rheumatoid arthritis and reversal by anti-TNFalpha therapy. *J Exp Med.* 2004;200:277–285.
21. Zanin-Zhorov A, Ding Y, Kumari S, et al. Protein kinase C-theta mediates negative feedback on regulatory T cell function. *Science.* 2010;328:372–376.
22. Nie H, Zheng Y, Li R, et al. Phosphorylation of FOXP3 controls regulatory T cell function and is inhibited by TNF-alpha in rheumatoid arthritis. *Nat Med.* 2013;19:322–328.
23. Wehrens EJ, Mijnheer G, Duurland CL, et al. Functional human regulatory T cells fail to control autoimmune inflammation due to PKB/c-akt hyperactivation in effector cells. *Blood.* 2011;118:3538–3548.
24. Cao D, van Vollenhoven R, Klareskog L, et al. CD25brightCD4+ regulatory T cells are enriched in inflamed joints of patients with chronic rheumatic disease. *Arthritis Res Ther.* 2004;6:R335–R346.
25. Jiao Z, Wang W, Jia R, et al. Accumulation of FoxP3-expressing CD4+CD25+ T cells with distinct chemokine receptors in synovial fluid of patients with active rheumatoid arthritis. *Scand J Rheumatol.* 2007;36:428–433.
26. Liu MF, Wang CR, Fung LL, et al. The presence of cytokine-suppressive CD4+CD25+ T cells in the peripheral blood and synovial fluid of patients with rheumatoid arthritis. *Scand J Immunol.* 2005;62:312–317.
27. van Amelsfort JM, Jacobs KM, Bijlsma JW, et al. CD4(+)CD25(+) regulatory T cells in rheumatoid arthritis: differences in the presence, phenotype, and function between peripheral blood and synovial fluid. *Arthritis Rheum.* 2004;50:2775–2785.
28. Krummel MF, Allison JP. CTLA-4 engagement inhibits IL-2 accumulation and cell cycle progression upon activation of resting T cells. *J Exp Med.* 1996;183:2533–2540.
29. Walunas TL, Bakker CY, Bluestone JA. CTLA-4 ligation blocks CD28-dependent T cell activation. *J Exp Med.* 1996;183:2541–2550.
30. Wing K, Onishi Y, Prieto-Martin P, et al. CTLA-4 control over Foxp3+ regulatory T cell function. *Science.* 2008;322:271–275.
31. Waterhouse P, Penninger JM, Timms E, et al. Lymphoproliferative disorders with early lethality in mice deficient in Ctla-4. *Science.* 1995;270:985–988.
32. Tivol EA, Borriello F, Schweitzer AN, et al. Loss of CTLA-4 leads to massive lymphoproliferation and fatal multiorgan tissue destruction, revealing a critical negative regulatory role of CTLA-4. *Immunity.* 1995;3:541–547.
33. Nadkarni S, Mauri C, Ehrenstein MR. Anti-TNF-alpha therapy induces a distinct regulatory T cell population in patients with rheumatoid arthritis via TGF-beta. *J Exp Med.* 2007;204:33–39.
34. Samson M, Audia S, Janikashvili N, et al. Brief report: inhibition of interleukin-6 function corrects Th17/Treg cell imbalance in patients with rheumatoid arthritis. *Arthritis Rheum.* 2012;64:2499–2503.

35. Takayanagi H. Osteoimmunology: shared mechanisms and crosstalk between the immune and bone systems. *Nat Rev Immunol.* 2007;7:292–304.
36. Takayanagi H. SnapShot: osteoimmunology. *Cell Metab.* 2015;21. e501–502.
37. Kim YG, Lee CK, Nah SS, et al. Human CD4+CD25+ regulatory T cells inhibit the differentiation of osteoclasts from peripheral blood mononuclear cells. *Biochem Biophys Res Commun.* 2007;357:1046–1052.
38. Zaiss MM, Axmann R, Zwerina J, et al. Treg cells suppress osteoclast formation: a new link between the immune system and bone. *Arthritis Rheum.* 2007;56:4104–4112.
39. Bozec A, Zaiss MM, Kagwiria R, et al. T cell costimulation molecules CD80/86 inhibit osteoclast differentiation by inducing the IDO/tryptophan pathway. *Sci Transl Med.* 2014;6:235ra260.
40. Takayanagi H. Osteoimmunology in 2014: two-faced immunology-from osteogenesis to bone resorption. *Nat Rev Rheumatol.* 2015;11:74–76.
41. Zaiss MM, Frey B, Hess A, et al. Regulatory T cells protect from local and systemic bone destruction in arthritis. *J Immunol.* 2010;184:7238–7246.
42. Zaiss MM, Sarter K, Hess A, et al. Increased bone density and resistance to ovariectomy-induced bone loss in FoxP3-transgenic mice based on impaired osteoclast differentiation. *Arthritis Rheum.* 2010;62:2328–2338.
43. Negishi-Koga T, Gober HJ, Sumiya E, et al. Immune complexes regulate bone metabolism through FcRgamma signalling. *Nat Commun.* 2015;6:6637.
44. Jordan MS, Boesteanu A, Reed AJ, et al. Thymic selection of CD4+CD25+ regulatory T cells induced by an agonist self-peptide. *Nat Immunol.* 2001;2:301–306.
45. Zhou X, Bailey-Bucktrout SL, Jeker LT, et al. Instability of the transcription factor Foxp3 leads to the generation of pathogenic memory T cells in vivo. *Nat Immunol.* 2009;10:1000–1007.
46. Rubtsov YP, Niec RE, Josefowicz S, et al. Stability of the regulatory T cell lineage in vivo. *Science.* 2010;329:1667–1671.
47. Miyao T, Floess S, Setoguchi R, et al. Plasticity of Foxp3(+) T cells reflects promiscuous Foxp3 expression in conventional T cells but not reprogramming of regulatory T cells. *Immunity.* 2012;36:262–275.
48. Komatsu N, Okamoto K, Sawa S, et al. Pathogenic conversion of Foxp3+ T cells into TH17 cells in autoimmune arthritis. *Nat Med.* 2014;20:62–68.
49. Danks L, Komatsu N, Guerrini MM, et al. RANKL expressed on synovial fibroblasts is primarily responsible for bone erosions during joint inflammation. *Ann Rheum Dis.* 2015. http://dx.doi.org/10.1136/annrheumdis-2014-207137. pii: annrheumdis-2014-207137.
50. Ohkura N, Hamaguchi M, Morikawa H, et al. T cell receptor stimulation-induced epigenetic changes and Foxp3 expression are independent and complementary events required for Treg cell development. *Immunity.* 2012;37:785–799.
51. Miyara M, Yoshioka Y, Kitoh A, et al. Functional delineation and differentiation dynamics of human CD4+ T cells expressing the FoxP3 transcription factor. *Immunity.* 2009;30:899–911.
52. Ito Y, Hashimoto M, Hirota K, et al. Detection of T cell responses to a ubiquitous cellular protein in autoimmune disease. *Science.* 2014;346:363–368.
53. Gagliani N, Vesely MC, Iseppon A, et al. Th17 cells transdifferentiate into regulatory T cells during resolution of inflammation. *Nature.* 2015;523:221–225.

CHAPTER TEN

The Immune Fulcrum: Regulatory T Cells Tip the Balance Between Pro- and Anti-inflammatory Outcomes upon Infection

Laura E. Richert-Spuhler*, Jennifer M. Lund*,†,1
*Vaccine and Infectious Disease Division, Fred Hutchinson Cancer Research Center, Seattle, Washington, USA
†Department of Global Health, University of Washington, Seattle, Washington, USA
[1]Corresponding author: e-mail address: jlund@fhcrc.org

Contents

Abstract

Regulatory T cells (Tregs) are indispensable for immune homeostasis and the prevention of autoimmunity. In the context of infectious diseases, Tregs are multidimensional. Here, we describe how they may potentiate effector responses by assisting in recruitment of T cells into the infection site to resolve infection, facilitate accelerated antigen-specific memory responses, limit pathology, and contribute to disease resolution and healing, to the great benefit of the host. We also explore the villainous functions of Tregs during infection by reviewing several diseases in which the depletion or reduction in Treg frequency allows for better generation of effector memory, and results in acute resolution of infection, as opposed to chronicity or severe long-term outcomes. We describe

Progress in Molecular Biology and Translational Science, Volume 136
ISSN 1877-1173
http://dx.doi.org/10.1016/bs.pmbts.2015.07.015

findings generated using mouse models of infection as well as experiments performed using human cells and tissues. We propose that Tregs represent an immunologic fulcrum, promoting both pathogen clearance and damage control by preventing excessive destruction of infected tissues though unchecked immune responses.

1. INTRODUCTION

Thymically selected regulatory T cells (Tregs), defined by the lineage marker forkhead box P3 (Foxp3), quintessentially promote self-tolerance and limit autoimmunity.[1] Thus, aberrant Treg function is extensively implicated in the manifestation of autoimmune syndromes such as inflammatory bowel disease, rheumatoid arthritis, and multiple sclerosis (reviewed in Ref. 2). Most severely, in the absence of Treg function, as exemplified by Treg-deficient scurfy mice, or individuals carrying nonfunctional or impaired versions of the *FOXP3* gene, life-threatening multiorgan autoimmunities and lymphoproliferative diseases manifest during early development.[3,4] In the context of infection, many studies have demonstrated that Tregs act to dampen overexuberant effector immune responses, yet accumulating evidence suggests that Treg function is in fact far more versatile than a simple provision of immune inhibition, as we will highlight here. Tregs thus participate in diverse mechanisms of immunomodulation in a context-dependent fashion. We herein specifically emphasize Treg responses to pathogens, with a focus on their abilities to potentiate appropriate immune responses, and propose that Tregs serve as a fulcrum, tipping the immune system between pathogen clearance and protection from collateral damage.

Tregs are broadly categorized into two groups, thymically derived Tregs (tTregs, or natural nTregs) and peripherally induced Tregs (pTregs, or iTregs). Both subsets are defined by their high expression of IL-2Rα (CD25) and Foxp3, although peripheral Tregs may be converted to this phenotype from CD25−Foxp3− precursors under the influence of IL-2 and TGF-β.[5,6] Efforts to distinguish the phenotype of tTregs from pTregs have identified Helios, a member of the Ikaros transcription factor family, as its expression tends to be enriched on tTregs.[6–9] However, upon certain *in vitro* culture conditions or immunization schemes, low-level induction of Helios expression from peripheral Tregs and Foxp3− T cells can be detected.[6,9,10] tTregs may also be differentiated by their increased expression of PD-1, neuropilin-1, and CD73 as compared to pTregs, whereas both subsets may express activation markers including GITR and CTLA-4.[6,11] Other types of peripheral Foxp3−CD4+ T cells may also exhibit regulatory functions, such as the secretion of IL-10 (Tr1) and TGF-β (Th3)[12–14]; furthermore, regulatory CD8+ T cells may express Foxp3+ and produce IL-10.[15,16] Taken together, peripheral Tregs likely represent a highly plastic population, prone to evolution in response to environmental cues or other impetus.[17] Here, we will focus on classic CD4+CD25+Foxp3+ Treg biology.

Human studies largely identify Tregs based on their expression of CD4, CD25, and FOXP3, with low expression of CD127 (IL-7R). Unfortunately, the necessity of intracellular staining for FOXP3 complicates functional assays from human cells, whereas in mice, *Foxp3^{gfp}* reporter mice provide an excellent mechanism of Treg distinction without cellular fixation.[18] Thus, surface markers must suffice for many functional human analyses. Furthermore, FOXP3 is not a definitive and reliable marker for human Tregs, as it can also be expressed on recently activated CD4+ effector T cells.[19–22] In mice, Tregs are definitively identified based on expression of CD4 and Foxp3, and both mouse and human Tregs may express CD39, CD73,[23] and activation and trafficking markers such as GITR, ICOS, and CTLA-4.[24] Activation and proliferation or polarization markers which are not necessarily specific to Tregs such as CD62L, CD69, CD103, CD44, Ki-67, and Tbet (among many others) may also be present, depending on the circumstance.

Murine models of Treg cell ablation have provided excellent tools for the examination of Treg function, and loss of function, which logistically cannot be achieved in human studies *in vivo*. Germline *Foxp3* dysfunction results in catastrophic autoimmunity, thus necessitating conditional knockout strategies for Foxp3+ Tregs. Foxp3-diphtheria toxin receptor (Foxp3-DTR) mice were created by inserting the human DT receptor under the control of the *Foxp3* locus, thus allowing for the selective ablation of Foxp3-expressing cells upon diphtheria toxin (DT) administration.[25] Similarly, bacterial artificial chromosome (BAC)-transgenic "depletion of regulatory T cells" (DEREG) mice were created, again reliant on DT-mediated Foxp3 ablation.[26] Both strategies allow for the specific and inducible depletion of Foxp3+ Tregs at desired times, although neonatal but not adult DEREG mice exhibit spontaneous autoimmunity upon DT-ablation of Tregs.[25–27] Thus, it appears that Treg elimination is incomplete in adult DEREG mice, or cannot be sustained due to the repopulation of Tregs lacking the DT-sensitive transgene.[27,28] Prior to the availability of Foxp3-DTR and

DEREG mouse models, it was also common to deplete Tregs using monoclonal antibodies against CD25. Whereas many of these studies laid important groundwork for the field, CD25 can be expressed on activated T cells, as well as Tregs, thereby complicating the clarity of these results.

2. ACUTE INFECTIONS IN DELICATE TISSUES

2.1 The Lung

The lung is a site of significant antigen exposure, where inhaled allergens and pathogens are constantly introduced, and must be quickly controlled to maintain homeostasis and avoid airway consolidation. In this regard, sentinel alveolar macrophages maintain the airways in a tolerogenic state, partially via active TGF-β, tethered to the surface of neighboring alveolar epithelial cells.[29,30] However, upon the introduction of a pathogen, a rapid and tightly regulated immune response must be generated where, ideally, the infection can be cleared with minimal collateral damage, since overly exuberant responses can compromise air exchange and are therefore dangerous to the host.

2.1.1 Influenza Virus

Influenza virus infection elicits a strong proinflammatory cytokine response with abundant lymphocyte and innate cell infiltration, and whereas T cell and antibody responses are generally effective at controlling the virus, serious sequelae may result from acute infection.[31] Such morbidities become especially pronounced in more susceptible hosts, or during seasons when emerging or highly pathogenic strains are introduced. Tregs thus provide an irreplaceable mechanism for dampening influenza-associated pathology, controlling the magnitude of the immune response, and promoting healing. These dampening Treg functions have been well elucidated in animal models.[32–35] During primary influenza infection of mice, Tregs accumulate in the lung prior to the influx of antigen-specific CD4+ and CD8+ T cells and suppress effector T cell responses.[32–34] These responses are retained *ex vivo*, and the adoptive transfer of influenza-exposed Tregs provides protection to infected recipients. More dramatically, the adoptive transfer of Tregs ameliorates influenza virus-associated morbidity and prolongs survival in immunodeficient *Rag1*$^{-/-}$ mice, whereas effector CD4+ T cells do not facilitate the same protection.[32] As *Rag1*−/− mice are significantly immunocompromised, the repletion of Tregs is insufficient to control viral load, yet Treg introduction delays the accumulation and organizes infiltrating monocytes, providing protective effects for tissue preservation, potentially through CCL8 suppression.[32]

Treg-mediated suppression appears to be antigen specific, as several groups have identified influenza-specific Tregs that acquire functional specialization by secreting IL-10, and upregulating activation markers CTLA-4, GITR, CD44, CD69 and ICOS, and the T_H1-associated transcription factor, Tbet.[33,34] In line with this evidence, Brincks *et al.* have shown that influenza-specific Tregs additionally acquire a memory phenotype, and are uniquely adept at trafficking to the lung and draining lymph nodes (dLNs) upon rechallenge with a heterologous virus strain.[36] Memory Tregs then control the intensity of memory T cell responses and protect the host from collateral damage in an MHC II-dependent, antigen-specific manner. These results cannot be recapitulated by replacing memory Tregs with naïve Tregs, further implicating long-term immune-shaping effects of Tregs on both primary and memory responses to influenza virus.[36]

While the importance of Treg-mediated immune suppression should not be overlooked, particularly in delicate lung tissue, unexpected non-suppressive roles for Tregs during influenza infection are beginning to be elucidated as well. For example, León *et al.* demonstrated that Treg depletion impairs the germinal center (GC) response to influenza virus challenge by impairing T follicular helper (T_{FH}) cell function.[37] In this context, Tregs compete for IL-2, which is inhibitory to T_{FH} function, and in their absence, an overabundance of IL-2 impedes the correct organization of the effector response in the lung and dLN. This work highlights the multidimensional role for Treg function in the lung, placing emphasis on the ability of Tregs to orchestrate prudent effector responses by encouraging GC formation, and regulating primary and memory responses to influenza virus.

2.1.2 Respiratory Syncytial Virus

Respiratory syncytial virus (RSV) infection is another prevalent respiratory illness and constitutes a major concern in infants due to its potential to result in severe outcomes, such as bronchiolitis.[38] Additionally, early childhood RSV infections are known to drive strong T_H2-dominated responses, placing affected children at a higher risk for asthma later in childhood.[39,40] Attempts at developing a safe vaccine against RSV have failed to date, most famously in the case of a formalin-inactivated RSV vaccine candidate that unexpectedly enhanced the acquisition and severity of natural infection in vaccinees.[41] It is thought that the generation of inappropriate antibody epitopes and a lack of local CD8+ T cell responses played a role in this failure, though studies in mice using a similar vaccine demonstrated that a "crash" in Treg presence may also be contributory.[42] In this regard, immune responses to RSV are enigmatic in that natural infection drives incomplete immunity, allowing for viral reinfection, and while passive antibody transfer can be protective, natural antibody titers do not always correspond with protection.[38] Effector CD8+ T cell-mediated immunity appears to better correlate with protection and viral control (although some caveats exist for these rules), but paradoxically, T cell responses can contribute to enhanced immunopathology and disease severity.[38] These scenarios strongly imply a necessity for regulatory mechanisms.

Many groups have shown that RSV infection elicits the expansion and activation of Tregs in the lung parenchyma, airway spaces (as assessed by bronchoalveolar lavage), and dLN.[43–46] Upon Treg ablation, animals lose significantly more body weight and experience a protracted recovery period.[43,44,46] Perhaps not surprisingly, the peak of weight loss correlates with disease severity and has been reported to be accompanied by innate immune cell infiltrate, and activated NK cells, and T cells which express high levels of CD69 and secrete IFN-γ.[43] In other reports, Treg depletion results in strong T_H2-driven immune responses, with IFN-γ and IL-13-producing CD4+ T cell recruitment, and abundant eosinophilia.[46] There exists some disagreement regarding the effect of Treg ablation on viral load, although the timing of analysis post-RSV challenge seems to impact these conclusions.[43–47] To distill several reports, it appears likely that whereas the peak titer of RSV may remain unaffected, viral clearance may be delayed (although not inhibited) by Treg ablation. Emphasis on the timing of response has been further highlighted by several groups who report that the depletion of Treg cells prior to RSV challenge impairs the early egress of RSV-specific CD8+ T cells from the dLN into the lung.[44,45] Such a delay in the antiviral response permitted a protracted infection course, despite a later influx of polyfunctional CD8+ T cells producing IFN-γ and TNF-α.[44,45] Thus, Tregs serve an important role during acute RSV infection by facilitating the early trafficking of antigen-specific CD8+ T cells into the effector site. Moreover, Ruckwardt *et al.* report that not only are CD8+ antigen-specific T cell responses delayed in accumulating in the lung in Treg-depleted mice, but the viral epitopes to which the responses are directed are also altered, tipping the balance toward an increased frequency of immunodominant epitope-specific CD8+ T cells.[45]

Interestingly, Treg ablation apparently exerts an opposite effect on memory responses to RSV, as memory CD8+ RSV-specific T cell responses are increased in frequency in mice depleted of Tregs during acute RSV infection.[45] The explanation of acute versus memory discrepancies has yet to be formally revealed. However, it seems reasonable to assume that since the antigen-specific CD8+ T cell responses normalize at later timepoints during acute infection, and that an increased production of cytokines and associated damage ensues, stronger, and possibly dysregulated, CD8+ T cell responses in Treg-ablated mice could lead to an increased frequency of memory T cells. Taken together, we conclude that Tregs provide essential control of early effector cell recruitment, and later limitation of excessive cellular infiltrate and cytokine production in the lung, thereby promoting efficient and appropriate effector immunity, and aiding the resolution of inflammation.

2.1.3 *Mycobacterium tuberculosis*

Mycobacterium tuberculosis (Mtb) infects tolerogenic alveolar macrophages, where it slowly replicates until the bacteria kill the overwhelmed cell, or T cell responses are finally generated to command the phagocyte to mount an antibacterial response.[48] Mtb is a master at manipulating immune responses, and acute exposure may result in chronic infection. In this regard, the exact requirements for successful bacterial clearance and disease resolution remain enigmatic, although many lines of evidence indicate that robust, early effector T cell responses are critical for disease resolution.[48–50]

Upon infection of mice, Tregs (including recently identified Mtb-specific Tregs[51]) have been shown to expand in the dLN and lung early during infection and upregulate activation markers such as CTLA-4, GITR, and ICOS.[51,52] Human peripheral and pleural Tregs similarly respond and expand upon exposure to *M. tuberculosis*,[53–55] and several groups have reported a correlation between chronically elevated frequencies of Tregs in tuberculosis patients with active disease, compared to far lower frequencies in individuals who resolve infection, or who were never infected.[53,56,57] The theme from these publications would implicate Treg responses as inhibitory to pathogen clearance and disease resolution. In contrast, Wergeland *et al.* reveal that increased frequencies of Tregs are accompanied by increased frequencies of activated (HLA-DR+CD38+) T cells in the blood of Mtb-infected individuals during the acute versus latent stage of infection, which were both elevated over frequencies in naïve individuals.[58] These data may simultaneously argue for Treg-mediated protection against the detrimental effects of chronic immune activation, which notoriously predisposes individuals to unrelated health complications.

Investigations using mouse models of Treg ablation have revealed additional complexities. Treg ablation has been reported to either reduce Mtb bacterial load in the lungs[52] or have little effect on bacterial clearance.[59] Ozeki *et al.* have reported that the timing of Treg depletion appears to be most important, as a transient, early Treg ablation results in superior bacterial clearance and decreased granuloma formation at 2 weeks postinfection, but elicits no effect on later bacterial load or lung pathology, as assessed at 3 and 5 weeks postinfection.[60] Moreover, Leepiyasakulchai *et al.* show that in the lungs (but not LNs) of highly susceptible DBA/2 mice, despite sustained high Mtb load, the frequency of Tregs decreases at 3 weeks postinfection, and this declination continues through chronicity.[61] Thus, expanded Treg frequency may not always be directly linked to the inhibition of pathogen clearance, and here the authors suggest a role for both CD103+ dendritic cells (DCs) and Tregs in promoting immunity toward Mtb challenge.[61]

2.1.4 Fungal Infections

Immunocompetent individuals rarely acquire pathogenic fungal infections of the lung; however, for susceptible individuals such as those who are immunosuppressed due to HIV infection, bone marrow transplantation, or chemotherapy, infections with organisms such as *Candida*, *Pneumocystis*, *Aspergillus*, *Cryptococcus*, and *Coccidioides* are more common. Our discussion of the fungal-Treg axis will be very brief, as current literature regarding a role for Tregs is fairly limited. There are, however, some lines of evidence for both Treg-mediated mitigation of pulmonary inflammation[62,63] and Treg-mediated inhibition of antifungal responses.[64] Using a mouse model of *Pneumocystis* pneumonia and immune reconstitution inflammatory syndrome (IRIS), McKinley *et al.* report that both anti-CD25-mediated depletion of regulatory T cells in wild-type mice, and the reconstitution of *Pneumocystis*-infected SCID mice with CD4+CD25− T cells elicit proinflammatory cytokine production, whereas the reconstitution of SCID mice with Tregs has no effect on disease exacerbation, but augment *Pneumocystis* burden.[62] As *Pneumocystis* infection remains a health concern, especially for HIV+ individuals with low CD4+ T cell counts, the authors propose that Treg dysfunction may partially underlie the observed pathogenic responses of IRIS.[62] Schulze *et al.* report a more dramatic phenotype of Treg ablation in mice infected with *Cryptococcus neoformans*.[63] Tregs expand and accumulate in the lung during *C. neoformans* infection to control T_H2-mediated inflammation, and in their absence, eosinophilia, overt mucus production, T_H2-skewed cytokine production, and fungal burden are exacerbated.[63] In contrast, the early depletion of Tregs during *Paracoccidioides brasiliensis* infection reduced tissue pathology and fungal burden in the lungs, liver, and spleen of two genetically diverse strains of mice, representing variation in resistance to *P. brasiliensis* infection.[64] Finally, Tregs have been shown to promote acute T_H17 cell responses to *Candida albicans*, allowing for enhanced fungal clearance, while at later timepoints they had suppressive roles.[65] These functions were mediated by IL-2 competition, especially at early timepoints, again indicating the importance of appropriate timing in response profiles.

2.2 The Brain

2.2.1 West Nile Virus

Studies which have interrogated the role for Tregs in clearing neurotropic viruses from the central nervous system (CNS) and brain similarly highlight the necessity for a regulated balance of effector and regulatory T cell responses. For example, sterilizing immunity against West Nile virus (WNV) requires the recruitment of CD8+ T cells into the CNS,[66] although these responses must be regulated to avoid excessive damage to nonrenewable neurons. In agreement with this hypothesis, Lanteri *et al.* demonstrate that individuals who remain asymptomatic during West Nile infection have an increased frequency of Treg cells in peripheral blood, as compared to individuals with more severe disease.[67] Their findings are further corroborated in a mouse model of Treg ablation in which case Treg-deficient mice similarly suffer an increase in disease severity, weight loss, and lethality.[67] A link between these complex and pendulous mechanisms has been recently reported by Graham *et al.* who show that Tregs work to limit both the proliferation and inflammatory cytokine secretion of T cells in the CNS, while simultaneously encouraging the retention and maintenance of protective resident memory T cells in the brain.[68] Moreover, an early Treg presence in the brain is important to this protection.[68] Such a dual role for Tregs during infectious disease is thus becoming a repetitive theme where both the timing and the calibration of Treg responses are essential to a positive outcome, particularly in tissues with unique requirements or delicate architecture.

3. ACUTE SYSTEMIC VIRAL INFECTIONS

3.1 Dengue Virus

Dengue virus, similar to WNV, is another example of a vector-borne acute flaviviral infection which can present with a spectrum of symptom severity and pathology. Instances of extreme disease severity are characterized by cytokine storm, life-threatening vascular leakage syndrome, and hemorrhagic fever or shock.[69] Immune control of dengue virus, particularly in the context of a heterologous strain challenge, can be enigmatic as cross-strain reactive antibody and memory T cell responses often amplify viral replication and associated pathology.[69] Interestingly, children suffering more severe outcomes during acute dengue virus infection harbor reduced frequencies of Treg cells in peripheral blood.[70] However, Treg function in terms of cytokine secretion and suppressive abilities remain unaltered despite the infection course.[70] Immunohistological analysis of liver biopsies from adults with severe dengue hemorrhagic fever similarly shows high expression of proinflammatory markers, TLR2 and TLR3, iNOS, IL-6, IL-18, TGF-β, and granzyme B, whereas Foxp3 was notably undetectable.[71] Along these lines, mouse models have shown a lack of Treg expansion in the spleen in response to dengue viral challenge.[72] A conclusive role for Tregs in response to dengue virus infection remains elusive, but again appears to depend on appropriate effector timing. This may be particularly important in the context of reinfection with a disparate strain of dengue virus.

3.2 Lymphocytic Choriomeningitis Virus

Lymphocytic choriomeningitis virus (LCMV) of mice is a widely used model for both acute and chronic systemic viral infection. Adult mice infected with LCMV strain Armstrong (Arm) undergo acute infection, which is resolved by robust T cell responses roughly 8–10 days after challenge, whereas mice infected with the strain clone 13 (Cl13) remain chronically infected.[73,74] Akin to other viral infections, chronic LCMV infection results in the exhaustion of effector T cells which upregulate PD-1 and further exhibit impairment in cytokine secretion and killing.[75] Moreover, mice chronically infected with LCMV elicit an increased frequency of splenic Tregs bearing a heightened activation marker expression profile.[76] Thus, it has been suggested that Tregs actively constrain CD8+ T cell responses during chronic infection, allowing for viral persistence. Upon Treg ablation of chronically LCMV-infected mice, CD8+ T cells rapidly rebound in both number and function including proliferation, production of IFN-γ, TNF-α, CD107a/b and granzyme B, and target cell killing after antigen-specific *ex vivo* stimulation.[76] Interestingly, whereas rescued CD8+ T cells from Treg-depleted mice showed restored functionality *ex vivo*, these responses have no effect on controlling the LCMV viral burden. The authors go on to show that Treg depletion elicited an upregulation of PD-L1, in particular by LCMV-infected cells, providing an additional link between T cell exhaustion and Treg function. Viral control could be restored only when chronically infected mice were both Treg-depleted and dosed with PD-L1 blocking antibody.[76]

Early type one interferon (IFN) signaling is pivotal to infection outcome for both acute and chronic LCMV infections. As acute infection (with strain Arm) is resolved, IFN-α/β dissipates, whereas during chronic infection (with Clone 13) IFN-α/β remains elevated, as do other indicators of immune activation.[77] Systemically depleting IFN-α/β after chronic infection can improve disease outcome by ameliorating chronic immune activation and viral burden, and restoring lymphoid architecture.[77,78] Along these lines, within 2 days postacute infection with LCMV Arm, IFN-α/β is known to act directly on activated Treg cells to provide inhibitory signals.[79] Downstream, the effect of early IFN-α/β-mediated Treg inhibition is the allowance of optimal CD4+ and CD8+ effector T cell expansion, which act to clear the virus. Tellingly, Foxp3-DTR mice which had undergone DT-mediated Treg ablation, and were reconstituted with *Ifnar1* −/− Tregs, were more susceptible to LCMV infection severity due to poor production of IFN-γ, reduced killing capacity of CD8+ T cells, and an increased viral load at day 7 postinfection.[79] In this regard, the transient, early inhibition of Tregs appears to promote protection by encouraging the generation of effector immunity. Complete Treg ablation during early acute infection, however, is highly detrimental to viral clearance.[80] Treg-ablated mice harbor an increased liver viral load at day 5 post-LCMV Arm challenge as compared to intact mice.[80] Taken together, viral resolution appears to rely on the generation of timely effector T cell responses, which are allowed by transient IFN-α/β interference of Tregs. Paradoxically, Treg presence is necessary for viral clearance, implying additional functions of Tregs during LCMV infection.

4. GASTROINTESTINAL INFECTIONS

The gut and intestines are constantly challenged by a dense array of food antigens, normal microbiota, and potential pathogens, implicating Tregs as a prevalent mechanism of controlling tolerance and pathogen breach in these sites. During oral rotavirus challenge of mice, Tregs expand significantly in the spleen, mesenteric lymph nodes, and gut-associated lymphoid tissue.[81,82] However, the ablation of Tregs has no effect on measurements of disease severity including diarrheal incidence, viral shedding, or early IgA responses, despite an expansion of T and B cells, upregulation of activation markers such as CD69, and increased IFN-γ production, with concurrent loss of IL-10.[81,82] Similarly, during *Listeria monocytogenes* infection, the depletion of Tregs prior to challenge with rLM-OVA does not affect the ability of the mice to clear bacteria during primary challenge.[83] However, Treg ablation does impair antigen-specific CD8+ T cell responses in terms of number and avidity during the peak of the primary immune response. Such primary impairment may have grave effects on memory responses, and indeed, during secondary bacterial challenge, those mice which had been depleted of Tregs prior to primary infection were significantly less efficient at clearing bacteria. This could be attributed to a reduction in the overall numbers and affinity of antigen-specific T cells, as well as their ability to dually produce IFN-γ and granzyme B upon *ex vivo* stimulation.[83]

Further evidence for Treg potentiation of effector immunity during intestinal infection comes from investigations of *Citrobacter rodentium*. *C. rodentium* notoriously drives proinflammatory T_H17 responses for effective bacterial clearance, which must be tightly controlled to limit damage. T_H17 cells share a unique "reciprocal" and plastic[17] relationship with intestinal Tregs through their shared expression of ROR-γt and responsiveness to TGF-β,[84,85] indicating their cooperative balance. Wang *et al.* show that Treg depletion prior to and during acute *C. rodentium* infection significantly exacerbates morbidity, mortality, and systemic bacterial dissemination.[86] The mechanisms underlying exacerbated disease appear to be mediated by a loss of T_H17 cell and neutrophil frequency and function, specifically at the site of infection. Moreover, neutralizing IL-2 counteracted the exacerbated disease in Treg-deficient animals, indicating that Treg-mediated control of IL-2 availability in the mucosa promotes appropriate T_H17 cell responses. Conversely, during *Salmonella* infection, Treg ablation was beneficial to the host in accelerating bacterial clearance, as well as leading to timely activation of effector T cells.[87] Again, the role of Tregs in infection, with gastrointestinal infections highlighted here, proves to be dynamic in nature, with Treg responses and cooperation with effector cells in a fluid state of change, a theme in common for many infectious disease scenarios.

5. CHRONIC INFECTIONS

5.1 Herpes Simplex Virus

Infection with herpes simplex virus (HSV) type-1 and type-2 elicits a dominant T cell-mediated response, in which successful resolution of primary disease requires a carefully orchestrated balance of IFN and toll-like receptor signaling, and antigen-presenting cell and T cell trafficking.[88,89] Importantly, Tregs serve an indispensable function in orchestrating these effector responses. Upon intravaginal challenge of mice with HSV-2, CD11b+ submucosal DCs and plasmacytoid DCs provide early antigen presentation,[90] and IFN,[91] respectively. T cell responses subsequently follow, specifically leading to both effector T cell and Treg activation, expansion, production of effector cytokines, and trafficking to the vagina and dLN.[80] In the absence of Tregs, HSV-2-infected mice succumb to infection rapidly, whereas Treg-sufficient mice succumb more slowly to fatal infection. Specifically, following Treg ablation, viral replication, and dissemination into spinal cord tissue is augmented, concurrent with clinical disease severity and mortality.[80] In the absence of Tregs, strong effector responses and the production of proinflammatory cytokines (TNF-α, IL-2, and IFN-α/β and IFN-γ) ensue in the dLN. However, early local cytokine signaling and T cell trafficking into the vagina are notably impaired, due to dysregulated retention of effector lymphocytes in the dLN.[80] Cellular egress into the vagina relies on correct and complex chemokine gradients, which were found to be altered upon Treg ablation, and therefore linked to the impairment of cellular trafficking in the absence of Tregs.[80] Thus, Tregs coordinate the timely balance of effector cell recruitment and trafficking into the vaginal mucosa, control the chemokine gradient, and suppress overly exuberant proinflammatory responses to the great benefit of the host.[80,92]

During ocular infection with HSV-1, adept viral control similarly relies on strong effector immunity, and again, associated collateral damage can be highly detrimental in the eye. Specifically, recurrent effector immune responses to viral recrudescence are a prevalent cause of blindness.[88] In this regard, Tregs act to ameliorate disease pathology and severity by limiting neutrophil and T cell migration into the delicate site, and secreting IL-10 and TGF-β.[93,94] Conversely, when HSV-1 is delivered as a subcutaneous challenge, where collateral damage may not be as costly, Treg-mediated suppression of effector T cell responses is arguably detrimental to the intensity of the response.[95,96] In this scenario, Treg ablation enhances both primary and memory vaccine-amnestic T cell functions after CD8+ T cell epitope immunization, and as a result, primed effector T cells in Treg-ablated mice are more successful in clearing virus.[97,98] Both of these scenarios implicate Treg function as pivotal to well-balanced immunological outcomes where viral clearance and the preservation of tissue architecture and function may both be attained.

5.2 Human Immunodeficiency Virus

Human immunodeficiency virus (HIV) uniquely infects and is maintained in CD4+ T cells and susceptible APCs, posing a major hurdle for the generation of effective antiviral immunity. HIV-infected individuals remain infected for life, and most suffer consequent health complications including chronic immune activation and increased susceptibility to unrelated pathogens, cancers, and inflammatory disorders.[99] Correlates of protection against HIV acquisition and cure remain unknown, though some hints derived from rare individuals who are refractory to HIV acquisition or progress significantly more slowly than average remain an intensive area of interest.[100–104] In this regard, a role for Tregs in HIV acquisition and disease progression is multifaceted. Akin to other infections reviewed here, the timing and phase of immune responses to HIV pivotally impact disease outcome.

Highly HIV-exposed seronegative (HESN) individuals provide a unique opportunity to investigate natural resistance to HIV acquisition, which has been proposed to be partially facilitated by unique Treg responses. For example, commercial sex works that have remained uninfected for at least 7 years, despite high HIV exposure, uniformly show heightened frequencies of Tregs, and a concurrent reduction in the frequencies CD69+-activated T cells.[105] Similar results have been shown in terms of reduction in immune activation for other cohorts of HESN,[102,106,107] potentially implying a contributory role for Tregs in protection from HIV infection. Along these lines, seronegative infants born to HIV-positive mothers, who were not receiving antiretroviral treatment during pregnancy, were found to have enhanced frequencies of Tregs and a reduction in the frequencies of activated effector T cells in cord blood.[108] Thematically, these studies implicate a state of low immune activation, expanded Treg numbers, and overall immune quiescence in resistance to HIV acquisition. Conversely, Pattacini *et al.* demonstrate that a subset of HESN individuals from a cohort of heterosexual serodiscordant couples have impaired Treg-mediated suppression of HIV-specific CD4+ T cell responses.[109] Interestingly, MIP-1β was found to be upregulated in these individuals, and since the autocrine signaling of MIP-1β is known to block the dominant HIV coreceptor, CCR5, Treg allowance of enhanced MIP-1β signaling could play a role in protection from HIV acquisition.[109]

HIV disease progression is also impacted by Treg frequency and function. Studies examining the frequencies of Tregs in long-term nonprogressors and elite controllers, who have significant delay in disease progression and/or exert some control over HIV infection, have found that Treg frequencies are similar to, or less than, those of uninfected control individuals.[110,111] Importantly, whereas a reduction in the frequency of Tregs may allow for better viral control, immune activation and associated comorbidities were found to be heightened in these cohorts.[110–112] Moreover, others have shown that a dominant Treg presence during HIV infection may better allow viral persistence by detrimentally inhibiting effector T cell responses.[113,114] Taken together, the role for Tregs during HIV infection is central, and extremely context dependent, reliant upon virus–host interaction, and disease phase.

5.3 Hepatitis B and C Viruses

Hepatitis B (HBV) and hepatitis C viruses (HCV) are members of disparate viral families, Hepadnaviridae and Flaviviridae, respectively; however both are hepatotropic and require similar T_H1 and CTL-dominated immunity for successful disease resolution.[115] Infection with HBV or HCV results in one of two outcomes—viral clearance or chronicity, where the relative vigor of the acute immune response to infection controls disease outcome. A robust and early T_H1 response is thought to be sufficient to clear acute infection; however decreases in T_H1 intensity or T_H2 skewing tend to result in chronicity.[116,117] Interestingly, 95% of HBV infections spontaneously resolve, whereas less than 30% of individuals are able to acutely control HCV infection.[115] Some have hypothesized that CD8+ T cell exhaustion, and the loss of CD4+ T cell help may together be responsible for the inability to clear virus,[117] although a second body of evidence suggests that a shift to favor Treg induction promotes viral persistence for both HBV and HCV.[118–123] In support of this argument is the observation that the proportion of Tregs is increased in chronically infected individuals in both the liver and periphery.[117–127]

Functionally, CD4+CD25+ Tregs isolated from chronically infected HCV patients inhibit peptide-specific CD8+ T cell proliferation, IFN-γ production, and perforin expression reliant on cell-to-cell contact.[125,126,128] Li and colleagues further show that CD25 and Foxp3 are both upregulated upon antigenic restimulation *in vitro*, suggesting that these cells are HCV specific, and responsive.[129] IL-10-producing CD4+ (Tr1) T cells are additionally found to be upregulated during chronic HCV infection and some groups suggest that regulatory Tr1 cells are the subset responsible for effector

T cell interference during chronic HCV infection.[119,130,131] Along these lines, Eckels *et al.* noted that a lack of IL-2, and low levels of IFN-γ, in combination with high levels of IL-10, corresponds to HCV viral chronicity, whereas the production of IFN-γ with little to no IL-10 corresponds to a self-limited infection.[132]

Viral factors, such as an HCV-associated promotion signal, may engage Treg expansion, thus encouraging immune tolerance to the virus and allowing a prolonged infection.[133,134] Alternatively, the body may induce the shift to Treg cell production in an attempt to mitigate cellular damage caused by the caustic primary antiviral immune response. The liver is an inherently tolerogenic environment where DCs secrete high levels of IL-10, and the surrounding hepatocytes naturally produce TGF-β in order to sustain these liver DCs. Moreover, recent evidence additionally implicates upregulation of the healing and T cell-suppressive factor amphiregulin in response to chronic HBV infection.[135] Thus, environmental cues in the liver may additionally contribute to the propensity for a host-mediated induction of Tregs during infection.[119,131] Taken together, Tregs arguably provide the host both benefit and detriment, by permitting long-term chronicity, but also protecting from excessive liver damage and hepatocellular carcinoma.[119]

5.4 Parasitic Infections

Parasitic disease is a major global health concern, especially in areas where climate and poverty synergize to allow for transmission. The complex lifecycle of parasites combined with vector involvement adds delay and difficulty in generating sterilizing immunity and durable memory responses.[136] Moreover, immunity to these types of infections is additionally convoluted in that natural infection often induces T_H2-driven tolerance rather than clearance. A role for Tregs is equally enigmatic. For instance, the nematode *Heligmosomoides polygyrus* can drive the differentiation of Tregs from naïve CD4+ T cells through the engagement TGF-β,[137] and *Litomosoides sigmodontis* requires Treg induction to establish long-lived infection.[138] In contrast, for African trypanosomiasis, Tregs are known to limit host and parasite-derived pathology, without compromising parasite clearance,[139] and the transfer of Tregs improves pregnancy outcomes in mice infected with *Toxoplasma gondii*.[140] Liver disease attributed to *Schistosoma mansoni* is limited by Tregs, and upon their removal, liver granulomas are augmented.[141] Intriguingly, *Leishmania major* infection elicits an expansion of Tregs to establish chronic infection, yet the long-term exposure of antigen paradoxically allows for the generation of strong protective memory responses upon subsequent reexposure.[142] In these studies, mice which generated complete sterilizing immunity during primary infection were not protected against secondary challenge, implicating a role for host–pathogen coevolution.[142]

Efficacious immunity to malaria infection is thematically different from other parasitic infections in that T_H1-dominated responses prevail, and protect, with the caveat that reliable memory responses often remain absent. Upon acute malaria infection, IFN-γ-production and antibody responses are essential to parasite control.[143] Perhaps not surprisingly, *Plasmodium* infection (of various species) in both mice and humans elicits the expansion of Tregs (reviewed extensively in Ref. 144). Importantly, the ratio of effector T cells to Tregs seems to be most indicative of outcome, with increased frequencies of Tregs correlating to an increased parasite burden, although these ratios also appear to be a "moving target," in that T cell frequencies react fluidly to disease severity and stage. Moreover, adults and children show conflicting results in that Gambian children with both uncomplicated and severe malaria disease show increased Treg frequencies, which are correlated with high parasite burden, whereas increased Treg frequencies in Papuan adults could only be correlated with an increased parasite burden during severe disease.[144–146] Importantly, in the cohort of children, parasite loads exceeded 10^5 parasites per microliter of blood in both severe and uncomplicated disease, whereas infection of adults only reached such extreme concentrations of parasitemia upon complicated infections, potentially indicating a parasite burden threshold for Treg expansion.[144–146] Along these lines, using a mouse model of cerebral malaria, Amante *et al.* show that delivery of therapeutic anti-CD25 protects against neurologic disease and parasite burden in the brain, even after the establishment of blood-stage infection.[147] Interestingly, these mice did not show increased neurologic damage. In this regard, the distillation of Treg biology during parasite infection is convoluted, where discrepancies in consensus details appear to rely on the model, ethnicity, age, and timing and burden of parasitemia in relation to effector and regulatory immune responses. However, it seems reasonable to propose that similar to many other infectious diseases, an appropriate and well-timed Treg response provides important protection by limiting pathology, and in certain scenarios, potentiating protective memory.

6. CONCLUSIONS AND FUTURE DIRECTIONS

In conclusion, Tregs are an indispensable subset of T cells that not only provide governance of self-tolerance but also critically impact disease outcomes during infection. As we have extensively discussed, depending on the timing, the pathogen, and the severity of disease, Tregs may potentiate protective responses via recruiting mechanisms, dampen overexuberant responses, diminish tissue pathology, and aid in resolution and memory responses upon secondary challenge. Alternatively, various published reports ascribe a more villainous component to Treg cell function, as effector T cell responses may be pathologically diminished by Tregs, negating protective effects for T cell responses.

Whereas Treg interactions controlling effector T cell responses have been explored in many infection models, a role for Treg control of B cells and humoral responses is less well defined. It is known that the loss of function of Foxp3 in mice and humans causes dysfunctional humoral immunity with spontaneous GC formation, recruitment of T_{FH} cells, and the overproduction of IgE, implicating an undeniable role for Tregs in management of B cell responses.[148–151] In line with these findings, the short-term depletion of Tregs increases the frequency of B cells of several phenotypes and states of maturation.[152] Phenotypic data additionally shows that Foxp3+ Tregs are known to upregulate the expression of the T_{FH} markers CXCR5 and Bcl6, allowing them to enter the GC and exert control over GC B cell expansion.[150,153,154] It has been suggested that CTLA-4-dependent suppression is important for these results.[152] However, contrasting evidence from León *et al.*, who utilized an influenza model in mice, shows that Treg depletion prevents the formation of GCs.[37] Thus, a role for Treg-mediated governance of B cell responses in terms of autoimmunity, innocuous antigen exposure, and infectious disease is likely disparate and will necessitate more in-depth analysis.

Finally, we have described several examples of antigen-specific Tregs presence and function during infectious disease.[36,51] However, it remains unknown whether Tregs are always antigen specific, or if global bystander functions may also be in play, and the circumstance for each response. Moreover, it remains unknown whether infection drives the conversion of conventional CD4+ T cells into pTregs. Gagliani *et al.* have begun to elucidate some of these mechanisms by fate tracking of peripherally induced Tregs, which they report to have transdifferentiated from T_H17 cells.[17] However, very few transdifferentiated cells expressed Foxp3, despite abundant detection of IL-10-producing Tr1 cells. These data are exciting, and such models could be pursued for a variety of infections and lineages beyond T_H17, as to date the accurate distinction of thymically derived Tregs from peripherally induced Tregs has been hindered by overlapping phenotypic signatures in both subsets.

REFERENCES

1. Belkaid Y, Tarbell K. Regulatory T cells in the control of host-microorganism interactions (*). *Annu Rev Immunol.* 2009;27:551–589.
2. Buckner JH. Mechanisms of impaired regulation by CD4(+)CD25(+)FOXP3(+) regulatory T cells in human autoimmune diseases. *Nat Rev Immunol.* 2010;10(12):849–859.
3. Gambineri E, Torgerson TR, Ochs HD. Immune dysregulation, polyendocrinopathy, enteropathy, and X-linked inheritance (IPEX), a syndrome of systemic autoimmunity caused by mutations of FOXP3, a critical regulator of T-cell homeostasis. *Curr Opin Rheumatol.* 2003;15(4):430–435.
4. Brunkow ME, Jeffery EW, Hjerrild KA, et al. Disruption of a new forkhead/winged-helix protein, scurfin, results in the fatal lymphoproliferative disorder of the scurfy mouse. *Nat Genet.* 2001;27(1):68–73.
5. Horwitz DA, Zheng SG, Gray JD. Natural and TGF-beta-induced Foxp3(+)CD4(+) CD25(+) regulatory T cells are not mirror images of each other. *Trends Immunol.* 2008;29(9):429–435.
6. Lin X, Chen M, Liu Y, et al. Advances in distinguishing natural from induced Foxp3(+) regulatory T cells. *Int J Clin Exp Pathol.* 2013;6(2):116–123.
7. Sugimoto N, Oida T, Hirota K, et al. Foxp3-dependent and -independent molecules specific for CD25+CD4+ natural regulatory T cells revealed by DNA microarray analysis. *Int Immunol.* 2006;18(8):1197–1209.
8. Kelley CM, Ikeda T, Koipally J, et al. Helios, a novel dimerization partner of Ikaros expressed in the earliest hematopoietic progenitors. *Curr Biol.* 1998;8(9):508–515.
9. Shevach EM, Thornton AM. tTregs, pTregs, and iTregs: similarities and differences. *Immunol Rev.* 2014;259(1):88–102.
10. Akimova T, Beier UH, Wang L, Levine MH, Hancock WW. Helios expression is a marker of T cell activation and proliferation. *PLoS One.* 2011;6(8):e24226.
11. Yadav M, Louvet C, Davini D, et al. Neuropilin-1 distinguishes natural and inducible regulatory T cells among regulatory T cell subsets *in vivo*. *J Exp Med.* 2012;209(10):1713–1722. S1711–1719.
12. Groux H, O'Garra A, Bigler M, et al. A CD4+ T-cell subset inhibits antigen-specific T-cell responses and prevents colitis. *Nature.* 1997;389(6652):737–742.
13. Chen ZM, O'Shaughnessy MJ, Gramaglia I, et al. IL-10 and TGF-beta induce alloreactive CD4+CD25- T cells to acquire regulatory cell function. *Blood.* 2003;101(12):5076–5083.
14. Gagliani N, Magnani CF, Huber S, et al. Coexpression of CD49b and LAG-3 identifies human and mouse T regulatory type 1 cells. *Nat Med.* 2013;19(6):739–746.
15. Sun J, Madan R, Karp CL, Braciale TJ. Effector T cells control lung inflammation during acute influenza virus infection by producing IL-10. *Nat Med.* 2009;15(3):277–284.
16. Zou Q, Wu B, Xue J, et al. CD8+ Treg cells suppress CD8+ T cell-responses by IL-10-dependent mechanism during H5N1 influenza virus infection. *Eur J Immunol.* 2014;44(1):103–114.

17. Gagliani N, Vesely MC, Iseppon A, et al. Th17 cells transdifferentiate into regulatory T cells during resolution of inflammation. *Nature*. 2015;523:221–225.
18. Fontenot JD, Rasmussen JP, Williams LM, Dooley JL, Farr AG, Rudensky AY. Regulatory T cell lineage specification by the forkhead transcription factor foxp3. *Immunity*. 2005;22(3):329–341.
19. Allan SE, Passerini L, Bacchetta R, et al. The role of 2 FOXP3 isoforms in the generation of human CD4+ Tregs. *J Clin Invest*. 2005;115(11):3276–3284.
20. Morgan ME, van Bilsen JH, Bakker AM, et al. Expression of FOXP3 mRNA is not confined to CD4+CD25+ T regulatory cells in humans. *Hum Immunol*. 2005;66(1):13–20.
21. Walker MR, Kasprowicz DJ, Gersuk VH, et al. Induction of FoxP3 and acquisition of T regulatory activity by stimulated human CD4+CD25- T cells. *J Clin Invest*. 2003;112(9):1437–1443.
22. Wang J, Ioan-Facsinay A, van der Voort EI, Huizinga TW, Toes RE. Transient expression of FOXP3 in human activated nonregulatory CD4+ T cells. *Eur J Immunol*. 2007;37(1):129–138.
23. Deaglio S, Dwyer KM, Gao W, et al. Adenosine generation catalyzed by CD39 and CD73 expressed on regulatory T cells mediates immune suppression. *J Exp Med*. 2007;204(6):1257–1265.
24. Wing K, Onishi Y, Prieto-Martin P, et al. CTLA-4 control over Foxp3+ regulatory T cell function. *Science*. 2008;322(5899):271–275.
25. Kim JM, Rasmussen JP, Rudensky AY. Regulatory T cells prevent catastrophic autoimmunity throughout the lifespan of mice. *Nat Immunol*. 2007;8(2):191–197.
26. Lahl K, Loddenkemper C, Drouin C, et al. Selective depletion of Foxp3+ regulatory T cells induces a scurfy-like disease. *J Exp Med*. 2007;204(1):57–63.
27. Rudensky AY. Regulatory T cells and Foxp3. *Immunol Rev*. 2011;241(1):260–268.
28. Christiaansen AF, Boggiatto PM, Varga SM. Limitations of Foxp3(+) Treg depletion following viral infection in DEREG mice. *J Immunol Methods*. 2014;406:58–65.
29. Westphalen K, Gusarova GA, Islam MN, et al. Sessile alveolar macrophages communicate with alveolar epithelium to modulate immunity. *Nature*. 2014;506(7489):503–506.
30. Kopf M, Schneider C, Nobs SP. The development and function of lung-resident macrophages and dendritic cells. *Nat Immunol*. 2015;16(1):36–44.
31. Chiu C, Openshaw PJ. Antiviral B cell and T cell immunity in the lungs. *Nat Immunol*. 2015;16(1):18–26.
32. Antunes I, Kassiotis G. Suppression of innate immune pathology by regulatory T cells during Influenza A virus infection of immunodeficient mice. *J Virol*. 2010;84(24):12564–12575.
33. Betts RJ, Prabhu N, Ho AW, et al. Influenza A virus infection results in a robust, antigen-responsive, and widely disseminated Foxp3+ regulatory T cell response. *J Virol*. 2012;86(5):2817–2825.
34. Bedoya F, Cheng GS, Leibow A, et al. Viral antigen induces differentiation of Foxp3+ natural regulatory T cells in influenza virus-infected mice. *J Immunol*. 2013;190(12):6115–6125.
35. Moser EK, Hufford MM, Braciale TJ. Late engagement of CD86 after influenza virus clearance promotes recovery in a FoxP3+ regulatory T cell dependent manner. *PLoS Pathog*. 2014;10(8):e1004315.
36. Brincks EL, Roberts AD, Cookenham T, et al. Antigen-specific memory regulatory CD4+Foxp3+ T cells control memory responses to influenza virus infection. *J Immunol*. 2013;190(7):3438–3446.

37. León B, Bradley JE, Lund FE, Randall TD, Ballesteros-Tato A. FoxP3+ regulatory T cells promote influenza-specific Tfh responses by controlling IL-2 availability. *Nat Commun*. 2014;5:3495.
38. Openshaw PJ, Chiu C. Protective and dysregulated T cell immunity in RSV infection. *Curr Opin Virol*. 2013;3(4):468–474.
39. Krishnamoorthy N, Khare A, Oriss TB, et al. Early infection with respiratory syncytial virus impairs regulatory T cell function and increases susceptibility to allergic asthma. *Nat Med*. 2012;18(10):1525–1530.
40. Sigurs N, Bjarnason R, Sigurbergsson F, Kjellman B. Respiratory syncytial virus bronchiolitis in infancy is an important risk factor for asthma and allergy at age 7. *Am J Respir Crit Care Med*. 2000;161(5):1501–1507.
41. Kapikian AZ, Mitchell RH, Chanock RM, Shvedoff RA, Stewart CE. An epidemiologic study of altered clinical reactivity to respiratory syncytial (RS) virus infection in children previously vaccinated with an inactivated RS virus vaccine. *Am J Epidemiol*. 1969;89(4):405–421.
42. Loebbermann J, Durant L, Thornton H, Johansson C, Openshaw PJ. Defective immunoregulation in RSV vaccine-augmented viral lung disease restored by selective chemoattraction of regulatory T cells. *Proc Natl Acad Sci USA*. 2013;110(8):2987–2992.
43. Lee DC, Harker JA, Tregoning JS, et al. CD25+ natural regulatory T cells are critical in limiting innate and adaptive immunity and resolving disease following respiratory syncytial virus infection. *J Virol*. 2010;84(17):8790–8798.
44. Fulton RB, Meyerholz DK, Varga SM. Foxp3+ CD4 regulatory T cells limit pulmonary immunopathology by modulating the CD8 T cell response during respiratory syncytial virus infection. *J Immunol*. 2010;185(4):2382–2392.
45. Ruckwardt TJ, Bonaparte KL, Nason MC, Graham BS. Regulatory T cells promote early influx of CD8+ T cells in the lungs of respiratory syncytial virus-infected mice and diminish immunodominance disparities. *J Virol*. 2009;83(7):3019–3028.
46. Durant LR, Makris S, Voorburg CM, Loebbermann J, Johansson C, Openshaw PJ. Regulatory T cells prevent Th2 immune responses and pulmonary eosinophilia during respiratory syncytial virus infection in mice. *J Virol*. 2013;87(20):10946–10954.
47. Loebbermann J, Thornton H, Durant L, et al. Regulatory T cells expressing granzyme B play a critical role in controlling lung inflammation during acute viral infection. *Mucosal Immunol*. 2012;5(2):161–172.
48. Orme IM, Robinson RT, Cooper AM. The balance between protective and pathogenic immune responses in the TB-infected lung. *Nat Immunol*. 2015;16(1):57–63.
49. Urdahl KB, Shafiani S, Ernst JD. Initiation and regulation of T-cell responses in tuberculosis. *Mucosal Immunol*. 2011;4(3):288–293.
50. Kursar M, Koch M, Mittrucker HW, et al. Cutting edge: regulatory T cells prevent efficient clearance of Mycobacterium tuberculosis. *J Immunol*. 2007;178(5):2661–2665.
51. Shafiani S, Dinh C, Ertelt JM, et al. Pathogen-specific Treg cells expand early during mycobacterium tuberculosis infection but are later eliminated in response to Interleukin-12. *Immunity*. 2013;38(6):1261–1270.
52. Scott-Browne JP, Shafiani S, Tucker-Heard G, et al. Expansion and function of Foxp3-expressing T regulatory cells during tuberculosis. *J Exp Med*. 2007;204(9):2159–2169.
53. Garg A, Barnes PF, Roy S, et al. Mannose-capped lipoarabinomannan- and prostaglandin E2-dependent expansion of regulatory T cells in human Mycobacterium tuberculosis infection. *Eur J Immunol*. 2008;38(2):459–469.
54. Feruglio SL, Tonby K, Kvale D, Dyrhol-Riise AM. Early dynamics of T helper cell cytokines and T regulatory cells in response to treatment of active Mycobacterium tuberculosis infection. *Clin Exp Immunol*. 2015;179(3):454–465.

55. Toor JS, Singh S, Sharma A, Arora SK. Mycobacterium tuberculosis modulates the gene interactions to activate the HIV replication and faster disease progression in a co-infected host. *PLoS One*. 2014;9(9):e106815.
56. Chen X, Zhou B, Li M, et al. CD4(+)CD25(+)FoxP3(+) regulatory T cells suppress Mycobacterium tuberculosis immunity in patients with active disease. *Clin Immunol*. 2007;123(1):50–59.
57. Beiranvand E, Abediankenari S, Rezaei MS, Khani S, Sardari S, Beiranvand B. Increased expression of forkhead box protein 3 gene of regulatory T cells in patients with active tuberculosis. *Inflamm Allergy Drug Targets*. 2014;13(5):330–334.
58. Wergeland I, Assmus J, Dyrhol-Riise AM. T regulatory cells and immune activation in Mycobacterium tuberculosis infection and the effect of preventive therapy. *Scand J Immunol*. 2011;73(3):234–242.
59. Quinn KM, McHugh RS, Rich FJ, et al. Inactivation of CD4+ CD25+ regulatory T cells during early mycobacterial infection increases cytokine production but does not affect pathogen load. *Immunol Cell Biol*. 2006;84(5):467–474.
60. Ozeki Y, Sugawara I, Udagawa T, et al. Transient role of CD4+CD25+ regulatory T cells in mycobacterial infection in mice. *Int Immunol*. 2010;22(3):179–189.
61. Leepiyasakulchai C, Ignatowicz L, Pawlowski A, Kallenius G, Skold M. Failure to recruit anti-inflammatory CD103+ dendritic cells and a diminished CD4+ Foxp3+ regulatory T cell pool in mice that display excessive lung inflammation and increased susceptibility to Mycobacterium tuberculosis. *Infect Immun*. 2012;80(3):1128–1139.
62. McKinley L, Logar AJ, McAllister F, Zheng M, Steele C, Kolls JK. Regulatory T cells dampen pulmonary inflammation and lung injury in an animal model of pneumocystis pneumonia. *J Immunol*. 2006;177(9):6215–6226.
63. Schulze B, Piehler D, Eschke M, et al. CD4(+) FoxP3(+) regulatory T cells suppress fatal T helper 2 cell immunity during pulmonary fungal infection. *Eur J Immunol*. 2014;44(12):3596–3604.
64. Felonato M, Pina A, de Araujo EF, et al. Anti-CD25 treatment depletes Treg cells and decreases disease severity in susceptible and resistant mice infected with Paracoccidioides brasiliensis. *PLoS One*. 2012;7(11):e51071.
65. Pandiyan P, Conti HR, Zheng L, et al. CD4(+)CD25(+)Foxp3(+) regulatory T cells promote Th17 cells *in vitro* and enhance host resistance in mouse Candida albicans Th17 cell infection model. *Immunity*. 2011;34(3):422–434.
66. Shrestha B, Diamond MS. Role of CD8+ T cells in control of West Nile virus infection. *J Virol*. 2004;78(15):8312–8321.
67. Lanteri MC, O'Brien KM, Purtha WE, et al. Tregs control the development of symptomatic West Nile virus infection in humans and mice. *J Clin Invest*. 2009;119(11):3266–3277.
68. Graham JB, Da Costa A, Lund JM. Regulatory T cells shape the resident memory T cell response to virus infection in the tissues. *J Immunol*. 2014;192(2):683–690.
69. Rothman AL. Immunity to dengue virus: a tale of original antigenic sin and tropical cytokine storms. *Nat Rev Immunol*. 2011;11(8):532–543.
70. Luhn K, Simmons CP, Moran E, et al. Increased frequencies of CD4+ CD25(high) regulatory T cells in acute dengue infection. *J Exp Med*. 2007;204(5):979–985.
71. Pagliari C, Quaresma JA, Fernandes ER, et al. Immunopathogenesis of dengue hemorrhagic fever: contribution to the study of human liver lesions. *J Med Virol*. 2014;86(7):1193–1197.
72. Yauch LE, Prestwood TR, May MM, et al. CD4+ T cells are not required for the induction of dengue virus-specific CD8+ T cell or antibody responses but contribute to protection after vaccination. *J Immunol*. 2010;185(9):5405–5416.
73. Sullivan BM, Emonet SF, Welch MJ, et al. Point mutation in the glycoprotein of lymphocytic choriomeningitis virus is necessary for receptor binding, dendritic cell infection, and long-term persistence. *Proc Natl Acad Sci USA*. 2011;108(7):2969–2974.

74. Matloubian M, Concepcion RJ, Ahmed R. CD4+ T cells are required to sustain CD8+ cytotoxic T-cell responses during chronic viral infection. *J Virol.* 1994;68(12):8056–8063.
75. Wherry EJ. T cell exhaustion. *Nat Immunol.* 2011;12(6):492–499.
76. Penaloza-MacMaster P, Kamphorst AO, Wieland A, et al. Interplay between regulatory T cells and PD-1 in modulating T cell exhaustion and viral control during chronic LCMV infection. *J Exp Med.* 2014;211(9):1905–1918.
77. Wilson EB, Yamada DH, Elsaesser H, et al. Blockade of chronic type I interferon signaling to control persistent LCMV infection. *Science.* 2013;340(6129):202–207.
78. Teijaro JR, Ng C, Lee AM, et al. Persistent LCMV infection is controlled by blockade of type I interferon signaling. *Science.* 2013;340(6129):207–211.
79. Srivastava S, Koch MA, Pepper M, Campbell DJ. Type I interferons directly inhibit regulatory T cells to allow optimal antiviral T cell responses during acute LCMV infection. *J Exp Med.* 2014;211(5):961–974.
80. Lund JM, Hsing L, Pham TT, Rudensky AY. Coordination of early protective immunity to viral infection by regulatory T cells. *Science.* 2008;320(5880):1220–1224.
81. Kim B, Feng N, Narvaez CF, et al. The influence of CD4+ CD25+ Foxp3+ regulatory T cells on the immune response to rotavirus infection. *Vaccine.* 2008;26(44):5601–5611.
82. Miller AD, Blutt SE, Conner ME. FoxP3+ regulatory T cells are not important for rotavirus clearance or the early antibody response to rotavirus. *Microbes Infect.* 2014;16(1):67–72.
83. Pace L, Tempez A, Arnold-Schrauf C, et al. Regulatory T cells increase the avidity of primary CD8+ T cell responses and promote memory. *Science.* 2012;338(6106):532–536.
84. Weaver CT, Hatton RD. Interplay between the TH17 and TReg cell lineages: a (co-) evolutionary perspective. *Nat Rev Immunol.* 2009;9(12):883–889.
85. Bettelli E, Carrier Y, Gao W, et al. Reciprocal developmental pathways for the generation of pathogenic effector TH17 and regulatory T cells. *Nature.* 2006;441(7090):235–238.
86. Wang Z, Friedrich C, Hagemann SC, et al. Regulatory T cells promote a protective Th17-associated immune response to intestinal bacterial infection with C. rodentium. *Mucosal Immunol.* 2014;7(6):1290–1301.
87. Johanns TM, Ertelt JM, Rowe JH, Way SS. Regulatory T cell suppressive potency dictates the balance between bacterial proliferation and clearance during persistent Salmonella infection. *PLoS Pathog.* 2010;6(8):e1001043.
88. Chentoufi AA, Benmohamed L. Mucosal herpes immunity and immunopathology to ocular and genital herpes simplex virus infections. *Clin Dev Immunol.* 2012;2012:149135.
89. Iwasaki A. Antiviral immune responses in the genital tract: clues for vaccines. *Nat Rev Immunol.* 2010;10(10):699–711.
90. Zhao X, Deak E, Soderberg K, et al. Vaginal submucosal dendritic cells, but not Langerhans cells, induce protective Th1 responses to herpes simplex virus-2. *J Exp Med.* 2003;197(2):153–162.
91. Lund JM, Linehan MM, Iijima N, Iwasaki A. Cutting edge: plasmacytoid dendritic cells provide innate immune protection against mucosal viral infection in situ. *J Immunol.* 2006;177(11):7510–7514.
92. Diaz GA, Koelle DM. Human CD4+ CD25 high cells suppress proliferative memory lymphocyte responses to herpes simplex virus type 2. *J Virol.* 2006;80(16):8271–8273.
93. Suvas S, Azkur AK, Kim BS, Kumaraguru U, Rouse BT. CD4+CD25+ regulatory T cells control the severity of viral immunoinflammatory lesions. *J Immunol.* 2004;172(7):4123–4132.
94. Veiga-Parga T, Suryawanshi A, Mulik S, et al. On the role of regulatory T cells during viral-induced inflammatory lesions. *J Immunol.* 2012;189(12):5924–5933.

95. Fernandez MA, Yu U, Zhang G, et al. Treg depletion attenuates the severity of skin disease from ganglionic spread after HSV-2 flank infection. *Virology*. 2013;447(1–2):9–20.
96. Fernandez MA, Puttur FK, Wang YM, Howden W, Alexander SI, Jones CA. T regulatory cells contribute to the attenuated primary CD8+ and CD4+ T cell responses to herpes simplex virus type 2 in neonatal mice. *J Immunol*. 2008;180(3):1556–1564.
97. Toka FN, Suvas S, Rouse BT. CD4+ CD25+ T cells regulate vaccine-generated primary and memory CD8+ T-cell responses against herpes simplex virus type 1. *J Virol*. 2004;78(23):13082–13089.
98. Suvas S, Kumaraguru U, Pack CD, Lee S, Rouse BT. CD4+CD25+ T cells regulate virus-specific primary and memory CD8+ T cell responses. *J Exp Med*. 2003;198(6):889–901.
99. Lucas S, Nelson AM. HIV and the spectrum of human disease. *J Pathol*. 2015;235(2):229–241.
100. Poropatich K, Sullivan Jr DJ. Human immunodeficiency virus type 1 long-term non-progressors: the viral, genetic and immunological basis for disease non-progression. *J Gen Virol*. 2011;92(pt 2):247–268.
101. Moore PL, Williamson C, Morris L. Virological features associated with the development of broadly neutralizing antibodies to HIV-1. *Trends Microbiol*. 2015;23(4):204–211.
102. Poudrier J, Thibodeau V, Roger M. Natural immunity to HIV: a delicate balance between strength and control. *Clin Dev Immunol*. 2012;2012:875821.
103. Autran B, Descours B, Avettand-Fenoel V, Rouzioux C. Elite controllers as a model of functional cure. *Curr Opin HIV AIDS*. 2011;6(3):181–187.
104. Card CM, Ball TB, Fowke KR. Immune quiescence: a model of protection against HIV infection. *Retrovirology*. 2013;10:141.
105. Card CM, McLaren PJ, Wachihi C, Kimani J, Plummer FA, Fowke KR. Decreased immune activation in resistance to HIV-1 infection is associated with an elevated frequency of CD4(+)CD25(+)FOXP3(+) regulatory T cells. *J Infect Dis*. 2009;199(9):1318–1322.
106. Koning FA, van der Vorst TJ, Schuitemaker H. Low levels of human immunodeficiency virus type 1 DNA in high-risk seronegative men. *J Virol*. 2005;79(10):6551–6553.
107. Jennes W, Evertse D, Borget MY, et al. Suppressed cellular alloimmune responses in HIV-exposed seronegative female sex workers. *Clin Exp Immunol*. 2006;143(3):435–444.
108. Legrand FA, Nixon DF, Loo CP, et al. Strong HIV-1-specific T cell responses in HIV-1-exposed uninfected infants and neonates revealed after regulatory T cell removal. *PLoS One*. 2006;1:e102.
109. Pattacini L, Murnane PM, Kahle EM, et al. Differential regulatory T cell activity in HIV type 1-exposed seronegative individuals. *AIDS Res Hum Retroviruses*. 2013;29(10):1321–1329.
110. Hunt PW, Landay AL, Sinclair E, et al. A low T regulatory cell response may contribute to both viral control and generalized immune activation in HIV controllers. *PLoS One*. 2011;6(1):e15924.
111. Shaw JM, Hunt PW, Critchfield JW, et al. Increased frequency of regulatory T cells accompanies increased immune activation in rectal mucosae of HIV-positive noncontrollers. *J Virol*. 2011;85(21):11422–11434.
112. Chevalier MF, Weiss L. The split personality of regulatory T cells in HIV infection. *Blood*. 2013;121(1):29–37.

113. Andersson J, Boasso A, Nilsson J, et al. The prevalence of regulatory T cells in lymphoid tissue is correlated with viral load in HIV-infected patients. *J Immunol.* 2005;174(6):3143–3147.
114. Aandahl EM, Michaelsson J, Moretto WJ, Hecht FM, Nixon DF. Human CD4+ CD25+ regulatory T cells control T-cell responses to human immunodeficiency virus and cytomegalovirus antigens. *J Virol.* 2004;78(5):2454–2459.
115. Rehermann B, Nascimbeni M. Immunology of hepatitis B virus and hepatitis C virus infection. *Nat Rev Immunol.* 2005;5(3):215–229.
116. Tsai SL, Liaw YF, Chen MH, Huang CY, Kuo GC. Detection of type 2-like T-helper cells in hepatitis C virus infection: implications for hepatitis C virus chronicity. *Hepatology.* 1997;25(2):449–458.
117. Shoukry NH, Cawthon AG, Walker CM. Cell-mediated immunity and the outcome of hepatitis C virus infection. *Annu Rev Microbiol.* 2004;58:391–424.
118. Bolacchi F, Sinistro A, Ciaprini C, et al. Increased hepatitis C virus (HCV)-specific CD4+CD25+ regulatory T lymphocytes and reduced HCV-specific CD4+ T cell response in HCV-infected patients with normal versus abnormal alanine aminotransferase levels. *Clin Exp Immunol.* 2006;144(2):188–196.
119. Manigold T, Racanelli V. T-cell regulation by CD4 regulatory T cells during hepatitis B and C virus infections: facts and controversies. *Lancet Infect Dis.* 2007;7(12):804–813.
120. Wang JH, Layden TJ, Eckels DD. Modulation of the peripheral T-cell response by CD4 mutants of hepatitis C virus: transition from a Th1 to a Th2 response. *Hum Immunol.* 2003;64(7):662–673.
121. Stoop JN, van der Molen RG, Baan CC, et al. Regulatory T cells contribute to the impaired immune response in patients with chronic hepatitis B virus infection. *Hepatology.* 2005;41(4):771–778.
122. Xu D, Fu J, Jin L, et al. Circulating and liver resident CD4+CD25+ regulatory T cells actively influence the antiviral immune response and disease progression in patients with hepatitis B. *J Immunol.* 2006;177(1):739–747.
123. Yang G, Liu A, Xie Q, et al. Association of CD4+CD25+Foxp3+ regulatory T cells with chronic activity and viral clearance in patients with hepatitis B. *Int Immunol.* 2007;19(2):133–140.
124. Sugimoto K, Ikeda F, Stadanlick J, Nunes FA, Alter HJ, Chang KM. Suppression of HCV-specific T cells without differential hierarchy demonstrated *ex vivo* in persistent HCV infection. *Hepatology.* 2003;38(6):1437–1448.
125. Boettler T, Spangenberg HC, Neumann-Haefelin C, et al. T cells with a CD4+CD25 + regulatory phenotype suppress *in vitro* proliferation of virus specific CD8+ T cells during chronic hepatitis C virus infection. *J Virol.* 2005;79(12):7860–7867.
126. Rushbrook SM, Ward SM, Unitt E, et al. Regulatory T cells suppress *in vitro* proliferation of virus-specific CD8+ T cells during persistent hepatitis C virus infection. *J Virol.* 2005;79(12):7852–7859.
127. Cabrera R, Tu Z, Xu Y, et al. An immunomodulatory role for CD4(+)CD25(+) regulatory T lymphocytes in hepatitis C virus infection. *Hepatology.* 2004;40(5): 1062–1071.
128. Godkin A, Ng WF, Gallagher K, Betts G, Thomas HC, Lechler RI. Expansion of hepatitis C-specific CD4+CD25+ regulatory T cells after viral clearance: a mechanism to limit collateral damage? *J Allergy Clin Immunol.* 2008;121(5):1277–1284. e1273.
129. Li S, Jones KL, Woollard DJ, et al. Defining target antigens for CD25+ FOXP3+IFN-gamma- regulatory T cells in chronic hepatitis C virus infection. *Immunol Cell Biol.* 2007;85(3):197–204.
130. Belkaid Y. Regulatory T cells and infection: a dangerous necessity. *Nat Rev Immunol.* 2007;7(11):875–888.

131. MacDonald AJ, Duffy M, Brady MT, et al. CD4 T helper type 1 and regulatory T cells induced against the same epitopes on the core protein in hepatitis C virus-infected persons. *J Infect Dis.* 2002;185(6):720–727.
132. Eckels DD, Wang H, Bian TH, Tabatabai N, Gill JC. Immunobiology of hepatitis C virus (HCV) infection: the role of CD4 T cells in HCV infection. *Immunol Rev.* 2000;174:90–97.
133. Gremion C, Cerny A. Hepatitis C virus and the immune system: a concise review. *Rev Med Virol.* 2005;15(4):235–268.
134. Racanelli V, Rehermann B. Hepatitis C virus infection: when silence is deception. *Trends Immunol.* 2003;24(8):456–464.
135. Dai K, Huang L, Chen J, Yang L, Gong Z. Amphiregulin promotes the immunosuppressive activity of intrahepatic $CD4^+$ regulatory T cells to impair $CD8^+$ T cell immunity against hepatitis B virus infection. *Immunology.* 2014;144:506–517.
136. Yazdanbakhsh M, Sacks DL. Why does immunity to parasites take so long to develop? *Nat Rev Immunol.* 2010;10(2):80–81.
137. Wilson MS, Taylor MD, Balic A, Finney CA, Lamb JR, Maizels RM. Suppression of allergic airway inflammation by helminth-induced regulatory T cells. *J Exp Med.* 2005;202(9):1199–1212.
138. Taylor JJ, Krawczyk CM, Mohrs M, Pearce EJ. Th2 cell hyporesponsiveness during chronic murine schistosomiasis is cell intrinsic and linked to GRAIL expression. *J Clin Invest.* 2009;119(4):1019–1028.
139. Guilliams M, Oldenhove G, Noel W, et al. African trypanosomiasis: naturally occurring regulatory T cells favor trypanotolerance by limiting pathology associated with sustained type 1 inflammation. *J Immunol.* 2007;179(5):2748–2757.
140. Liu Y, Zhao M, Xu X, et al. Adoptive transfer of Treg cells counters adverse effects of Toxoplasma gondii infection on pregnancy. *J Infect Dis.* 2014;210(9):1435–1443.
141. Hesse M, Piccirillo CA, Belkaid Y, et al. The pathogenesis of schistosomiasis is controlled by cooperating IL-10-producing innate effector and regulatory T cells. *J Immunol.* 2004;172(5):3157–3166.
142. Belkaid Y, Piccirillo CA, Mendez S, Shevach EM, Sacks DL. CD4+CD25+ regulatory T cells control Leishmania major persistence and immunity. *Nature.* 2002;420(6915):502–507.
143. Doolan DL, Hoffman SL. The complexity of protective immunity against liver-stage malaria. *J Immunol.* 2000;165(3):1453–1462.
144. Scholzen A, Minigo G, Plebanski M. Heroes or villains? T regulatory cells in malaria infection. *Trends Parasitol.* 2010;26(1):16–25.
145. Minigo G, Woodberry T, Piera KA, et al. Parasite-dependent expansion of TNF receptor II-positive regulatory T cells with enhanced suppressive activity in adults with severe malaria. *PLoS Pathog.* 2009;5(4):e1000402.
146. Walther M, Jeffries D, Finney OC, et al. Distinct roles for FOXP3 and FOXP3 CD4 T cells in regulating cellular immunity to uncomplicated and severe Plasmodium falciparum malaria. *PLoS Pathog.* 2009;5(4):e1000364.
147. Amante FH, Stanley AC, Randall LM, et al. A role for natural regulatory T cells in the pathogenesis of experimental cerebral malaria. *Am J Pathol.* 2007;171(2):548–559.
148. Wildin RS, Smyk-Pearson S, Filipovich AH. Clinical and molecular features of the immunodysregulation, polyendocrinopathy, enteropathy, X linked (IPEX) syndrome. *J Med Genet.* 2002;39(8):537–545.
149. Lin W, Truong N, Grossman WJ, et al. Allergic dysregulation and hyperimmunoglobulinemia E in Foxp3 mutant mice. *J Allergy Clin Immunol.* 2005;116(5):1106–1115.
150. Chung Y, Tanaka S, Chu F, et al. Follicular regulatory T cells expressing Foxp3 and Bcl-6 suppress germinal center reactions. *Nat Med.* 2011;17(8):983–988.

151. Wing JB, Sakaguchi S. Foxp3(+) T(reg) cells in humoral immunity. *Int Immunol.* 2014;26(2):61–69.
152. Wing JB, Ise W, Kurosaki T, Sakaguchi S. Regulatory T cells control antigen-specific expansion of Tfh cell number and humoral immune responses via the coreceptor CTLA-4. *Immunity.* 2014;41(6):1013–1025.
153. Wollenberg I, Agua-Doce A, Hernandez A, et al. Regulation of the germinal center reaction by Foxp3+ follicular regulatory T cells. *J Immunol.* 2011;187(9):4553–4560.
154. Linterman MA, Pierson W, Lee SK, et al. Foxp3+ follicular regulatory T cells control the germinal center response. *Nat Med.* 2011;17(8):975–982.

CHAPTER ELEVEN

Regulatory T Cells in Autoimmune Diabetes: Mechanisms of Action and Translational Potential

Vitalijs Ovcinnikovs[1], Lucy S.K. Walker
Institute of Immunity & Transplantation, Division of Infection & Immunity, University College London, London, United Kingdom
[1]Corresponding author: e-mail address: vitalijs.ovcinnikovs.13@ucl.ac.uk

Contents

Abstract

Since the discovery of specialized T cells with regulatory function, harnessing the power of these cells to ameliorate autoimmunity has been a major goal. Here we collate the evidence that regulatory T cells (Treg) can inhibit Type 1 diabetes in animal models and humans. We discuss the anatomical sites and molecular mechanisms of Treg suppressive function in the Type 1 diabetes setting, citing evidence that Treg can function in both the pancreatic lymph nodes and within the pancreatic lesion. Involvement of the CTLA-4 pathway, as well as TGF-β and IL-2 deprivation will be considered. Finally, we summarize current efforts to manipulate Treg therapeutically in individuals with Type 1 diabetes. The translation of this research area from bench to bedside is still in its infancy, but the remarkable therapeutic potential of successfully manipulating Treg populations is clear to see.

Progress in Molecular Biology and Translational Science, Volume 136
ISSN 1877-1173
http://dx.doi.org/10.1016/bs.pmbts.2015.08.004

1. INTRODUCTION

Type 1 diabetes (T1D), also known as early-onset or juvenile diabetes, is an autoimmune disease, caused by the destruction of insulin-producing beta-cells in the pancreatic islets of Langerhans. It manifests as a development of inflammatory infiltrates in the pancreas accompanied by a loss of blood glucose homeostasis, eventually leading to life-long dependency on exogenous insulin for the affected individual. T1D is considered a result of breakdown of tolerance to self-antigens of the pancreas; however, the precise disease-causing mechanisms are multifactorial. The ~50% concordance in identical twins suggests a significant environmental influence in addition to the strong genetic links implicating mechanisms of central and peripheral immune tolerance.[1] Considerable insight into the genetics of T1D has derived from analysis of the insulin-dependent diabetes (*idd*) loci of nonobese diabetic mice (NOD), an established animal model of T1D research. Importantly, human genome-wide association studies reveal striking similarity in genetic susceptibility regions between species.[2]

Development of central tolerance takes place in the thymus, where T cell precursors are selected on the strength of their T cell receptor (TCR) interaction with peptide presented on the major histocompatibility complex (MHC) proteins (HLA in humans). Thymocytes that express TCR with high affinity for self-peptides are eliminated via the process of negative selection, which prevents highly autoreactive T cells from arising. In some cases, self-reactive T cells can differentiate into regulatory T cells (Treg). The process of central tolerance is, however, imperfect and allows some autoreactive T cells to escape into the periphery.[3,4] Fortunately, in healthy individuals the autoreactive conventional T cell (Tconv) pool is effectively controlled by peripheral tolerance, most notably via Treg populations, whose suppressive mechanisms, orchestrated by the master transcription factor FoxP3, serve to prevent autoimmunity from arising.[5]

1.1 Genetic Control of Autoimmunity

Given that HLA molecules determine what T cells can "see," it is not surprising that alleles at the HLA locus represent a major genetic susceptibility to T1D and indeed influence susceptibility to multiple autoimmune diseases.[6] In the case of T1D, variations in HLAII-DQ and -DR loci, affecting specific peptide-binding pockets, may negatively affect self-antigen presentation in the thymus, thus affecting the selection of the T cell repertoire.[7,8]

For instance, insulin-specific T cells in NOD mice recognize a suboptimal MHC-binding register of insulin that is insufficient to trigger apoptosis due to weak affinity.[9–11] A similar HLA-DQ8 restricted insulin-specific population of T cells has been identified in humans.[12] In addition to T1D, the abnormal topology of TCR binding that leads to suboptimal MHC interactions also features in multiple sclerosis and its mouse models suggesting that altered antigen presentation may be a common factor in autoimmunity.[13,14] In addition, variation in the magnitude of self-antigen expression may contribute to autoimmune susceptibility. For instance, the IDDM2 polymorphism, which maps to the promoter region of the insulin gene has been linked to reduced insulin mRNA in the thymus, possibly influencing central tolerance development via impaired deletion of insulin-specific T cells; or peripheral tolerance via reduced development of Treg.[15,16]

PTPN22, a nonreceptor tyrosine phosphatase implemented in TCR signaling, is also strongly associated with susceptibility to autoimmune disease.[17,18] Several studies have suggested that the R620W variant of this gene lowers TCR signaling and allows for more autoreactive T cells to escape thymic selection that would otherwise develop into Treg.[19,20] Maine[21] and Brownlie[22] report an increase in thymic Treg output in PTPN22-deficient mice, pointing to a possible shift in selection thresholds. Conversely, others report PTPN22-R620W as a gain-of-function variant that is unable to bind Csk, a negative regulator of TCR activation. Thus by reducing the threshold of T cells activation R620W is responsible for the overactive responses from lower affinity TCR T cells in the periphery.[23] A mouse model of the human R620W variant supports this by showing enhanced TCR signaling as a cause of increased effector and memory T cell generation alongside general features of systemic autoimmunity, while the Treg function remains unaffected.[24] Work from the Zamoyska lab demonstrated that the increase in T cell activation in PTPN22 knockout mice was due to enhanced responses to low-affinity antigens in the periphery, indicating that PTPN22 acts as a brake on responses to weak antigens.[25] Dampening of weak-to-medium TCR signals by phosphatases has recently been shown to be critical for selection of weakly self-reactive clones to occur in the thymus.[26] It is therefore possible that controlled responses to weak antigens may play a role in both thymic selection and later activation in the periphery.

Several mutations within the interleukin-(IL-)2 pathway are also found to associate strongly with T1D, including IL-2 itself as well as IL-2R-α (CD25) and IL-2R-β (CD25 and CD122) receptor susceptibility loci.[1,27,28]

IL-2 is central to the development and homeostasis of Treg that constitutively express CD25, the α-chain of the high-affinity IL-2 receptor. At higher doses it acts as a growth factor for effector CD4 and CD8 cells, therefore sensing and removing IL-2 from the local environment is an important mechanism of Treg.[29] Whether the contribution of IL-2-related polymorphisms is causative for T1D is unclear. This is due to low disease penetrance despite common occurrence of these polymorphisms in humans.[30] In NOD mice, who harbor all of these mutations, low IL-2 levels, reduced CD25, and increased apoptosis are observed on Treg within the pancreatic lesion, suggesting altered Treg homeostasis. Furthermore, the absence of IL-2, as an inhibitor of T follicular helper (Tfh) development, might also lead to the predominant Tfh cell phenotype found in T1D diabetes.[31,32]

Lastly, cytotoxic T lymphocyte antigen 4 (CTLA-4), a major suppressive mechanism of Treg cells, has been identified as a susceptibility locus in T1D.[33] CTLA-4 plays a central role in preventing T cell activation by depriving them of costimulatory ligands CD80 and CD86 (B7) from the antigen-presenting cells (APC).[34] The role of costimulation is to dramatically enhance the TCR-MHCII signal and amplify T cell activation, leading to subsequent clonal expansion and differentiation aided by the cytokine environment. Importantly, B7/CD28 pathway is carefully balanced by CTLA-4, a close homologue of CD28 with greater affinity to its ligands that does not relay a costimulatory signal.[35] CTLA-4 is dramatically upregulated upon TCR-MHCII binding on Tconv and to an even greater extent on Treg.[36] Levels of CTLA-4 expression in humans[37] as well as various mouse isoforms have been connected to T1D susceptibility.[38,39]

Taken together, the genetic association data now implicate considerable numbers of loci in contributing to autoimmune susceptibility. Collectively, these data firmly indicate a role for T cells and increasingly point toward variability in immune tolerance mechanisms that are required to keep self-reactive T cells under control. From an immunological perspective, it is increasingly clear that such self-tolerance is in large part maintained by specialized populations of Treg.

1.2 Control of Autoimmune Diabetes by Regulatory T Cells

Multiple lines of evidence support a key role for Treg in suppressing the development of T1D. Perhaps most striking is the observation that T1D manifests in approximately 80% of individuals suffering from IPEX due to deficient Treg development.[40] This suggests a nonredundant role for

Foxp3-expressing Treg in preventing islet autoimmunity in humans. Substantial evidence from mouse models also supports this conclusion. BDC2.5 TCR (specific for an islet autoantigen) transgenic NOD mice rendered Treg deficient by the scurfy mutation developed rapid and aggressive diabetes, with 100% incidence by 21 days post-birth.[41] Similarly, punctual ablation of Treg in BDC2.5/NOD mice, using the transgenic diphtheria toxin receptor system, leads to fulminant diabetes within 3 days.[42] Consistent with this idea, diabetes caused by adoptive transfer of pancreas-specific conventional T cells can be controlled by coinjection of $CD25^+$ Treg.[43]

In T1D patients, despite initial reports to the contrary,[44] it appears that the frequency of Treg in peripheral blood is probably unchanged compared to healthy controls,[45–47] and Treg from patients typically exhibit normal suppressive capability *in vitro*.[48] Nevertheless, careful longitudinal analysis during the 12 months following T1D onset suggested a transient decrease in Treg function that manifested between 2 weeks and 6 months post-diagnosis and had resolved by 9 months.[49] Some evidence also suggests that certain individuals with longstanding T1D may exhibit reduced Treg function,[50] for example, as a result of alterations in the capacity of their Treg to respond to IL-2.[51,52]

Thus, Treg appear to be an important defense against the development of autoimmune diabetes in both mice and humans. Understanding the mechanisms by which Treg control autoimmunity offers an opportunity to gain insight into the molecular control of disease development and ultimately to potentially develop better treatments.

2. SITES OF TREG ACTION IN T1D

An important question concerning the regulation of autoimmune diabetes by Treg is that of anatomical location: do Treg exhibit their suppressive effects in the pancreas-draining lymph node or within the pancreas itself? Addressing this issue has implications for our understanding of the biological processes that naturally inhibit diabetes in healthy individuals, as well as highlighting migration considerations that are pertinent to therapeutic Treg manipulation.

2.1 Pancreatic LN

The priming and activation of naïve T cells occurs in the draining lymph node where antigen-bearing dendritic cells (DCs) migrate from the tissue. When islet-specific BDC2.5 TCR transgenic CD4 T cells were adoptively transferred into NOD recipients, they proliferated specifically in the pancreatic lymph node (PanLN), and this occurred at timepoints significantly earlier than insulitis could be detected.[53] Transgenic antigens (e.g., Ovalbumin, Hen Egg Lysozyme) expressed under the insulin promoter are also presented to T cells in the PanLN.[53,54] Elegant studies using intrapancreatic injection of CFSE-labeled cells[55] or mice transgenically expressing GFP under the control of the insulin promoter[56] have permitted the DCs trafficking antigen to the PanLN to be visualized, indicating that they bear a CD11c+CD11b+ CD8α-phenotype. The importance of T cell activation that takes place in the PanLN is illustrated by its excision, which leads to protection from disease in NOD mice.[57,58] The interaction between T cells and DCs within lymph nodes has been demonstrated *in vivo* by numerous studies using two-photon microscopy.[59–61] There, T cells actively scan the DC network in search of their cognate antigen and upon recognition exhibit arrest and form stable interactions with DCs.[62] Celli *et al.*[63] estimate that a 6 h duration of the arrest phase is required to trigger proliferation and clonal expansion. The ability of multiple transient interactions (swarming) to initiate T cell proliferation has also been reported,[64] and abundance and quality of antigen[65] as well as the state of DC maturation[66] are key factors in determining the size and nature of the T cell response.

Tadokoro *et al.*[67] demonstrate that the arrest duration of Tconv upon recognition of antigen delivered by immunization is reduced twofold when Treg are also present. This phenomenon of Treg influencing the nature of T cell:DC interactions is also observed in response to self-antigen as demonstrated by Tang *et al.*[68] In this study, the authors demonstrated that islet-specific (BDC2.5) Treg formed stable interactions with the APCs in the PanLN, while denying stable contact formation by BDC2.5 Tconv and promoting swarming behavior. This was associated with reduced Tconv proliferation. Importantly, Treg did not form contacts with Tconv directly and did not appear to compete for antigen on the antigen-bearing DC as BDC2.5 TCR Treg could suppress conventional T cells specific for a different antigen (BDC12-4.1 TCR Tconv). This carries an important implication for therapeutic use of Treg since it implies that a single specificity Treg population can suppress self-reactive T cell clones of multiple epitope specificities provided they are interacting with antigen in the same microenvironment. Using a model involving diabetogenic TCR transgenic CD8 T cells (8.3-CD8), Serra *et al.* generated evidence that Treg were able to modulate the phenotype of APC in the PanLN of NOD mice and suppress CD8 T cell responses.[69] The effect of Treg in this system was to lower expression of costimulatory ligands on the PanLN-resident DCs, particularly CD80, as well as altering expression of CD40 and CD11c.

In vitro assays of Treg function traditionally measure the inhibition of conventional T cell proliferation, a readout of suppression which is robust and dose dependent. However, the effects of Treg *in vivo* in diabetes models have been less clear. Indeed our own experiments investigating the consequences of Treg in an adoptive transfer model of diabetes revealed a surprisingly modest effect on Tconv proliferation and IL-2 production, but a dramatic inhibition of Tconv IFN-γ production and islet infiltration.[43] Decreased IFN-γ production was linked with impaired upregulation of the chemokine receptor CXCR3,[43] consistent with a potential role for this chemokine receptor in mediating pancreas entry. In other models, the presence of Treg robustly inhibited the proliferation of Tconv in the PanLN.[68] These conflicting observations were largely reconciled by the findings of Tang *et al.*[68] who reported that if Treg were transferred 2 days prior to Tconv, they were able to completely abrogate Tconv proliferation in the PanLN. In contrast, when transferred at the same time, proliferation was unaffected but IFN-γ production was markedly reduced. Thus, the balance of Treg:Tconv within the local microenvironment likely dictates whether Treg suppression targets Tconv cytokines alone or also shuts down Tconv proliferation. Interestingly, Treg appear to selectively inhibit IFN-γ rather than orchestrating blanket inhibition of Th1 differentiation since induction of Tbet and TNFα production are spared.[70] Thus, the actions of Treg in the pancreatic LN can involve both effects on Tconv proliferation and on effector cytokine production.

A role for Treg suppression within the draining LN is consistent with the observation that CCR7 −/− Treg, that lack the capacity to home to LN, fail to suppress contact hypersensitivity or IBD despite effectively migrating to inflamed skin or gut, respectively.[71] However, evidence from T1D models suggests that the draining LN is not the only site of Treg action as will be discussed below.

2.2 Pancreatic Islets of Langerhans

Treg are known to be present within the infiltrated pancreas in mouse models of T1D raising the possibility that they may also act at the site of the autoimmune attack (Fig. 1). In BDC2.5/NOD mice, $CD4^{+}CD25^{+}CD69^{-}$ Treg are found alongside conventional CD4 cells within the lesion even before the onset of disease. Infiltrating T cells appear to arrive at around 2–3 weeks of

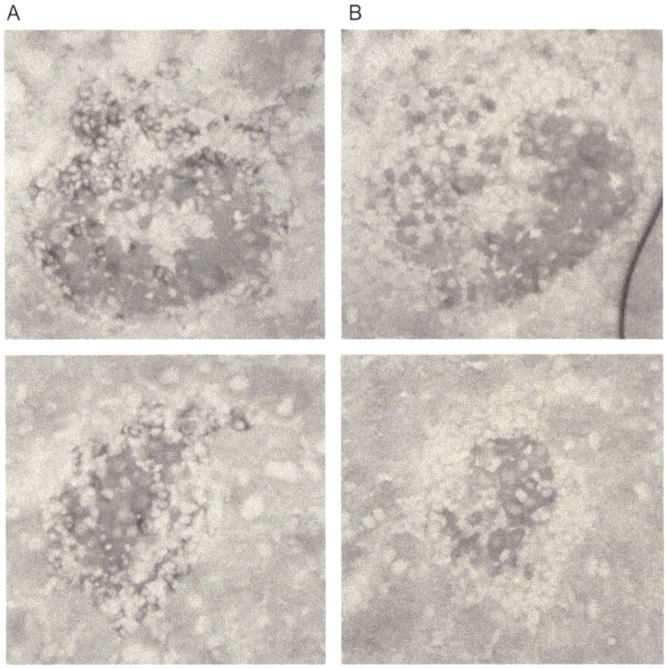

Figure 1 Treg are present in the pancreatic islet infiltrate of diabetic mice. Images show serial sections from frozen acetone-fixed pancreas of diabetic DO11xripOVA mice. (A) CD4+ T cells (blue; dark gray in the print version) infiltrating the islet of Langerhans marked by insulin-producing (brown; dark gray in the print version) beta-cells. (B) FoxP3+ Treg cells (blue; dark gray in the print version) are present within the infiltrate. Insulin-producing beta-cells are stained in brown (dark gray in the print version).

age, more than 2 months before the onset of disease.[72] Indeed live imaging of diabetogenic CD8 T cells suggests that they enter the lesion via postcapilliary vessels directly adjacent to the islet site that appear to become "leaky" during the onset of prediabetes.[73]

Whether such "leakiness" permits entry of bystander T cells to the islet has been a contentious issue, with some studies suggesting islet entry is tightly restricted to antigen-specific T cells.[74,75] However, more recently it has been suggested that T cell infiltration is beta-cell antigen specific in the early stages but progresses to become more open to cells with wider specificity after the onset of islet inflammation, with IFN-γ signaling and induced expression of CXCR3 chemokines playing a key role in the recruitment of bystander T cells.[76] Recent work using the Kaede transgenic mouse model to track fluorescently labeled immune cells from a nondraining LN supports the open status of the diabetic lesion at all stages of disease with many infiltrating T cells bearing a naïve phenotype.[77] Treg were also shown to access the pancreatic infiltrate in this model, although they exhibited less active migration dynamics both in terms of exiting the labeled LN and appearing in the pancreas.

The presence of Treg in the pancreas of humans with T1D is far harder to assess. A significant factor here is that insulitis is mainly thought to be present close to T1D diagnosis[78] such that even pancreas tissue obtained from individuals with "new onset" disease may exhibit very little lymphocytic infiltration. Welcome developments in early disease detection and management have all but eradicated fatal ketoacidosis at T1D presentation. However, prior to these developments, a small number of unfortunate cases meant that pancreas tissue was available to study very close to the time of T1D diagnosis. In one such instance, tissue was obtained from a 12-year-old girl who died within 24 h of diagnosis and had insulitis in 24% of islets. Most of the infiltrating T cells were positive for HLA-DR, suggestive of activation, and a proportion expressed CD25 raising the possibility that they were Treg.[79] Willcox and colleagues further demonstrated the presence of FoxP3+ cells within the inflamed islets of recent onset T1D patients, although these cells were understandably rare given the scarcity of lymphocytes in general.[80] In this regard, it should be noted that the highest frequency of FoxP3+ cells within infiltrated islets was observed prior to the onset of diabetes in NOD mice,[31] a timeframe that is understandable difficult to study in humans.

Assuming then that Treg are present within the pancreatic infiltrate in mice and humans with T1D, do they elicit suppressive function at this site? Evidence suggesting this is the case comes from careful comparative analysis of diabetes development in either the presence or absence of Treg in the BDC2.5/NOD model. Abrogation of Treg development by the scurfy mutation did not alter the activation of T cells in the pancreatic LN in this model, but instead had a profound impact on the nature of the insulitis rendering it immediately aggressive.[41] These data suggest that Treg operate within the pancreas itself to moderate the capacity of conventional T cells to penetrate the islet cell mass and elicit damage to the beta-cells. A similar conclusion was reached in experiments where Treg were targeted for deletion by expression of the diphtheria toxin receptor under the control of the FoxP3-promoter: within hours of diphtheria toxin-mediated Treg depletion, the islet-resident cells mounted a dramatic response, including large amounts of IFN-γ production by islet-resident NK cells.[42]

Further support for the idea of intrapancreatic regulation comes from the observation that the capacity of cyclophosphamide treatment to promote diabetes was associated with a sustained depletion of Treg in the pancreas, but not the PanLN of NOD mice.[81] In addition, progression to diabetes in NOD mice was associated with a decrease in the Treg:Teff ratio in the pancreas, and not in the PanLN, suggesting a loss of regulation *in situ*.[31] During disease development, islet-infiltrating Treg displayed a decreased expression of CD25, coupled with lower expression of FoxP3 and survival factor Bcl-2, suggesting the decline in frequency is due to increased apoptosis. This, in turn can be a result of defective IL-2 production, a cytokine known to be essential for Treg homeostasis.

Additional insight has been provided by Mahne *et al.*,[82] who demonstrated that Treg reduced the accumulation of islet-specific CD8 T cells. Furthermore, Treg abrogated the production of IFN-γ by both CD8 and CD4 T cells through the reduction of mTOR pathway activation by depriving effector cells of IL-2. Local APC phenotype was also modestly affected in these experiments, with decreased expression of CD86 and CD40, although T cell:DC interactions did not appear to be disrupted.[82]

Thus, considerable evidence suggests that Treg are present within inflamed islets and are able to exert suppressive function at this location.

3. MECHANISMS OF TREG SUPPRESSION IN T1D

The importance of Treg in preventing autoimmunity is extremely well demonstrated in the scurfy mice[83,84] that have a frame shift mutation in FoxP3, the master transcription factor responsible for the Treg programme.[85–87] These mice succumb to an early death from a lymphoproliferative syndrome with multiorgan inflammation. To achieve regulation, Treg are believed to employ a broad array of suppressive mechanisms, including use of soluble mediators and cell-to-cell contact.[88,89] Some of these have been demonstrated to contribute to immune regulation in a T1D autoimmune setting.

3.1 Role of CTLA-4

CTLA-4 is an essential negative regulator of T cell activation, deficiency of which causes lethal autoimmunity in mice.[90,91] CTLA-4 binds to the same ligands as the T cell costimulatory receptor CD28 (CD80 and CD86, previously called B7-1 and B7-2, respectively). Several mechanisms explaining the action of CTLA-4 have been proposed. Early models focused on the idea that CTLA-4 delivers a negative signal when bound to its ligands. However, evidence for this model is incomplete,[92] and within this framework there is currently no explanation of how CTLA-4 can suppress Tconv, but not Treg cells that express it at even higher levels. Furthermore, mouse bone marrow chimeras that produce a mixture of wild-type T cells and ctla4 −/− T cells are healthy, and show no evidence of lymphoproliferation or autoimmune tissue infiltration.[93,94] The elegant bone marrow chimera experiment serves as strong evidence for an extrinsic mode of CTLA-4 action, in which CTLA-4 on wild-type T cells effectively controls the activation of *ctla4−/−* cells.[93,94] This idea is consistent with a role of CTLA-4 in Treg since Treg are the archetypal mediators of cell extrinsic regulation.[95] It has been known for many years that CTLA-4 can compete with CD28 for its ligands and thus diminish costimulation and prevent T cell activation.[96] Indeed, costimulation of CD28 by CD80 and CD86 is the driver of pathology in *ctla4−/−* mice, as *ctla/cd28* double knockout or *ctla4/cd80/cd86* triple knockout animals show no evidence of T cell overactivation.[97,98] Thus, the biological role of CTLA-4 appears to be to regulate signals delivered through CD28. Interestingly, in numerous studies downregulation of CD80 and CD86 expression on APCs has been observed as the consequence of CTLA-4 action.[34,69,99–103] More recently, Qureshi *et al.*[104] have demonstrated that CTLA-4 can physically remove B7 from the surface of an APC, internalize and drive it down the endocytic compartment for eventual degradation by the process termed trans-endocytosis.

CTLA-4 is clearly necessary for the maintenance of tolerance to islet antigens (Fig. 2), as *ctla4−/−* BDC2.5 TCR mice develop diabetes at higher incidence and an accelerated rate compared to their CTLA-4 sufficient counterparts.[105] Diabetes is also enhanced in mice treated with a blocking anti-CTLA-4 antibody.[106] Furthermore, CTLA-4-deficient DO11 TCR transgenic T cells exhibit an increased capacity to induce diabetes in an adoptive transfer model compared with CTLA-4 sufficient T cells.[107] Whether CTLA-4 acts in the Treg compartment to suppress diabetes appears to depend on the particular *in vivo* model under study. In one TCR transgenic adoptive transfer system, Treg-expressed CTLA-4 was crucial for suppression of diabetes caused by adoptive transfer of islet autoantigen-specific CD25- T cells, as, unlike wild-type Treg, CTLA-4-deficient Treg failed to control the disease.[101] However, in another system, TCR transgenic Treg lacking CTLA-4 were as effective as their CTLA-4-expressing counterparts at inhibiting diabetes.[108] Importantly, in the latter model the recipient animals were CD28 −/− NOD mice, meaning that the pathogenic immune response was not driven by CD28 costimulation. If CTLA-4 serves to regulate CD28 engagement, CTLA-4 function would not be predicted to control a CD28-independent response. It therefore makes sense that CTLA-4 was not required for regulation in this setting

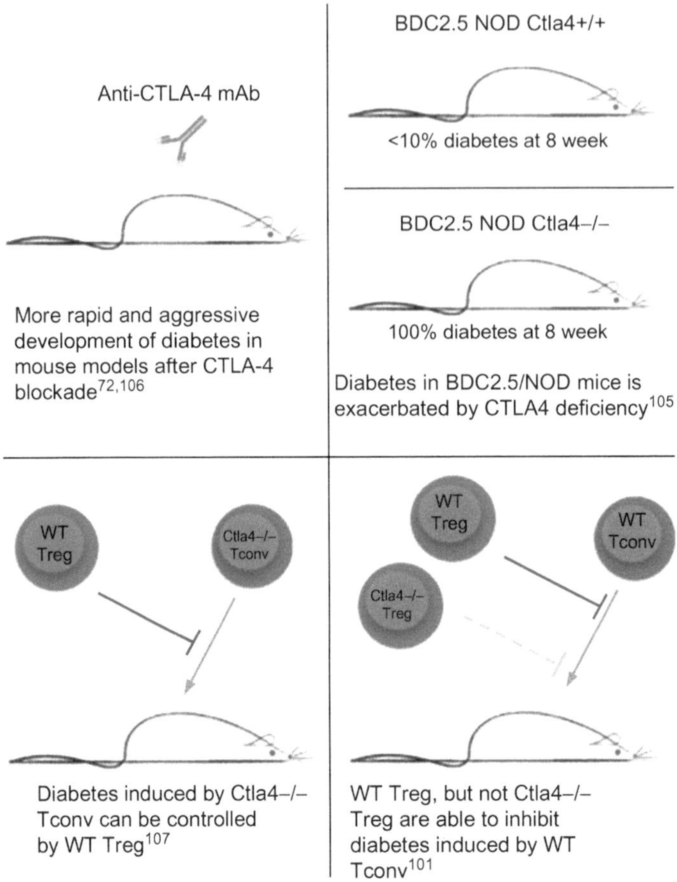

Figure 2 Summary of the evidence that supports a role for the CTLA-4 pathway in the regulation of autoimmune diabetes. This was demonstrated in a series of studies that utilized antibody blockade, genetic deficiency, and adoptive transfer methods.

and that alternative Treg mechanisms were at play. In this regard, it is known that Treg from CTLA-4-deficient mice express higher levels of TGF-β than wild-type Treg[101,109] and that this can partially compensate for the lack of CTLA-4 in *in vitro* suppression assays.[109]

Whether CTLA-4 serves as a Treg mechanism to regulate T1D in humans is yet to be revealed. However, heterozygous mutations in CTLA-4, resulting in defective Treg function, have recently been described in humans[110,111] and notably two studies suggest that T1D is among the autoimmune manifestations associated with this condition.[111,112]

3.2 Role of Cytokine Deprivation

Treg express high levels of CD25 (IL-2R-α) which in conjunction with IL-2R-β and the common-γ chain forms the high affinity IL-2 receptor. It has therefore been suggested that one facet of Treg suppression may be to consume IL-2 locally, thereby depriving conventional T cells of this well-recognized growth factor.[113,114] This notion was challenged by the observation that Treg from CD25-deficient mice, isolated on the basis of FoxP3-GFP expression, showed intact suppressive function *in vitro*.[115] Nevertheless, evidence that local cytokine deprivation can contribute to Treg function has been steadily growing. Pandiyan *et al.*[116] demonstrated that Treg were able to consume IL-2 produced by conventional T cells, taking advantage of an IL-2-GFP reporter to preclude the possibility that the Treg were instead inhibiting IL-2 transcription. Deprivation of IL-2 was associated with a fraction of the Tconv undergoing apoptosis by a pathway involving the Bcl2 family proteins Bad and Bim. The capacity of Treg to deplete their local microenvironment of IL-2 could potentially support bystander suppression of Tconv with distinct specificities that are present in the vicinity. In models of cytokine deprivation, the spatial separation of Treg and Tconv becomes a significant factor in determining the degree of suppression[117] as the cells embark on a "tug-of-war" over local IL-2.[118] Conclusions about the contact-dependence of Treg suppression based on transwell assays, where Treg and Tconv are separated by distances of an order of 100 times the diameter of a lymphocyte, may not be sound considering that IL-2 competition may require a proximity in the region of 1–2 cell diameters.[119]

Assuming then that cytokine deprivation can represent one mechanism of action of Treg, is this mechanism at play in the regulation of diabetes? In this regard, the loss of CD25 expression on islet-infiltrating Treg that precedes the onset of diabetes in NOD mice[31] could render them less able to sequester IL-2 from conventional T cells in addition to cutting them off from critical survival signals. Furthermore, within 8 h of Treg depletion in BDC2.5/NOD mice, a signature of IL-2-induced genes was detectable in the islet-resident NK cells and flow cytometry revealed a striking increase in STAT5 phosphorylation.[120] This suggests that the capacity of Treg to limit bioavailability of IL-2 within the islet lesion is a key factor controlling the immune response at this site. Consistent with the findings of Pandiyan *et al.*, this study found that the capacity of CD4 T cells to synthesize IL-2 at early timepoints was not altered by Treg, implying that IL-2 consumption rather than transcriptional inhibition was responsible.[120]

Thus, Treg appear to regulate ongoing anti-islet responses by competing for local IL-2: the defects in IL-2 signaling associated with T1D[51,52,121–122] may be relevant to this in ways yet to be elucidated.

If sequestering IL-2 represents a mechanism by which Treg control anti-islet T cell responses, then factors that alter IL-2 production may have an impact on immune control. In this regard, data from mouse models have suggested that intra-islet T cells produce the cytokine IL-21.[32,123,124] Interestingly, IL-21 was able to substitute for the lack of IL-2 in conventional T cells but not Treg,[125] thereby selectively impairing the regulatory arm of the immune system. Furthermore, IL-21 increased CD25 expression in Tconv rendering them better able to compete for residual IL-2. Therefore, factors that decrease T cell production of IL-2 have the capacity to interfere with immune regulation by impairing Treg homeostasis and interfering with their capacity to dominantly sequester this cytokine.

3.3 Role of TGF-β

It is now established that Treg are able to use TGF-β to suppress conventional T cell responses in some settings.[126–128] Indeed high expression of the integrin αvβ8 on effector Treg, enabling them to activate latent TGF-β, has recently shown to be important for their capacity to regulate active inflammation.[129] In the setting of diabetes, local TGF-β may act to positively regulate Treg homeostasis within the islet lesion.[130] The contribution of TGF-β to Treg control of diabetes has been tested in adoptive transfer models: diabetes induced by adoptive transfer of NOD splenocytes to NOD-SCID recipients was inhibited by transfer of CD25+ Treg, however, protection was abrogated in the presence of anti-TGF-β antibody.[131] In a different diabetes model, it was shown islet-reactive CD8 T cells that were unable to receive TGF-β signals were refractory to control by Treg.[132] On the other hand, data obtained from the BDC2.5/NOD diabetes model indicated that TGF-β neutralization failed to recapitulate the effects of Treg depletion,[120] despite internal controls indicating that the antibody was bioactive. This emphasizes the challenge of pinpointing which particular suppressive mechanism is dominant at any one time in a cell type (Treg) that is a "jack of all trades."[89] Assuming that many suppressive mechanisms will be at play simultaneously (CTLA-4 function, IL-2 deprivation, inhibitory cytokines), blocking one will only abrogate Treg function if the others fail to compensate. Small differences in the particular diabetes model under study, or the timing of an experimental intervention, may lead to differences in the cellular composition of the lesion, and shift the balance between which regulatory mechanism is dominant. In short, blocking an individual suppressive mechanism and looking for Treg failure is a blunt tool for surmising whether that mechanism is in use. Nevertheless, on balance the evidence suggests that TGF-β can contribute to regulation of the anti-islet immune response.

4. THERAPEUTIC MANIPULATION OF TREG IN T1D

4.1 Therapeutic Treg Expansion

The current understanding of etiology of autoimmune disease and the molecular pathways that underlie the maintenance or breakdown of immune tolerance offer the potential for a range of therapeutic interventions. Such interventions can broadly act to dampen immune activation, promote immune tolerance or, perhaps for the largest effect, utilize a combination of both. With respect to Treg, this can be achieved by altering the Treg:Tconv ratio through boosting Treg numbers or enhancing their function (Fig. 3).

4.1.1 Anti-CD3

Monoclonal antibodies have shown promise in depleting pathogenic T cells and creating a favorable Treg:Tconv ratio. FcR nonbinding anti-CD3-ε treatment promoted tolerance in NOD mice by selectively depleting pathogenic T cells, while sparing the Treg population.[133] In addition, the intervention appeared to induce the generation or expansion of Treg that were capable of suppressing in a TGF-β-dependent manner.[131] In an experiment involving the treatment of overtly diabetic mice, coadministration of a

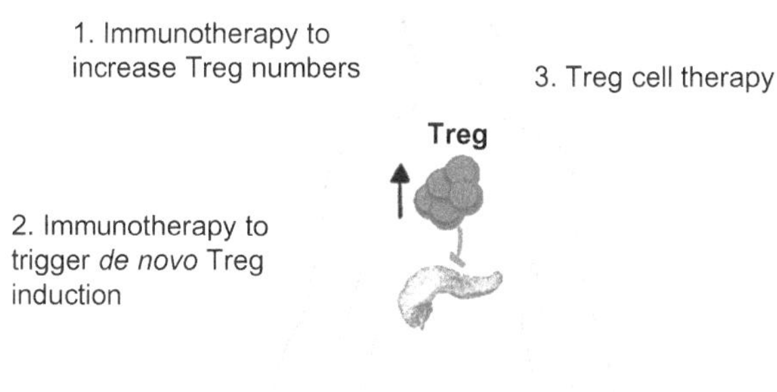

Figure 3 Therapeutic approaches centered around the enhancement of the Treg population for the treatment of Type 1 diabetes, with the goal of inhibiting autoimmune destruction of the pancreas.

blocking anti-CTLA-4 Ab abrogated the disease remission induced by anti-CD3 treatment[134]: this could suggest a role for CTLA-4-dependent Treg function, or could reflect the fact anti-CD3 induced tolerance mechanisms cannot compensate for the lack of the CTLA-4 pathway. In a humanized mouse model, anti-CD3 was shown to trigger IL-10 production from a subset of T cells that migrated to the gut.[135]

The exact route by which anti-CD3 can differentially impact effector T cells and Treg is still unknown. Valle *et al.* demonstrate that *in vivo* anti-CD3 effects correlate with heterogeneity of CD3 expression levels and therefore lower expressing Treg are preferentially spared.[136] It has also been shown that anti-CD3 treatment downregulates IL-7R on Tconv while upregulating it on Treg, suggesting a model in which IL-7 signals may contribute to the maintenance of Treg homeostasis.[137]

Following the success of this therapy in animal models, the anti-CD3 treatment has also been tested in clinical trials of human T1D. A single treatment course of anti-CD3-ε (Teplizumab) in newly diagnosed individuals led to significant improvements in C-peptide response 2 years post-treatment.[138] The improvement correlated with a reduction of IFN-γ production by CD8 T cells as well as higher levels of IL-10. In a previous trial, however, anti-CD3 was found to cause increased adverse effects with elevated dosing and did not prevent the decline in C-peptide production, although there was a trend for the fall in production to be slower.[139] Glycosylation modified anti-CD3-ε (Otelixizumab) at a total dose of 48–65 mg preserved beta-cell function and reduced insulin requirements maintaining the effect at 3 years post-trial.[140] Reactivation of EBV was a prominent side effect that is resolved by using lower drug dosing, which may come at the expense of efficacy.[141,142]

Thus, anti-CD3 antibodies have shown considerable promise in modulating the diabetogenic immune response in mouse models and T1D patients, with the actions of Treg most likely contributing the protective mechanisms. However, challenges remain regarding the choice of endpoints for clinical trials (with composite endpoints proving problematic[143]) and the selection of doses which preserve efficacy while avoiding adverse effects.

4.1.2 Low-Dose IL-2

It is well known that IL-2 is a critical survival factor for Treg. This is exemplified by the lymphoproliferation and autoimmunity observed after blockade[144] or genetic deficiency[145,146] of IL-2 in mice. Harnessing the IL-2 pathway to promote Treg homeostasis therefore offers huge therapeutic potential.[147] For instance in diabetic mice, low-dose IL-2 promoted homeostasis of Treg in the pancreas and it was further effective at preventing and reversing T1D.[31,148,149]

In humans, low-dose IL-2 has been trialed in GvHD patients in a dose escalation study that showed protective effect of Treg at lower doses. Furthermore, patients demonstrated increased thymic output and proliferation of Treg with improved resistance to apoptosis.[150] The results were in line with the role of IL-2 for Treg development and homeostasis. A similar boost in Treg function was found in patients with autoimmune HCV vasculitis[151]; and low-dose IL-2 was generally well tolerated.[150]

With regard to diabetes, the therapeutic expansion of Treg by IL-2 has also been tested in clinical trials. In the first trial, IL-2 was used in conjunction with rapamycin, the mTOR inhibitor known to inhibit the activation of effector T cells.[152] Although an increase in Treg number was observed and persisted for some time after the trial, a drop in C-peptide was seen, signaling a deleterious effect of the treatment. Moreover, an expansion in NK cell population was also noted in addition to marked eosinophilia. In a second IL-2-only dose-range estimation trial, no negative effect on beta-cell function was observed, pointing to the deleterious contribution of rapamycin in the previous study.[153,154] Interestingly, further studies in the NOD model suggested that although rapamycin augmented IL-2-induced Treg expansion it impaired the capacity of those Treg to suppress T cell cytokine production.[149] Subsequent immunomonitoring of the second trial revealed a dose range of 0.3–1 MIU/day that does not initiate the NK cell transcriptional profile while expanding and activating Treg, as measured by enhanced GITR, CD25, and CTLA-4 expression.[155]

Overall, despite the caution warranted by the prospect of IL-2 altering additional immune cell populations, harnessing the IL-2 pathway to promote Treg homeostasis appears a promising strategy in T1D and other autoimmune diseases. A meticulous approach to dose-range estimation will be important for success, and relevant dose response trials are already underway.[156] The modification of IL-2 to alter its half-life and binding preferences also offers exciting opportunities to maximize the immune-regulatory effects of this therapy.[157]

4.2 Therapeutic Treg Induction

In addition to polyclonal Treg expansion, attempts have been made to influence tolerance directly *in vivo* in an antigen-dependent fashion. Tolerance to an antigen can be established at mucosal sites, most notably, the gut, via the induction of Treg from naïve conventional T cells.[158] In NOD mice, oral,[159] nasal, and subcutaneous[160] routes of insulin administration all showed some disease protection. The following human trials have despite initial promise[161] largely presented with negative results.[162–164] Nevertheless, TrialNet oral insulin T1D prevention study is currently underway and with more detailed immunological analysis may shed light on the translational challenges.[165]

In humans, where insulin injection is a standard therapy to control blood glucose levels, patients typically enter a short remission phase, shortly after the start of treatment. It has been suggested that this may be associated with the induction of insulin-specific Treg.[166] Attempts to mimic this phenomenon in NOD mice showed early promise, where administration of exogenous insulin has led to protection[167] or delayed onset of T1D,[168] suggesting a potential benefit of insulin immunization. Once again, translation has been difficult, as intramuscular insulin in adjuvant was ineffective at preserving the C-peptide response, however, Treg development was observed and these newly induced cells persisted 2 years post-study.[169] Immunization with the 65 kDa isoform of glutamic acid decarboxylase, another major T1D antigen, showed mixed results, with a potential preservation of stimulated C-peptide level, but no difference were observed in other measurements when compared to the placebo group.[170–172]

Despite limited success of Treg induction approaches, new methods are still being researched. For instance, in a recent study by the Roncaloro group a novel approach at Treg induction was tested.[173] By targeting non-integrating lentiviral vector-mediated expression of insulin B chain 9-23 to hepatocytes, induction of both B9-23 specific Tconv and Treg was achieved in NOD mice. The induced Treg were able to inhibit islet infiltration and in combination with a single suboptimal dose of anti-CD3 this was able to reverse T1D. This study suggests that an antigen-specific non-integrating gene therapy that offers low risk of genotoxicity[174] could be combined with other immunomodulatory approaches to potentially become an effective T1D treatment.

4.3 Treg Cell Therapy

A different approach to harness the multitude of tolerance mechanisms offered by Treg is by treating patients with *in vitro*-expanded Treg cells. Similar to therapeutic Treg expansion *in vivo*, the goal is to modulate the deleterious immune response to islet auto-antigens without causing global immunosuppression and increased susceptibility to infection or cancer. Furthermore, the tolerogenic environment created by such therapy could conceivably be long-lasting without the need for future cell reinfusions or for lifetime drug treatment.

In mice, ample evidence demonstrates that adoptively transferred Treg are able to effectively prevent[43,175–177] and even reverse[178,179] autoimmune diabetes. Although antigen-specific Treg are more potent than polyclonal Treg populations, there are indications that functional suppression can extend beyond the TCR specificity of the injected Treg population.[178] Accordingly, injection of BDC2.5 TCR transgenic Treg, which recognize a single islet antigen, is sufficient to control a polyclonal autoimmune response in NOD mice, that presumably encompasses multiple antigenic targets.[177,178] This is consistent with the idea of "infectious tolerance" as proposed by Waldmann and colleagues[180] whereby "linked suppression" can be generated against antigens presented by the same APC.[181] In the case of T1D, this is epitomized by BDC2.5 Treg exhibiting the capacity to suppress the proliferation of 4.1 TCR transgenic T cells that are located in the same draining PanLN but presumably recognize a different islet antigen.[68]

A certain constraint when it comes to humans is the difficulty of obtaining sufficient numbers of cells from peripheral blood. Isolation can be performed on $CD4^{+}CD127^{lo}CD25^{+}$ surface markers, which enriches for the $FoxP3^{+}$ positive cells to >95% purity.[45] The following expansion with anti-CD3- and anti-CD28-coated bead stimulation in the presence of high concentrations of IL-2 results in a 3000-fold increase in cell numbers after 2 weeks and a 90% FoxP3 purity with a stable Treg phenotype.[48] Clinical grade protocols have been developed and are utilized in human trials.[182]

A second issue relates to the stability of adoptively transferred Treg populations. While the $FoxP3^{+}$ subset is relatively stable, Treg display a degree of phenotypic plasticity under the influence of inflammatory conditions.[183] Reprogramming of $FoxP3^{+}$ BDC2.5 T cells was observed in the NOD setting by Zhou *et al.*[184] in mice that permanently expressed a YFP reporter following FoxP3 expression. The frequency of "exTreg" cells, that had once expressed FoxP3 but were not currently doing so, was small; however, this population was slightly larger in the pancreas of diabetic mice. FoxP3 expression is sustained by methylation and is reliant on the presence of IL-2 signaling.[185,52] One study has suggested that IL-6 can remethylate a suppressive CpG motif and abrogate FoxP3 expression.[186] It is therefore possible that under conditions of low IL-2 and increased proinflammatory cytokines, Treg can acquire plasticity through the loss of FoxP3 expression. Data from Battaglia group offer partial supports for the notion of Treg plasticity in T1D by demonstrating that cells isolated from human PanLNs include a population of epigenetically imprinted TSDR^{+} "former" Treg that are FoxP3 negative.[187] FoxP3 is also known to be upregulated in activated T cells, and this could offer an alternative explanation for the induction of reporter expression.

A number of human trials investigating the safety and efficacy of Treg cell therapy have been initiated or completed (summarized in Ref. 188). Safety results appear encouraging, as no adverse off-target suppression or increased risk of cancer has been reported in either T1D or GvHD trials.[189–191] Moreover, a large proportion of injected cells persisted for over 12 months *in vivo*.[182] In a study with 10 T1D children that received polyclonal expanded Treg within 2 months of diagnosis preliminary suggestions of efficacy were seen.[192] Four months following therapy, C-peptide levels had declined in the control group but remained significantly higher in recipients of Treg cells.[192] Eight out of the 10 recipients of Treg cells were still in partial remission (informally referred to as the honeymoon phase) compared with only 4 out of the 10 patients in the comparison group. Follow up at the 1 year timepoint (with the recruitment of two additional patients to the Treg arm) showed that 8 out of 12 recipients of Treg cells were still in partial remission with two children insulin-independent.[193] In contrast, only 2 out of 10 patients from the untreated control group remained in partial remission and this group exhibited lower C-peptide levels and a higher insulin requirement compared to those treated with Treg cells. The purity of injected Treg cells was high in this trial, with FoxP3 > 90% being one of the release criteria for the Treg preparation so it is possible that this, combined with the timing of intervention so close to diagnosis created favorable conditions for efficacy. However, it should be noted that the length of the honeymoon phase is notoriously variable so caution must be used in interpreting the results of trials involving small numbers of participants. Moreover, it appears the rate of decline in C-peptide levels is not constant over the 2 years following T1D diagnosis, further complicating interpretation of such data.[194] Nevertheless, the fact that Treg therapy had a detectable impact on circulating Treg numbers in the short term and did not raise safety issues is highly encouraging.

To improve the efficacy of Treg therapy in T1D, further benefit can potentially be gained from the use of antigen-specific Treg.[178] Several approaches exist. First, antigen-specific Treg could theoretically be expanded from polyclonal populations *in vitro* by stimulation with recombinant MHC class II presenting islet peptide mimotope as demonstrated in mice.[195] This approach however has the disadvantage of very low cell numbers. Alternatively, induced Treg can be generated by culturing bulk antigen-specific CD4 T cells in the presence of polarizing conditions (i.e., TGF-β and vitamin D).[196] Here, however, lineage stability may become an issue, and this could potentially increase the pool of diabetogenic T cells thereby posing a safety concern. Another approach is to redirect antigen specificity by introducing transgenic TCRs.[197] Their dual specificity may, however, represent a caveat, risking off-target action due to the endogenous TCR, suboptimal homing to target tissues or hampered activity due to insufficient CD3 availability.[198] To address these possible issues, chimeric antigen receptors may be used instead, taking advantage of added activation or homing signals as well as suicide triggers as an invaluable safety mechanism should adverse reactions take place.[199]

5. CONCLUDING REMARKS

Despite initial problems in defining the Treg phenotype, the field has progressed rapidly and is now at the point where clinical application has become a realistic possibility.[200] The capacity of Treg to prevent or even reverse autoimmune diabetes has been amply demonstrated in mouse models, highlighting the exciting potential that harnessing the power of these cells offers. The ability of Treg to suppress CD4 and CD8 T cell responses, as well as B cell responses[201–203] and even innate immunity[204] means that they are well placed to regulate the complex multicellular events that lead to T1D. There is evidence to support the view that Treg can act in both the pancreas and in its draining lymph nodes, suggesting the prospect of inhibiting both early initiating events and the ongoing immune-mediated damage. Whether the balance of suppressive mechanisms employed by Treg differs between these two sites remains open to speculation.

There are several areas worthy of consideration regarding the translation of Treg-directed therapies to the clinic. For example, it is important to consider factors that may counteract the activity of Treg. Using the DO11xripOVA model Clough *et al.*[205] showed that Tconv from diabetic mice were less sensitive to Treg suppression than Tconv from nondiabetic animals. This result is consistent with reports from the NOD model,[206] where age-dependent development of Tconv resistance to Treg suppression was observed.[176,207] The Tconv resistance in the DO11xripOVA diabetes model was correlated with elevated levels of the cytokine IL-21 in the PanLN. In humans, IL-21 can also interfere with Treg suppression,[208] and there is some evidence that Tconv from the peripheral blood of T1D patients exhibit resistance to Treg suppression.[50,209] Intriguingly, it recently transpired that Tfh cells, which produce IL-21 and can coexpress IFN-γ and TNFα, are overrepresented in the effector T cell pool in DO11xripOVA mice and peripheral blood of T1D patients.[32,210] Thus, it is possible that IL-21 or other mediators[211] could counteract the suppressive effects of Treg in T1D. Although, there is evidence that suppression can be restored if Treg are present in high enough numbers,[205] an additional strategy might to accompany Treg cell therapy with the simultaneous blockade of pathways shown to interfere with suppression.

Another consideration pertinent to immunotherapy aimed at expanding Treg *in vivo* is the subtle difference in CD25 expression between mouse and human systems. Although the broad principles are conserved between species (Treg being CD25hi and naive T cells expressing the lowest levels), the distinction between CD25 levels on Treg versus other T cell populations is far clearer in mice than in humans. A key factor here is likely to be the presence of a larger (CD25+) memory T cell population in humans, who have faced years of immunological challenge unlike mice that are typically maintained in a more restricted antigenic environment. One consequence of this is that IL-2 immunotherapy is more likely to act selectively on Treg in mouse models. Thus careful dose response analyses are required to establish the optimal delivery strategy for IL-2 immunotherapy in humans.[156]

Finally, while we have focused on CD25+FoxP3+ Treg in this review, it is clear that additional regulatory populations also exist (reviewed in Ref. 213). For example, IL-10 can potently regulate diabetes in BDC2.2/NOD mice, and its main cellular source does not appear to be CD4+CD25+ Treg.[213] There is evidence that the balance between production of IL-10 and proinflammatory cytokines is also relevant to T1D disease pathogenesis in humans.[214,215] Thus, strategies aimed at augmenting or inducing additional types of regulatory populations may also be worthwhile.

Overall, it is an exciting time for immunotherapy in the setting of T1D and therapeutic manipulation of Treg populations is gathering momentum as a credible strategy. Increasing refinements are likely to result in the combination of this approach with additional interventions in order to maximize therapeutic benefit.

ACKNOWLEDGMENTS

LSKW holds a Medical Research Council Senior Non-Clinical Fellowship.

REFERENCES

1. Todd JA. Etiology of type 1 diabetes. *Immunity*. 2010;32(4):457–467.
2. Maier LM, Wicker LS. Genetic susceptibility to type 1 diabetes. *Curr Opin Immunol*. 2005;17(6):601–608.
3. Walker LS, Abbas AK. The enemy within: keeping self-reactive T cells at bay in the periphery. *Nat Rev Immunol*. 2002;2(1):11–19.
4. Reijonen H, Kwok WW, Nepom GT. Detection of CD4(+) autoreactive T cells in T1D using HLA class II tetramers. *Ann N Y Acad Sci*. 2003;1005:82–87. Immunology of Diabetes II: Pathogenesis from Mouse to Man.
5. Sakaguchi S, Wing K, Onishi Y, Prieto-Martin P, Yamaguchi T. Regulatory T cells: how do they suppress immune responses? *Int Immunol*. 2009;21(10):1105–1111.
6. Gough SCL, Simmonds MJ. The HLA region and autoimmune disease: associations and mechanisms of action. *Curr Genomics*. 2007;8(7):453–465.
7. Cucca F, Lampis R, Congia M, et al. A correlation between the relative predisposition of MHC class II alleles to type 1 diabetes and the structure of their proteins. *Hum Mol Genet*. 2001;10(19):2025–2037.
8. Johansson S, Lie BA, Todd JA, et al. Evidence of at least two type 1 diabetes susceptibility genes in the HLA complex distinct from HLA-DQB1,-DQA1 and-DRB1. *Genes Immun*. 2003;4(1):46–53.
9. Crawford F, Stadinski B, Jin NY, et al. Specificity and detection of insulin-reactive CD4(+) T cells in type 1 diabetes in the nonobese diabetic (NOD) mouse. *Proc Natl Acad Sci USA*. 2011;108(40):16729–16734.
10. Stadinski BD, Zhang L, Crawford F, Marrack P, Eisenbarth GS, Kappler JW. Diabetogenic T cells recognize insulin bound to IA(g7) in an unexpected, weakly binding register. *Proc Natl Acad Sci USA*. 2010;107(24):10978–10983.
11. Nayak DK, Calderon B, Vomund AN, Unanue ER. ZnT8-reactive T cells are weakly pathogenic in NOD mice but can participate in diabetes under inflammatory conditions. *Diabetes*. 2014;63(10):3438–3448.
12. Yang J, Chow IT, Sosinowski T, et al. Autoreactive T cells specific for insulin B:11-23 recognize a low-affinity peptide register in human subjects with autoimmune diabetes. *Proc Natl Acad Sci USA*. 2014;111(41):14840–14845.
13. Kawamura K, McLaughlin KA, Weissert R, Forsthuber TG. Myelin-reactive type B T cells and T cells specific for low-affinity MHC-binding myelin peptides escape tolerance in HLA-DR transgenic mice. *J Immunol*. 2008;181(5):3202–3211.
14. Wucherpfennig KW, Call MJ, Deng L, Mariuzza R. Structural alterations in peptide-MHC recognition by self-reactive T cell receptors. *Curr Opin Immunol*. 2009;21(6):590–595.
15. Barratt BJ, Payne F, Lowe CE, et al. Rentapping the insulin gene/IDDM2 locus in type 1 diabetes. *Diabetes*. 2004;53(7):1884–1889.
16. Bennett ST, Lucassen AM, Gough SCL, et al. Susceptibility to human type-1 diabetes at iddm2 is determined by tandem repeat variation at the insulin gene minisatellite locus. *Nat Genet*. 1995;9(3):284–292.
17. Bottini N, Peterson EJ. Tyrosine phosphatase PTPN22: multifunctional regulator of immune signaling, development, and disease. *Annu Rev Immunol*. 2014;32:83–119.
18. Bottini N, Musumeci L, Alonso A, et al. A functional variant of lymphoid tyrosine phosphatase is associated with type I diabetes. *Nat Genet*. 2004;36(4):337–338.

19. Bottini N, Vang T, Cucca F, Mustelin T. Role of PTPN22 in type 1 diabetes and other autoimmune diseases. *Semin Immunol.* 2006;18(4):207–213.
20. Fiorillo E, Orru V, Stanford SM, et al. Autoimmune-associated PTPN22 R620W variation reduces phosphorylation of lymphoid phosphatase on an inhibitory tyrosine residue. *J Biol Chem.* 2010;285(34):26506–26518.
21. Maine CJ, Hamilton-Williams EE, Cheung J, et al. PTPN22 alters the development of regulatory T cells in the thymus. *J Immunol.* 2012;188(11):5267–5275.
22. Brownlie RJ, Miosge LA, Vassilakos D, Svensson LM, Cope A, Zamoyska R. Lack of the phosphatase PTPN22 increases adhesion of murine regulatory T cells to improve their immunosuppressive function. *Sci Signal.* 2012;5(252):ra87.
23. Vang T, Congia M, Macis MD, et al. Autoimmune-associated lymphoid tyrosine phosphatase is a gain-of-function variant. *Nat Genet.* 2005;37(12):1317–1319.
24. Dai X, James RG, Habib T, et al. A disease-associated PTPN22 variant promotes systemic autoimmunity in murine models. *J Clin Invest.* 2013;123(5):2024–2036.
25. Salmond RJ, Brownlie RJ, Morrison VL, Zamoyska R. The tyrosine phosphatase PTPN22 discriminates weak self peptides from strong agonist TCR signals. *Nat Immunol.* 2014;15(9):875–883.
26. Fu G, Rybakin V, Brzostek J, Paster W, Acuto O, Gascoigne NRJ. Fine-tuning T cell receptor signaling to control T cell development. *Trends Immunol.* 2014;35(7):311–318.
27. Vella A, Cooper JD, Lowe CE, et al. Localization of a type 1 diabetes locus in the IL2RA/CD25 region by use of tag single-nucleotide polymorphisms. *Am J Hum Genet.* 2005;76(5):773–779.
28. Dendrou CA, Plagnol V, Fung E, et al. Cell-specific protein phenotypes for the autoimmune locus IL2RA using a genotype-selectable human bioresource. *Nat Genet.* 2009;41(9):1011–1015.
29. Liao W, Lin JX, Leonard WJ. Interleukin-2 at the crossroads of effector responses, tolerance, and immunotherapy. *Immunity.* 2013;38(1):13–25.
30. Hulme MA, Wasserfall CH, Atkinson MA, Brusko TM. Central role for interleukin-2 in type 1 diabetes. *Diabetes.* 2012;61(1):14–22.
31. Tang Q, Adams JY, Penaranda C, et al. Central role of defective interleukin-2 production in the triggering of islet autoimmune destruction. *Immunity.* 2008;28(5):687–697.
32. Kenefeck R, Wang CJ, Kapadi T, et al. Follicular helper T cell signature in type 1 diabetes. *J Clin Invest.* 2015;125(1):292–303.
33. Ueda H, Howson JMM, Esposito L, et al. Association of the T-cell regulatory gene CTLA4 with susceptibility to autoimmune disease. *Nature.* 2003;423(6939):506–511.
34. Cederbom L, Hall H, Ivars F. CD4(+)CD25(+) regulatory T cells down-regulate co-stimulatory molecules on antigen-presenting cells. *Eur J Immunol.* 2000;30(6):1538–1543.
35. Collins AV, Brodie DW, Gilbert RJC, et al. The interaction properties of costimulatory molecules revisited. *Immunity.* 2002;17(2):201–210.
36. Linsley PS, Bradshaw J, Greene J, Peach R, Bennett KL, Mittler RS. Intracellular trafficking of CTLA-4 and focal localization towards sites of TCR engagement. *Immunity.* 1996;4(6):535–543.
37. Atabani SF, Thio CL, Divanovic S, et al. Association of CTLA4 polymorphism with regulatory T cell frequency. *Eur J Immunol.* 2005;35(7):2157–2162.
38. Araki M, Chung D, Liu S, et al. Genetic evidence that the differential expression of the ligand-independent isoform of CTLA-4 is the molecular basis of the Idd5.1 type 1 diabetes region in nonobese diabetic mice. *J Immunol.* 2009;183(8):5146–5157.
39. Vijayakrishnan L, Slavik JM, Illes Z, et al. An autoimmune disease-associated CTLA-4 splice variant lacking the B7 binding domain signals negatively in T cells. *Immunity.* 2004;20(5):563–575.

40. Cheng MH, Anderson MS. Insights into type 1 diabetes from the autoimmune polyendocrine syndromes. *Curr Opin Endocrinol Diabetes Obes*. 2013;20(4):271–278.
41. Chen Z, Herman AE, Matos M, Mathis D, Benoist C. Where CD4+CD25+ T reg cells impinge on autoimmune diabetes. *J Exp Med*. 2005;202(10):1387–1397.
42. Feuerer M, Shen Y, Littman DR, Benoist C, Mathis D. How punctual ablation of regulatory T cells unleashes an autoimmune lesion within the pancreatic islets. *Immunity*. 2009;31(4):654–664.
43. Sarween N, Chodos A, Raykundalia C, Khan M, Abbas AK, Walker LS. CD4+CD25+ cells controlling a pathogenic CD4 response inhibit cytokine differentiation, CXCR-3 expression, and tissue invasion. *J Immunol*. 2004;173(5):2942–2951.
44. Kukreja A, Cost G, Marker J, et al. Multiple immuno-regulatory defects in type-1 diabetes. *J Clin Invest*. 2002;109(1):131–140.
45. Liu WH, Putnam AL, Xu-Yu Z, et al. CD127 expression inversely correlates with FoxP3 and suppressive function of human CD4(+) T reg cells. *J Exp Med*. 2006;203(7):1701–1711.
46. Brusko T, Wasserfall C, McGrail K, et al. No alterations in the frequency of FOXP3(+) regulatory T-cells in type 1 diabetes. *Diabetes*. 2007;56(3):604–612.
47. Tree TI, Roep BO, Peakman M. A mini meta-analysis of studies on CD4+CD25+ T cells in human type 1 diabetes: report of the Immunology of Diabetes Society T Cell Workshop. *Ann N Y Acad Sci*. 2006;1079:9–18.
48. Putnam AL, Brusko TM, Lee MR, et al. Expansion of human regulatory T-cells from patients with type 1 diabetes. *Diabetes*. 2009;58(3):652–662.
49. Hughson A, Bromberg I, Johnson B, Quataert S, Jospe N, Fowell DJ. Uncoupling of proliferation and cytokines from suppression within the CD4+CD25+Foxp3+ T-cell compartment in the 1st year of human type 1 diabetes. *Diabetes*. 2011;60(8):2125–2133.
50. Lawson JM, Tremble J, Dayan C, et al. Increased resistance to CD4(+)CD25(hi) regulatory T cell-mediated suppression in patients with type 1 diabetes. *Clin Exp Immunol*. 2008;154(3):353–359.
51. Garg G, Tyler JR, Yang JH, et al. Type 1 diabetes-associated IL2RA variation lowers IL-2 signaling and contributes to diminished CD4+CD25+ regulatory T cell function. *J Immunol*. 2012;188(9):4644–4653.
52. Long SA, Cerosaletti K, Bollyky PL, et al. Defects in IL-2R signaling contribute to diminished maintenance of FOXP3 expression in CD4(+)CD25(+) regulatory T-cells of type 1 diabetic subjects. *Diabetes*. 2010;59(2):407–415.
53. Hoglund P, Mintern J, Waltzinger C, Heath W, Benoist C, Mathis D. Initiation of autoimmune diabetes by developmentally regulated presentation of islet cell antigens in the pancreatic lymph nodes. *J Exp Med*. 1999;189(2):331–339.
54. Walker LS, Ausubel LJ, Chodos A, Bekarian N, Abbas AK. CTLA-4 differentially regulates T cell responses to endogenous tissue protein versus exogenous immunogen. *J Immunol*. 2002;169(11):6202–6209.
55. Turley S, Poirot L, Hattori M, Benoist C, Mathis D. Physiological beta cell death triggers priming of self-reactive T cells by dendritic cells in a type-1 diabetes model. *J Exp Med*. 2003;198(10):1527–1537.
56. Melli K, Friedman RS, Martin AE, et al. Amplification of autoimmune response through induction of dendritic cell maturation in inflamed tissues. *J Immunol*. 2009;182(5):2590–2600.
57. Gagnerault MC, Luan JJ, Lotton C, Lepault F. Pancreatic lymph nodes are required for priming of beta cell reactive T cells in NOD mice. *J Exp Med*. 2002;196(3):369–377.
58. Levisetti MG, Suri A, Frederick K, Unanue ER. Absence of lymph nodes in NOD mice treated with lymphotoxin-beta receptor immunoglobulin protects from diabetes. *Diabetes*. 2004;53(12):3115–3119.

59. Stoll S, Delon J, Brotz TM, Germain RN. Dynamic imaging of T cell-dendritic cell interactions in lymph nodes. *Science*. 2002;296(5574):1873–1876.
60. Bousso P, Robey E. Dynamics of CD8(+) T cell priming by dendritic cells in intact lymph nodes. *Nat Immunol*. 2003;4(6):579–585.
61. Moreau HD, Bousso P. Visualizing how T cells collect activation signals in vivo. *Curr Opin Immunol*. 2014;26:56–62.
62. Jacobelli J, Lindsay RS, Friedman RS. Peripheral tolerance and autoimmunity: lessons from in vivo imaging. *Immunol Res*. 2013;55(1-3):146–154.
63. Celli S, Lemaitre F, Bousso P. Real-time manipulation of T cell-dendritic cell interactions in vivo reveals the importance of prolonged contacts for CD4(+) T cell activation. *Immunity*. 2007;27(4):625–634.
64. Celli S, Garcia Z, Bousso P. CD4 T cells integrate signals delivered during successive DC encounters in vivo. *J Exp Med*. 2005;202(9):1271–1278.
65. Gottschalk RA, Hathorn MM, Beuneu H, et al. Distinct influences of peptide-MHC quality and quantity on in vivo T-cell responses. *Proc Natl Acad Sci USA*. 2012;109(3):881–886.
66. Thauland TJ, Koguchi Y, Dustin ML, Parker DC. CD28-CD80 interactions control regulatory T cell motility and immunological synapse formation. *J Immunol*. 2014;193(12):5894–5903.
67. Tadokoro CE, Shakhar G, Shen SQ, et al. Regulatory T cells inhibit stable contacts between CD4(+) T cells and dendritic cells in vivo. *J Exp Med*. 2006;203(3):505–511.
68. Tang QZ, Adams JY, Tooley AJ, et al. Visualizing regulatory T cell control of autoimmune responses in nonobese diabetic mice. *Nat Immunol*. 2006;7(1):83–92.
69. Serra P, Amrani A, Yamanouchi J, et al. CD40 ligation releases immature dendritic cells from the control of regulatory CD4+CD25+ T cells. *Immunity*. 2003;19(6):877–889.
70. Sojka DK, Fowell DJ. Regulatory T cells inhibit acute IFN-gamma synthesis without blocking T-helper cell type 1 (Th1) differentiation via a compartmentalized requirement for IL-10. *Proc Natl Acad Sci USA*. 2011;108(45):18336–18341.
71. Schneider MA, Meingassner JG, Lipp M, Moore HD, Rot A. CCR7 is required for the in vivo function of CD4+ CD25+ regulatory T cells. *J Exp Med*. 2007;204(4): 735–745.
72. Herman AE, Freeman GJ, Mathis D, Benoist C. CD4+CD25+ T regulatory cells dependent on ICOS promote regulation of effector cells in the prediabetic lesion. *J Exp Med*. 2004;199(11):1479–1489.
73. Coppieters K, Amirian N, von Herrath M. Intravital imaging of CTLs killing islet cells in diabetic mice. *J Clin Invest*. 2012;122(1):119–131.
74. Lennon GP, Bettini M, Burton AR, et al. T cell islet accumulation in type 1 diabetes is a tightly regulated, cell-autonomous event. *Immunity*. 2009;31(4):643–653.
75. Wang J, Tsai S, Shameli A, Yamanouchi J, Alkemade G, Santamaria P. In situ recognition of autoantigen as an essential gatekeeper in autoimmune CD8+ T cell inflammation. *Proc Natl Acad Sci USA*. 2010;107(20):9317–9322.
76. Calderon B, Carrero JA, Miller MJ, Unanue ER. Entry of diabetogenic T cells into islets induces changes that lead to amplification of the cellular response. *Proc Natl Acad Sci USA*. 2011;108(4):1567–1572.
77. Magnuson AM, Thurber GM, Kohler RH, Weissleder R, Mathis D, Benoist C. Population dynamics of islet-infiltrating cells in autoimmune diabetes. *Proc Natl Acad Sci USA*. 2015;112(5):1511–1516.
78. Gepts W. Pathologic anatomy of the pancreas in juvenile diabetes mellitus. *Diabetes*. 1965;14(10):619–633.
79. Bottazzo GF, Dean BM, McNally JM, Mackay EH, Swift PGF, Gamble DR. In situ characterization of autoimmune phenomena and expression of HLA molecules in the pancreas in diabetic insulitis. *N Engl J Med*. 1985;313(6):353–360.

80. Willcox A, Richardson SJ, Bone AJ, Foulis AK, Morgan NG. Analysis of islet inflammation in human type 1 diabetes. *Clin Exp Immunol.* 2009;155(2):173–181.
81. Brode S, Raine T, Zaccone P, Cooke A. Cyclophosphamide-induced type-1 diabetes in the NOD mouse is associated with a reduction of CD4+CD25+Foxp3+ regulatory T cells. *J Immunol.* 2006;177(10):6603–6612.
82. Mahne AE, Klementowicz JE, Chou A, Vinh N, Tang Q. Therapeutic regulatory T cells subvert effector T cell function in inflamed islets to halt autoimmune diabetes. *J Immunol.* 2015;194(7):3147–3155.
83. Brunkow ME, Jeffery EW, Hjerrild KA, et al. Disruption of a new forkhead/winged-helix protein, scurfin, results in the fatal lymphoproliferative disorder of the scurfy mouse. *Nat Genet.* 2001;27(1):68–73.
84. Godfrey VL, Wilkinson JE, Russell LB. X-linked lymphoreticular disease in the scurfy (sf) mutant mouse. *Am J Pathol.* 1991;138(6):1379–1387.
85. Hori S, Nomura T, Sakaguchi S. Control of regulatory T cell development by the transcription factor Foxp3. *Science.* 2003;299(5609):1057–1061.
86. Fontenot JD, Gavin MA, Rudensky AY. Foxp3 programs the development and function of CD4+CD25+ regulatory T cells. *Nat Immunol.* 2003;4(4):330–336.
87. Khattri R, Cox T, Yasayko SA, Ramsdell F. An essential role for Scurfin in CD4+ CD25+ T regulatory cells. *Nat Immunol.* 2003;4(4):337–342.
88. Vignali DA, Collison LW, Workman CJ. How regulatory T cells work. *Nat Rev Immunol.* 2008;8(7):523–532.
89. Tang Q, Bluestone JA. The Foxp3+ regulatory T cell: a jack of all trades, master of regulation. *Nat Immunol.* 2008;9(3):239–244.
90. Tivol EA, Borriello F, Schweitzer AN, Lynch WP, Bluestone JA, Sharpe AH. Loss of CTLA-4 leads to massive lymphoproliferation and fatal multiorgan tissue destruction, revealing a critical negative regulatory role of CTLA-4. *Immunity.* 1995;3(5):541–547.
91. Waterhouse P, Penninger JM, Timms E, et al. Lymphoproliferative disorders with early lethality in mice deficient in Ctla-4. *Science.* 1995;270(5238):985–988.
92. Walker LS, Sansom DM. Confusing signals: recent progress in CTLA-4 biology. *Trends Immunol.* 2015;36(2):63–70.
93. Bachmann MF, Kohler G, Ecabert B, Mak TW, Kopf M. Cutting edge: lymphoproliferative disease in the absence of CTLA-4 is not T cell autonomous. *J Immunol.* 1999;163(3):1128–1131.
94. Homann D, Dummer W, Wolfe T, et al. Lack of intrinsic CTLA-4 expression has minimal effect on regulation of antiviral T-cell immunity. *J Virol.* 2006;80(1):270–280.
95. Walker LSK. Treg and CTLA-4: two intertwining pathways to immune tolerance. *J Autoimmun.* 2013;45:49–57.
96. Thompson CB, Allison JP. The emerging role of CTLA-4 as an immune attenuator. *Immunity.* 1997;7(4):445–450.
97. Tai X, Van Laethem F, Sharpe AH, Singer A. Induction of autoimmune disease in CTLA-4(-/-) mice depends on a specific CD28 motif that is required for in vivo costimulation. *Proc Natl Acad Sci USA.* 2007;104(34):13756–13761.
98. Mandelbrot DA, McAdam AJ, Sharpe AH. B7-1 or B7-2 is required to produce the lymphoproliferative phenotype in mice lacking cytotoxic T lymphocyte-associated antigen 4 (CTLA-4). *J Exp Med.* 1999;189(2):435–440.
99. Oderup C, Cederbom L, Makowska A, Cilio CM, Ivars F. Cytotoxic T lymphocyte antigen-4-dependent down-modulation of costimulatory molecules on dendritic cells in CD4+ CD25+ regulatory T-cell-mediated suppression. *Immunology.* 2006;118(2): 240–249.
100. Kastenmuller W, Gasteiger G, Subramanian N, et al. Regulatory T cells selectively control CD8+ T cell effector pool size via IL-2 restriction. *J Immunol.* 2011;187(6): 3186–3197.

101. Schmidt EM, Wang CJ, Ryan GA, et al. CTLA-4 controls regulatory T cell peripheral homeostasis and is required for suppression of pancreatic islet autoimmunity. *J Immunol.* 2009;182(1):274–282.
102. Onishi Y, Fehervari Z, Yamaguchi T, Sakaguchi S. Foxp3+ natural regulatory T cells preferentially form aggregates on dendritic cells in vitro and actively inhibit their maturation. *Proc Natl Acad Sci USA.* 2008;105(29):10113–10118.
103. Wing K, Onishi Y, Prieto-Martin P, et al. CTLA-4 control over Foxp3(+) regulatory T cell function. *Science.* 2008;322(5899):271–275.
104. Qureshi OS, Zheng Y, Nakamura K, et al. Trans-endocytosis of CD80 and CD86: a molecular basis for the cell-extrinsic function of CTLA-4. *Science.* 2011;332(6029): 600–603.
105. Luhder F, Chambers C, Allison JP, Benoist C, Mathis D. Pinpointing when T cell costimulatory receptor CTLA-4 must be engaged to dampen diabetogenic T cells. *Proc Natl Acad Sci USA.* 2000;97(22):12204–12209.
106. Luhder F, Hoglund P, Allison JP, Benoist C, Mathis D. Cytotoxic T lymphocyte-associated antigen 4 (CTLA-4) regulates the unfolding of autoimmune diabetes. *J Exp Med.* 1998;187(3):427–432.
107. Eggena MP, Walker LS, Nagabhushanam V, Barron L, Chodos A, Abbas AK. Cooperative roles of CTLA-4 and regulatory T cells in tolerance to an islet cell antigen. *J Exp Med.* 2004;199(12):1725–1730.
108. Stumpf M, Zhou X, Bluestone JA. The B7-independent isoform of CTLA-4 functions to regulate autoimmune diabetes. *J Immunol.* 2013;190(3):961–969.
109. Tang Q, Boden EK, Henriksen KJ, Bour-Jordan H, Bi M, Bluestone JA. Distinct roles of CTLA-4 and TGF-beta in CD4+CD25+ regulatory T cell function. *Eur J Immunol.* 2004;34(11):2996–3005.
110. Kuehn HS, Ouyang W, Lo B, et al. Immune dysregulation in human subjects with heterozygous germline mutations in CTLA4. *Science.* 2014;345(6204):1623–1627.
111. Schubert D, Bode C, Kenefeck R, et al. Autosomal dominant immune dysregulation syndrome in humans with CTLA4 mutations. *Nat Med.* 2014;20(12):1410–1416.
112. Zeissig S, Petersen BS, Tomczak M, et al. Early-onset Crohn's disease and autoimmunity associated with a variant in CTLA-4. *Gut.* 2014. Epub ahead of print.
113. de la Rosa M, Rutz S, Dorninger H, Scheffold A. Interleukin-2 is essential for CD4+ CD25+ regulatory T cell function. *Eur J Immunol.* 2004;34(9):2480–2488.
114. Barthlott T, Moncrieffe H, Veldhoen M, et al. CD25+ CD4+ T cells compete with naive CD4+ T cells for IL-2 and exploit it for the induction of IL-10 production. *Int Immunol.* 2005;17(3):279–288.
115. Fontenot JD, Rasmussen JP, Gavin MA, Rudensky AY. A function for interleukin 2 in Foxp3-expressing regulatory T cells. *Nat Immunol.* 2005;6(11):1142–1151.
116. Pandiyan P, Zheng L, Ishihara S, Reed J, Lenardo MJ. CD4+CD25+Foxp3+ regulatory T cells induce cytokine deprivation-mediated apoptosis of effector CD4+ T cells. *Nat Immunol.* 2007;8(12):1353–1362.
117. Busse D, de la Rosa M, Hobiger K, et al. Competing feedback loops shape IL-2 signaling between helper and regulatory T lymphocytes in cellular microenvironments. *Proc Natl Acad Sci USA.* 2010;107(7):3058–3063.
118. Feinerman O, Jentsch G, Tkach KE, et al. Single-cell quantification of IL-2 response by effector and regulatory T cells reveals critical plasticity in immune response. *Mol Syst Biol.* 2010;6:437.
119. Scheffold A, Huhn J, Hofer T. Regulation of CD4+CD25+ regulatory T cell activity: it takes (IL-)two to tango. *Eur J Immunol.* 2005;35(5):1336–1341.
120. Sitrin J, Ring A, Garcia KC, Benoist C, Mathis D. Regulatory T cells control NK cells in an insulitic lesion by depriving them of IL-2. *J Exp Med.* 2013;210(6): 1153–1165.

121. Lowe CE, Cooper JD, Brusko T, et al. Large-scale genetic fine mapping and genotype-phenotype associations implicate polymorphism in the IL2RA region in type 1 diabetes. *Nat Genet.* 2007;39(9):1074–1082.
122. Todd JA, Walker NM, Cooper JD, et al. Robust associations of four new chromosome regions from genome-wide analyses of type 1 diabetes. *Nat Genet.* 2007;39(7):857–864.
123. Sutherland AP, Van Belle T, Wurster AL, et al. Interleukin-21 is required for the development of type 1 diabetes in NOD mice. *Diabetes.* 2009;58(5):1144–1155.
124. McGuire HM, Vogelzang A, Ma CS, et al. A subset of interleukin-21+ chemokine receptor CCR9+ T helper cells target accessory organs of the digestive system in autoimmunity. *Immunity.* 2011;34(4):602–615.
125. Attridge K, Wang CJ, Wardzinski L, et al. IL-21 inhibits T cell IL-2 production and impairs Treg homeostasis. *Blood.* 2012;119(20):4656–4664.
126. Nakamura K, Kitani A, Strober W. Cell contact-dependent immunosuppression by CD4(+)CD25(+) regulatory T cells is mediated by cell surface-bound transforming growth factor beta. *J Exp Med.* 2001;194(5):629–644.
127. Fahlen L, Read S, Gorelik L, et al. T cells that cannot respond to TGF-beta escape control by CD4(+)CD25(+) regulatory T cells. *J Exp Med.* 2005;201(5):737–746.
128. Li MO, Wan YY, Flavell RA. T cell-produced transforming growth factor-beta1 controls T cell tolerance and regulates Th1- and Th17-cell differentiation. *Immunity.* 2007;26(5):579–591.
129. Worthington JJ, Kelly A, Smedley C, et al. Integrin alphavbeta8-mediated TGF-beta activation by effector regulatory T cells is essential for suppression of T-cell-mediated inflammation. *Immunity.* 2015;42(5):903–915.
130. Peng Y, Laouar Y, Li MO, Green EA, Flavell RA. TGF-beta regulates in vivo expansion of Foxp3-expressing CD4+CD25+ regulatory T cells responsible for protection against diabetes. *Proc Natl Acad Sci USA.* 2004;101(13):4572–4577.
131. You S, Leforban B, Garcia C, Bach JF, Bluestone JA, Chatenoud L. Adaptive TGF-beta-dependent regulatory T cells control autoimmune diabetes and are a privileged target of anti-CD3 antibody treatment. *Proc Natl Acad Sci USA.* 2007;104(15):6335–6340.
132. Green EA, Gorelik L, McGregor CM, Tran EH, Flavell RA. CD4+CD25+ T regulatory cells control anti-islet CD8+ T cells through TGF-beta-TGF-beta receptor interactions in type 1 diabetes. *Proc Natl Acad Sci USA.* 2003;100(19):10878–10883.
133. Penaranda C, Tang Q, Bluestone JA. Anti-CD3 therapy promotes tolerance by selectively depleting pathogenic cells while preserving regulatory T cells. *J Immunol.* 2011;187(4):2015–2022.
134. Belghith M, Bluestone JA, Barriot S, Megret J, Bach JF, Chatenoud L. TGF-beta-dependent mechanisms mediate restoration of self-tolerance induced by antibodies to CD3 in overt autoimmune diabetes. *Nat Med.* 2003;9(9):1202–1208.
135. Waldron-Lynch F, Henegariu O, Deng S, et al. Teplizumab induces human gut-tropic regulatory cells in humanized mice and patients. *Sci Transl Med.* 2012;4(118):118ra112.
136. Valle A, Barbagiovanni G, Jofra T, et al. Heterogeneous CD3 expression levels in differing T cell subsets correlate with the in vivo anti-CD3-mediated T cell modulation. *J Immunol.* 2015;194(5):2117–2127.
137. Li L, Nishio J, van Maurik A, Mathis D, Benoist C. Differential response of regulatory and conventional CD4(+) lymphocytes to CD3 engagement: clues to a possible mechanism of anti-CD3 action? *J Immunol.* 2013;191(7):3694–3704.
138. Herold KC, Gitelman SE, Ehlers MR, et al. Teplizumab (anti-CD3 mAb) treatment preserves C-peptide responses in patients with new-onset type 1 diabetes in a randomized controlled trial metabolic and immunologic features at baseline identify a subgroup of responders. *Diabetes.* 2013;62(11):3766–3774.

139. Herold KC, Gitelman S, Greenbaum C, et al. Treatment of patients with new onset Type 1 diabetes with a single course of anti-CD3 mAb teplizumab preserves insulin production for up to 5 years. *Clin Immunol.* 2009;132(2):166–173.
140. Keymeulen B, Walter M, Mathieu C, et al. Four-year metabolic outcome of a randomised controlled CD3-antibody trial in recent-onset type 1 diabetic patients depends on their age and baseline residual beta cell mass. *Diabetologia.* 2010;53(4): 614–623.
141. Keymeulen B, Candon S, Fafi-Kremer S, et al. Transient Epstein-Barr virus reactivation in CD3 monoclonal antibody-treated patients. *Blood.* 2010;115(6):1145–1155.
142. Aronson R, Gottlieb PA, Christiansen JS, et al. Low-dose otelixizumab anti-CD3 monoclonal antibody DEFEND-1 study: results of the randomized phase III study in recent-onset human type 1 diabetes. *Diabetes Care.* 2014;37(10):2746–2754.
143. Sherry N, Hagopian W, Ludvigsson J, et al. Teplizumab for treatment of type 1 diabetes (Protege study): 1-year results from a randomised, placebo-controlled trial. *Lancet.* 2011;378(9790):487–497.
144. Setoguchi R, Hori S, Takahashi T, Sakaguchi S. Homeostatic maintenance of natural Foxp3(+) CD25(+) CD4(+) regulatory T cells by interleukin (IL)-2 and induction of autoimmune disease by IL-2 neutralization. *J Exp Med.* 2005;201(5):723–735.
145. Suzuki H, Kundig TM, Furlonger C, et al. Deregulated T cell activation and autoimmunity in mice lacking interleukin-2 receptor beta. *Science.* 1995;268(5216): 1472–1476.
146. Willerford DM, Chen J, Ferry JA, Davidson L, Ma A, Alt FW. Interleukin-2 receptor alpha chain regulates the size and content of the peripheral lymphoid compartment. *Immunity.* 1995;3(4):521–530.
147. Klatzmann D, Abbas AK. The promise of low-dose interleukin-2 therapy for autoimmune and inflammatory diseases. *Nat Rev Immunol.* 2015;15(5):283–294.
148. Grinberg-Bleyer Y, Baeyens A, You S, et al. IL-2 reverses established type 1 diabetes in NOD mice by a local effect on pancreatic regulatory T cells. *J Exp Med.* 2010;207(9):1871–1878.
149. Rabinovitch A, Suarez-Pinzon WL, Shapiro AM, Rajotte RV, Power R. Combination therapy with sirolimus and interleukin-2 prevents spontaneous and recurrent autoimmune diabetes in NOD mice. *Diabetes.* 2002;51(3):638–645.
150. Kosmaczewska A. Low-dose interleukin-2 therapy: a driver of an imbalance between immune tolerance and autoimmunity. *Int J Mol Sci.* 2014;15(10):18574–18592.
151. Saadoun D, Rosenzwajg M, Joly F, et al. Regulatory T-cell responses to low-dose interleukin-2 in HCV-induced vasculitis. *N Engl J Med.* 2011;365(22):2067–2077.
152. Long SA, Rieck M, Sanda S, et al. Rapamycin/IL-2 combination therapy in patients with type 1 diabetes augments Tregs yet transiently impairs beta-cell function. *Diabetes.* 2012;61(9):2340–2348.
153. Baeyens A, Perol L, Fourcade G, et al. Limitations of IL-2 and rapamycin in immunotherapy of type 1 diabetes. *Diabetes.* 2013;62(9):3120–3131.
154. Hartemann A, Bensimon G, Payan CA, et al. Low-dose interleukin 2 in patients with type 1 diabetes: a phase 1/2 randomised, double-blind, placebo-controlled trial. *Lancet Diabetes Endocrinol.* 2013;1(4):295–305.
155. Churlaud G, Pitoiset F, Jebbawi F, et al. Human and mouse CD8(+)CD25(+) FOXP3(+) regulatory T cells at steady state and during interleukin-2 therapy. *Front Immunol.* 2015;6:171.
156. Waldron-Lynch F, Kareclas P, Irons K, et al. Rationale and study design of the adaptive study of IL-2 dose on regulatory T cells in type 1 diabetes (DILT1D): a non-randomised, open label, adaptive dose finding trial. *BMJ Open.* 2014;4(6):e005559.
157. Bell CJ, Sun Y, Nowak UM, et al. Sustained in vivo signaling by long-lived IL-2 induces prolonged increases of regulatory T cells. *J Autoimmun.* 2015;56:66–80.

158. Pabst O, Mowat AM. Oral tolerance to food protein. *Mucosal Immunol.* 2012;5(3): 232–239.
159. Zhang ZJ, Davidson L, Eisenbarth G, Weiner HL. Suppression of diabetes in nonobese diabetic mice by oral administration of porcine insulin. *Proc Natl Acad Sci USA.* 1991;88(22):10252–10256.
160. Daniel D, Wegmann DR. Protection of nonobese diabetic mice from diabetes by intranasal or subcutaneous administration of insulin peptide B-(9-23). *Proc Natl Acad Sci USA.* 1996;93(2):956–960.
161. Keller RJ, Eisenbarth GS, Jackson RA. Insulin prophylaxis in individuals at high risk of type I diabetes. *Lancet.* 1993;341(8850):927–928.
162. Nanto-Salonen K, Kupila A, Simell S, et al. Nasal insulin to prevent type 1 diabetes in children with HLA genotypes and autoantibodies conferring increased risk of disease: a double-blind, randomised controlled trial. *Lancet.* 2008;372(9651):1746–1755.
163. Chaillous L, Lefevre H, Thivolet C, et al. Oral insulin administration and residual beta-cell function in recent-onset type 1 diabetes: a multicentre randomised controlled trial. Diabete insuline orale group. *Lancet.* 2000;356(9229):545–549.
164. Barker JM, McFann KK, Orban T. Effect of oral insulin on insulin autoantibody levels in the diabetes prevention trial type 1 oral insulin study. *Diabetologia.* 2007;50(8): 1603–1606.
165. von Herrath M, Peakman M, Roep B. Progress in immune-based therapies for type 1 diabetes. *Clin Exp Immunol.* 2013;172(2):186–202.
166. Tiittanen M, Huupponen JT, Knip M, Vaarala O. Insulin treatment in patients with type 1 diabetes induces upregulation of regulatory T-cell markers in peripheral blood mononuclear cells stimulated with insulin in vitro. *Diabetes.* 2006;55(12): 3446–3454.
167. Karounos DG, Bryson JS, Cohen DA. Metabolically inactive insulin analog prevents type I diabetes in prediabetic NOD mice. *J Clin Invest.* 1997;100(6):1344–1348.
168. Bowman MA, Campbell L, Darrow BL, Ellis TM, Suresh A, Atkinson MA. Immunological and metabolic effects of prophylactic insulin therapy in the NOD-scid/scid adoptive transfer model of IDDM. *Diabetes.* 1996;45(2):205–208.
169. Orban T, Farkas K, Jalahej H, et al. Autoantigen-specific regulatory T cells induced in patients with type 1 diabetes mellitus by insulin B-chain immunotherapy. *J Autoimmun.* 2010;34(4):408–415.
170. Ludvigsson J, Cheramy M, Axelsson S, et al. GAD-treatment of children and adolescents with recent-onset type 1 diabetes preserves residual insulin secretion after 30 months. *Diabetes Metab Res Rev.* 2014;30(5):405–411.
171. Ludvigsson J, Faresjo M, Hjorth M, et al. GAD treatment and insulin secretion in recent-onset type 1 diabetes. *N Engl J Med.* 2008;359(18):1909–1920.
172. Ludvigsson J, Krisky D, Casas R, et al. GAD65 antigen therapy in recently diagnosed type 1 diabetes mellitus. *N Engl J Med.* 2012;366(5):433–442.
173. Akbarpour M, Goudy KS, Cantore A, et al. Insulin B chain 9-23 gene transfer to hepatocytes protects from type 1 diabetes by inducing Ag-specific FoxP3+ Tregs. *Sci Transl Med.* 2015;7(289):289ra281.
174. Matrai J, Cantore A, Bartholomae CC, et al. Hepatocyte-targeted expression by integrase-defective lentiviral vectors induces antigen-specific tolerance in mice with low genotoxic risk. *Hepatology.* 2011;53(5):1696–1707.
175. Salomon B, Lenschow DJ, Rhee L, et al. B7/CD28 costimulation is essential for the homeostasis of the CD4(+)CD25(+) immunoregulatory T cells that control autoimmune diabetes. *Immunity.* 2000;12(4):431–440.
176. You S, Belghith M, Cobbold S, et al. Autoimmune diabetes onset results from qualitative rather than quantitative age-dependent changes in pathogenic T-cells. *Diabetes.* 2005;54(5):1415–1422.

177. Tarbell KV, Yamazaki S, Olson K, Toy P, Steinman RM. CD25+ CD4+ T cells, expanded with dendritic cells presenting a single autoantigenic peptide, suppress autoimmune diabetes. *J Exp Med.* 2004;199(11):1467–1477.
178. Tang QZ, Henriksen KJ, Bi MY, et al. In vitro-expanded antigen-specific regulatory T cells suppress autoimmune diabetes. *J Exp Med.* 2004;199(11):1455–1465.
179. Tarbell KV, Petit L, Zuo X, et al. Dendritic cell-expanded, islet-specific CD4+ CD25+ CD62L+ regulatory T cells restore normoglycemia in diabetic NOD mice. *J Exp Med.* 2007;204(1):191–201.
180. Qin S, Cobbold SP, Pope H, et al. "Infectious" transplantation tolerance. *Science.* 1993;259(5097):974–977.
181. Cobbold SP, Waldmann H. Regulatory cells and transplantation tolerance. *Cold Spring Harb Perspect Med.* 2013;3(6).
182. Tang Q, Bluestone JA. Regulatory T-cell therapy in transplantation: moving to the clinic. *Cold Spring Harb Perspect Med.* 2013;3(11).
183. Sakaguchi S, Vignali DAA, Rudensky AY, Niec RE, Waldmann H. The plasticity and stability of regulatory T cells. *Nat Rev Immunol.* 2013;13(6):461–467.
184. Zhou XY, Bailey-Bucktrout SL, Jeker LT, et al. Instability of the transcription factor Foxp3 leads to the generation of pathogenic memory T cells in vivo. *Nat Immunol.* 2009;10(9):1000–1104.
185. Lal G, Bromberg JS. Epigenetic mechanisms of regulation of Foxp3 expression. *Blood.* 2009;114(18):3727–3735.
186. Lal G, Zhang N, van der Touw W, et al. Epigenetic regulation of Foxp3 expression in regulatory T cells by DNA methylation. *J Immunol.* 2009;182(1):259–273.
187. Ferraro A, Socci C, Stabilini A, et al. Expansion of Th17 cells and functional defects in T regulatory cells are key features of the pancreatic lymph nodes in patients with type 1 diabetes. *Diabetes.* 2011;60(11):2903–2913.
188. Bluestone JA, Trotta E, Xu D. The therapeutic potential of regulatory T cells for the treatment of autoimmune disease. *Expert Opin Ther Targets.* 2015;19(8):1091–1103.
189. Trzonkowski P, Bieniaszewska M, Juscinska J, et al. First-in-man clinical results of the treatment of patients with graft versus host disease with human ex vivo expanded CD4+CD25+CD127-T regulatory cells. *Clin Immunol.* 2009;133(1):22–26.
190. Brunstein CG, Miller JS, Cao Q, et al. Infusion of ex vivo expanded T regulatory cells in adults transplanted with umbilical cord blood: safety profile and detection kinetics. *Blood.* 2011;117(3):1061–1070.
191. Di Ianni M, Falzetti F, Carotti A, et al. Tregs prevent GVHD and promote immune reconstitution in HLA-haploidentical transplantation. *Blood.* 2011;117(14): 3921–3928.
192. Marek-Trzonkowska N, Mysliwiec M, Dobyszuk A, et al. Administration of CD4+ CD25highCD127- regulatory T cells preserves beta-cell function in type 1 diabetes in children. *Diabetes Care.* 2012;35(9):1817–1820.
193. Marek-Trzonkowska N, Mysliwiec M, Dobyszuk A, et al. Therapy of type 1 diabetes with CD4(+)CD25(high) CD127-regulatory T cells prolongs survival of pancreatic islets—results of one year follow-up. *Clin Immunol.* 2014;153(1):23–30.
194. Greenbaum CJ, Beam CA, Boulware D, et al. Fall in C-peptide during first 2 years from diagnosis: evidence of at least two distinct phases from composite Type 1 Diabetes TrialNet data. *Diabetes.* 2012;61(8):2066–2073.
195. Masteller EL, Warner MR, Tang Q, Tarbell KV, McDevitt H, Bluestone JA. Expansion of functional endogenous antigen-specific CD4+CD25+ regulatory T cells from nonobese diabetic mice. *J Immunol.* 2005;175(5):3053–3059.
196. Kong N, Lan Q, Chen M, et al. Antigen-specific transforming growth factor beta-induced Treg cells, but not natural Treg cells, ameliorate autoimmune arthritis in mice by shifting the Th17/Treg cell balance from Th17 predominance to Treg cell predominance. *Arthritis Rheum.* 2012;64(8):2548–2558.

197. Brusko TM, Koya RC, Zhu S, et al. Human antigen-specific regulatory T cells generated by T cell receptor gene transfer. *PLoS One*. 2010;5(7):e11726.
198. Ahmadi M, King JW, Xue SA, et al. CD3 limits the efficacy of TCR gene therapy in vivo. *Blood*. 2011;118(13):3528–3537.
199. Jethwa H, Adami AA, Maher J. Use of gene-modified regulatory T-cells to control autoimmune and alloimmune pathology: is now the right time? *Clin Immunol*. 2014;150(1):51–63.
200. van der Net JB, Bushell A, Wood KJ, Harden PN. Regulatory T cells: first steps of clinical application in solid organ transplantation. *Transpl Int*. 2015. Epub ahead of print.
201. Lim HW, Hillsamer P, Banham AH, Kim CH. Cutting edge: direct suppression of B cells by CD4+ CD25+ regulatory T cells. *J Immunol*. 2005;175(7):4180–4183.
202. Zhao DM, Thornton AM, DiPaolo RJ, Shevach EM. Activated CD4+CD25+ T cells selectively kill B lymphocytes. *Blood*. 2006;107(10):3925–3932.
203. Iikuni N, Lourenco EV, Hahn BH, La Cava A. Cutting edge: regulatory T cells directly suppress B cells in systemic lupus erythematosus. *J Immunol*. 2009;183(3):1518–1522.
204. Maloy KJ, Salaun L, Cahill R, Dougan G, Saunders NJ, Powrie F. CD4+CD25+ T(R) cells suppress innate immune pathology through cytokine-dependent mechanisms. *J Exp Med*. 2003;197(1):111–119.
205. Clough LE, Wang CJ, Schmidt EM, et al. Release from regulatory T cell-mediated suppression during the onset of tissue-specific autoimmunity is associated with elevated IL-21. *J Immunol*. 2008;180(8):5393–5401.
206. Gregori S, Giarratana N, Smiroldo S, Adorini L. Dynamics of pathogenic and suppressor T cells in autoimmune diabetes development. *J Immunol*. 2003;171(8):4040–4047.
207. D'Alise AM, Auyeung V, Feuerer M, et al. The defect in T-cell regulation in NOD mice is an effect on the T-cell effectors. *Proc Natl Acad Sci USA*. 2008;105(50):19857–19862.
208. Peluso I, Fantini MC, Fina D, et al. IL-21 counteracts the regulatory T cell-mediated suppression of human CD4(+) T lymphocytes. *J Immunol*. 2007;178(2):732–739.
209. Schneider A, Rieck M, Sanda S, Pihoker C, Greenbaum C, Buckner JH. The effector T cells of diabetic subjects are resistant to regulation via CD4(+)FOXP3(+) regulatory T cells. *J Immunol*. 2008;181(10):7350–7355.
210. Ferreira RC, Simons HZ, Thompson WS, et al. IL-21 production by CD4(+) effector T cells and frequency of circulating follicular helper T cells are increased in type 1 diabetes patients. *Diabetologia*. 2015;58(4):781–790.
211. Walker LS. Regulatory T, cells overturned: the effectors fight back. *Immunology*. 2009;126(4):466–474.
212. Wong FS, Dayan CM. Regulatory T cells in autoimmune endocrine diseases. *Trends Endocrinol Metab*. 2008;19(8):292–299.
213. Phillips JM, Parish NM, Drage M, Cooke A. Cutting edge: interactions through the IL-10 receptor regulate autoimmune diabetes. *J Immunol*. 2001;167(11):6087–6091.
214. Arif S, Tree TI, Astill TP, et al. Autoreactive T cell responses show proinflammatory polarization in diabetes but a regulatory phenotype in health. *J Clin Invest*. 2004;113(3):451–463.
215. Petrich de Marquesini LG, Fu J, Connor KJ, et al. IFN-gamma and IL-10 islet-antigen-specific T cell responses in autoantibody-negative first-degree relatives of patients with type 1 diabetes. *Diabetologia*. 2010;53(7):1451–1460.

INDEX

Note: Page numbers followed by "*f*" indicate figures and "*t*" indicate tables.

N

O

P

R

S

T

V

W

X

Z

FSC
www.fsc.org
MIX
Papier | Fördert
gute Waldnutzung
FSC® C083411

Kolibri 360 GmbH
Ferdinand-Jühlke-Straße 7
99095 Erfurt, Deutschland
produktsicherheit@kolibri360.de